Deep Down Things

DEEP DOWN THINGS

The Breathtaking Beauty
of Particle Physics

Bruce A. Schumm

The Johns Hopkins University Press
Baltimore & London

The Johns Hopkins University Press
2715 North Charles Street
Baltimore, Maryland 21218-4363
www.press.jhu.edu

Library of Congress Cataloging-in-Publication Data

Schumm, Bruce A., 1958–
Deep down things : the breathtaking beauty of particle physics / Bruce A. Schumm.
p. cm.
Includes bibiliographical references and index.
ISBN 0-8018-7971-X (hardcover : acid-free paper)
1. Particles (Nuclear physics)   2. Quantum field theory   3. Gauge fields (Physics)
4. Symmetry (Physics)   I. Title.

QC793.2.S35 2004
539.7′2—dc22
2004050710

A catalog record for this book is available from the British Library.

Illustrations (except for fig. 4.12 on p. 90) are by Bill Rowe.

And for all this, nature is never spent;

There lives the dearest freshness deep down things

—Gerard Manley Hopkins, *God's Grandeur*

Contents

Preface

I didn't come up with the title for this book. For that, I can thank the people at the Johns Hopkins University Press, and in particular, Trevor Lipscombe, with whom I worked quite closely at various stages as the book came together. But this I will say: the title does about as good a job of capturing the essence of particle physics as any string of three monosyllabic words possibly can. Deep down within the atomic nucleus, deeply within the paradoxical richness of empty space, deep inside the synapses of the great scientific thinkers of the twentieth century—this is the territory of particle physics. The connection to the devotional poetry of Gerard Manley Hopkins is also well made; whether one is religious or not, to appreciate particle physics is to appreciate the miracle of nature. In fact, my only reservation about the title is that, as you will shortly see, it implies a degree of literacy to which I can't lay claim.

It will come as no surprise to anyone if I admit that writing a book is a massive and perhaps ill-advised undertaking. In this case, publication will have taken place more than four years since the time when, full of myself after having gotten tenure at the University of California, Santa Cruz, I first sat down to compose an explication of the Standard Model of particle physics for anyone interested enough to make a go of it. The students that gave me the first encouraging feedback on the early chapter drafts have long since moved on, pursuing advanced degrees, and I hope raising the intellectual Cain that we tried to train them to. For my part, when I think of all the late-night television I missed it almost makes me want to cry. But I guess that's all water under the bridge now.

In any regard, one doesn't emerge from a process such as this without favors to return, and it seems to be standard practice to put a few hasty acknowledgments at the front of the book and hope they stand as fair recompense.

First and foremost, I thank Trevor, fellow physicist and editor-in-chief at the Hopkins Press, whom I now consider, along with my seventh-grade and college-freshman English teachers, to be one of the three most effective writing instructors I have had. I also need to thank Trevor for letting me write the book the way I wanted to; for his wealth of clear and well-motivated suggestions, and for not being heavy-handed about them. In this vein, I also need to thank my astronomy colleague Stephen Thorsett for pointing me toward Trevor and the Hopkins Press.

I need to thank my good friend Bill Rowe, whose talent for illustration so ideally complements the content and tone of the book. Perhaps Bill and I will find another excuse to collaborate at some point in the future.

Debts of gratitude are owed to my physics department colleague Michael Riordan, himself a successful writer, and John Wilkes of our renowned science communication program. Their advice was invaluable as I negotiated my way through the unfamiliar world of publishers and literary agents.

And then there's the small army of theoreticians on whom I relied to get my thinking straight for me. Each of the following took time out to read a chapter of the book, confirming my conceptions and politely disabusing me of my misconceptions, playing out my scheme to avoid exposing the large and embarrassing gaps that tend to mottle experimentalists' understanding of their own work. From the Stanford Linear Accelerator Center: Tom Rizzo and Lance Dixon (JoAnne, you had your chance!). From the UCSC physics department: Michael Dine, Howie Haber, and Tom Banks. Thanks also to UCSC's Joel Primack for his bibliography of accessible books on cosmology. And I need to throw in another scientist: my mother (Margot Schumm, retired chemistry professor at Montgomery College in Montgomery County, Maryland), who gave me some particularly helpful comments on chapter 3, as well as one-half of the gift of life (for the other half, I need to thank my father, Richard Schumm, also a retired chemist).

And lastly, I need to thank you, the reader, for choosing to take this book on, rather than joining me in my quest to update my background in contemporary television programing. From my perspective, your abstinence will be well rewarded, for you are on the verge of a noble journey. Like Ulysses' slog through the perilous adventures of the Odyssey, the going may not be easy, but persistence will be rewarded. Honestly, I wish you the very best. Bon voyage!

Deep Down Things

1

Introduction

Within the past century, our understanding of the natural world has taken such a vast step forward that, looking back, the state of our scientific knowledge one hundred years ago seems almost laughably naive. Today, as we probe deeper and deeper into the workings of the universe, the responsibility for digging out the most fundamental and overarching of its operating principles has fallen, by and large, to a group of scientists known collectively as particle physicists. This is a book about the Standard Model of particle physics—the view of the natural world put forth by this community after a remarkable period of development stretching back fifty years or so.

As inhabitants of an astonishingly fertile island in the otherwise inhospitable reaches of interstellar space, we have developed a common-sense view of our world that is shaped by the microcosm in which we live. The laws of this world are known to everyone: find shelter from the elements; develop a livelihood that will enable easy access to food, to a means of transportation, and to whatever creature comforts might be available; and conduct life in a way that will allow others to do the same. These are the sorts of things that we need to know to make do in our corner of the universe.

Over the last few centuries, though, we have invented tools that allow us to extend our senses into previously unimaginable realms, and as a result, we have been forced to reformulate our conception of the rules that govern the physical world. The deeper we look, the more we find that our refined descriptions diverge from our original intuitive notions of natural law, and the more parochial those original notions seem.

The human body, we find, is not really a single object unto itself but rather an uncountable collection of individual and specialized cells, each almost with a life of its own, cooperating in comprehensible ways to enable the bodily functions that sustain life. Looking more deeply, we find that cells are composed of molecules, and those molecules are composed of atoms. As we attempt to understand and codify the rules of existence at this level, we enter the realm of quantum mechanics, with its jarring metaphysical implications.

Probing yet more deeply into the nugget of the atomic nucleus or to the core of the quantum-mechanical muck surrounding the electron, we enter the realm of particle physics. Particle physics is the study of the "within that lies within": of rules of order that take us into the world of the abstract mathematician, beyond the comparatively pedestrian realm of the quantum physics of atoms and molecules. Particle physicists attempt to understand the workings of the ultrasmall and ultraenergetic (a connection we'll come to understand in due time); within this realm lie many of the clues that are necessary to unravel the mystery of the nature and origin of the universe. What these clues suggest about that nature is, as we'll see, rather eye-opening.

As we undertake this fundamental exploration of the universe, we begin to discover that the natural world is a place of striking physical simplicity. This notion enters our blood as a philosophy and drives us toward the tantalizing goal of a precise formulation of the fundamental laws of nature in terms of a single, all-encompassing principle. At the same time, though, we begin to grasp the sophisticated mathematical underpinnings of natural order and how this mathematical structure acts to set the stage for the development of life. One of the most gratifying turns in the development of particle physics is the extraction of the abstract mathematician from the ivory tower of pure thought into the workaday world of physical science.

Einstein held that any physical theory worthy of respect must be explicable to any clear-thinking person. This book represents my attempt to elucidate the currently accepted theory of particle physics—the paradigm of the Standard Model—for the interested public. It is not, first and foremost, a story about the history of particle physics or of the lives of its protagonists. Nor is it a book of anecdotes about the culture and society of particle physicists. It *is* a book that presents, at a level well beyond the superficial, the conceptual ideas that underlie those physicists' view of the world.

To be deeply interested, however, does not mean to be steeped in the formal content, mathematically or scientifically, of physics. In particular, I presume little in the way of mathematical background—just some very basic notions from algebra and the concept of orders of magnitude.

It is impossible, though, to elicit the central notions of the theory of particle physics without an involved discussion of the mathematics that underlies it. The beautiful connection between the worlds of the mathematician and the physical scientist is one of the most interesting threads we'll follow and one for which there is precious little material available to nonprofessionals. That the abstractions of higher mathematics bear some relation to the physical world—in fact, seem to lie at the very heart of its order and operation—is one of the most profound revelations of the modern era.

The mathematics that we'll discuss, though, is not the calculation-mired pursuit that confronts one in introductory college-level courses but rather that of the abstract mathematician whose tools are more those of logic and generalization than the application of endless pages of arithmetic operation. It's a different sort of mathematics than most of us are used to—an evolved discipline that bears little resemblance to the practical, everyday manipulations from which it sprang. More so than numbers and equations, this mathematics is an exploration of structure and, in particular, the patterns exhibited by that structure. And it is through these patterns that, in the latter half of the twentieth century, the deep connection between abstract mathematics and the basic organizing principles of the physical universe were first recognized.

Before the mathematics, though, we'll need to ground ourselves in the backdrop of physics from which the Standard Model sprang in the fertile period of the 1960s. Illustrating the limits of the adjective *modern* in labeling any human development, this grounding begins with the quaint old "modern" physics of the early part of the twentieth century—Einstein's special relativity and conventional quantum mechanics. A discussion of the contemporary framework of causation that evolved from this base—the "quantum field theory" of Richard Feynman and others—then describes how the nineteenth-century notion of *action-at-a-distance* has been supplanted by the idea that physical forces are conveyed by the exchange of subatomic particles. Following this, we learn about the array of known subatomic particles, making note of the repeating patterns into which these particles fall, thus setting the stage for the ensuing discussion of abstract mathematics.

The Standard Model is the focus of the last three chapters of the book, with the introduction of gauge theory, the Higgs field, and spontaneously hidden symmetry. At last, we will be able to appreciate the development of this theory as the coalescence of the millennia-long quest to understand the fundamental nature of the universe into a paradigm whose success represents one of the most remarkable triumphs in the history of human thought. In the final chapter, we'll see why the years ahead hold great promise for further deepening our understanding of the natural world.

I hope that you enjoy the journey. Although the experimental and theoretical threads leading into the development of the Standard Model have had a telling effect on the way that we live our day-to-day lives, it is really the satisfaction of innate human curiosity and the attendant appreciation and awe of the workings of nature that stand as the historical and continuing motivation for particle physics research. While numerous practical spin-offs have arisen from this research, to date its funding decisions have been solely based on whether the proposed program stands to clarify and deepen our understanding of the order of the natural world. And while the program has been profoundly successful, garnering no fewer than twenty Nobel Prizes, our success at returning the fruits of particle physics research to its benefactors—the citizens of the many nations of the world that support it through their governments' science programs—has been more limited. I hope this book acts in some measure to redress this deficiency.

Viewed from this perspective, the list of patrons is lengthy. Particle physics research is a truly international and cooperative pursuit. The three traditional centers of the development of particle physics—Western Europe, Eastern Asia, and North America—have been roughly equal partners over the last fifty or so years. Increasingly, more and more nations and regions are being drawn into the effort, and all the continents (including Antarctica) are now represented. It is a pursuit, like many others in science, for which national allegiances play a relatively minor role, and collaborative professional and personal relationships spring forth quite independent of cultural background.

The support for particle physics research provided by this worldwide community does need to be acknowledged because it is not small. The annual global outlay for particle physics research is several billion U.S. dollars per year; for example, the United States' contribution to this total is about $800 million per year.

As we move into discussing the conceptual basis of particle physics, keep in mind that what follows is not fantasy, nor, for the most part, speculation. Instead, it is, to the best of our knowledge (and backed up, as we'll see, by impressive experimental verification), scientific fact. What it tells us about the world in which we live is an accurate and faithful representation of physical reality. The theory works, and as exacting tests performed in the decade of the 1990s demonstrated, it works remarkably well. It is as much of an absolute as any paradigm of the workings of nature could ever be.

Our world is an interesting place to live (in fact, is a place in which life itself is possible) because the world is rife with *causation*—the ability of things to influence other things. And the way that physical objects influence one another, whether as contiguous neurons in the taxed brain of a struggling physics student or as celestial bodies gliding through the depths of outer space, is through the exertion of forces. So it is here that we will begin, with a delineation and description of the four forces of nature.

In fact, there are neither four of them, nor is the notion of "force" quite the appropriate term for discussing the phenomenon of causation. But we do need to start somewhere, even if it's in the presentation of a point of view that we'll eventually need to refine substantially. Welcome to the world of particle physics!

2

The True Movers & Shakers

~~~~~~~~~~~~~~~~~~

The Forces of Nature

The world would be a dull place if the objects within it couldn't influence one another. Of what use is a can opener if it cannot open a can? Everything of interest to us in our everyday lives—conveying and receiving information, moving about through our surroundings, all of the biological functions that make life itself possible—is predicated on the ability of systems of matter to exert an influence on one another. The way things influence one another is by exerting *forces*.

It is often stated that there are four forces of nature—four fundamental and distinct ways in which objects can exert their influence on the world around them. These forces are gravity, electromagnetism, and the strong and weak nuclear forces. But how many forces are there really?

This remains an open question, providing a significant motivation for continued research into particle physics. One can ask how many of the four known forces are truly independent phenomena or merely different manifestations of a smaller number of underlying physical laws. Why should there be four separate laws of influence, each with its own peculiar set of rules, that somehow magically cooperate to underlie the myriad workings of the known universe? The fewer the laws, and the less arbitrary their rules, the easier it is to make sense of the whole picture.

Before the work of the Scottish physicist James Clerk Maxwell in the early 1860s, electricity and magnetism were thought to be separate forces. Maxwell developed a unified theory of electromagnetism, incorporating electricity and magnetism into a single theoretical framework and, as a dividend, explaining the phenomenon of light as wavelike disturbances in the electromagnetic force field. Even today, his work stands as one of the great triumphs of physics.

As we'll see, the Standard Model of particle physics reveals that two of the four forces of nature (electromagnetism and the weak nuclear force) are, similarly, manifestations of a single, unified "electroweak" force. So you might say that now we are down to three. However most particle physicists would consider it a bit of a defeat if, at some point, we don't reduce the number to just one.

Einstein spent a lot of effort trying to do just this—to merge the known forces into a single, "grand-unified theory of everything." Given the theoretical framework within which he worked in the middle of the twentieth century, this task lay beyond even his formidable intellectual powers. Today, we have much better tools at hand, and a large number of our most advanced theorists are still attempting to achieve this lofty goal. Success has so far eluded even these modern crusaders, although there is reason to be cautiously optimistic (note 2.1).

So, it's not clear that there really are four, or even three, separate forces of nature. However, it's convenient to begin our discussion of the nature of physical forces in terms of the traditional four forces.

There are a number of attributes that can be used to distinguish the fundamental forces among one another. Below we'll characterize each of the four forces in terms of three properties: the overall strength of the force; the range of the force; and the characteristic of any particular object, known as its "charge," that determines how severely the object is influenced by the given force.

The Electromagnetic Force

Atoms are composed of negatively charged electrons orbiting positively charged atomic nuclei. The negative electric charge of the electrons exactly cancels the positive electric charge of the nucleus so that, viewed from the outside, the atom has no net electric charge; we say that the atom as a whole is "electrically neutral." Inside the atom, though, the charge is separated between the stationary nucleus and orbiting electrons (see fig. 2.1). "Opposites attract," so the opposite charges of the orbiting electrons and stationary nucleus keep the whole system bound together in perpetuity. It is this

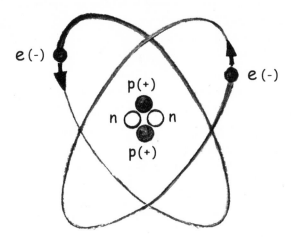

Fig. 2.1. The helium atom. The negative charges of the two electrons are of exactly the same magnitude as the positive charges of the two protons, so that the atom as a whole is electrically neutral (no net electric charge).

attractive electromagnetic force that keeps the orbiting electrons bound into their tight orbit around the nucleus.

If the weather is not too damp, and you have washed your hair within the last month or so, you can do an experiment that demonstrates this attractive force rather strikingly. Go to your bathroom sink, and turn the water on just enough to get a thin, steady stream rather than a rapid succession of drops. Comb your hair vigorously with a cheap plastic comb. When you place the comb close to the stream of water (without quite touching it) you should see a noticeable deflection of the stream. If conditions are right, this deflection can exceed 45 degrees.

The phenomenon you will observe is the electromagnetic attraction between the electrons you rubbed off your hair onto the previously uncharged comb and the nuclei of the water molecules. By the way, this works only because water is "polarizable," which means that the electrons in the water molecules like to move out of the way of the electrons on the comb, leaving the positively charged nuclei contained within the water molecules to be attracted by the electrons on the comb. The ready polarizability of water is directly related to it being a very good solvent, that is, ideal for washing, which you already knew, since you apparently washed your hair sometime in the last month. Let's consider this electromagnetic interaction between positively charged nuclei and negatively charged electrons.

First, the interaction is *strong*. Even after rubbing the comb through your hair, the comb is largely composed of neutral atoms. To a small fraction of those neutral atoms, an electron that formerly belonged to your hair is clinging. These few electrons, though, are able to exert a substantial force on the stream of water as it exits the faucet.

Second, the interaction is *long ranged*. The distance between the electrically charged comb and the stream of water may not seem far to a human, but it is huge compared with the width of an atom (one ten-millionth of a millimeter, with a millimeter being about one twenty-fifth of an inch), or the width of a nucleus (one-millionth of a millionth of a millimeter). Formally, the electromagnetic force is known as a $1/r^2$, or inverse square force, which means that the degree to which two charged objects influence each other decreases with the square of their separation—twice the separation, one quarter the force, and so on. This may not sound like a characteristic of a long-ranged force, but it's enough to allow electromagnetic forces to be explored with everyday apparatus, such as that found in the lab of an introductory physics class. As we'll see, not all of the forces are so amenable.

Third, the influence exerted by the comb on the stream of water depends on the electric charge of the comb—in other words, how many excess electrons happened to accumulate on the comb when you ran it through your hair. In general, the electromagnetic force is strong, but the precise strength of the electromagnetic influence between two objects depends on how much electric charge is accumulated on the objects. Note that electrons are not "electric charge"—they do *possess* electric charge, but then, so do the protons within atomic nuclei. As it turns out, most (but not all) fundamental particles possess electric charge—electric charge is an *attribute* of the particle, but particles are not electric charges in and of themselves.

Electrons and protons have a specific amount of electric charge associated with them ($-e$ and $+e$, respectively, where $e = 1.6021773 \times 10^{-19}$ coulombs, for what it's worth) (note 2.2). You can charge an object with any given multiple of this value ($+2e$, $-56e$, $12934582e$, etc.). You cannot charge it with a fraction of this value; for example, $12934582.4e$ is not possible. No exception to this rule has ever been observed (later on, when quarks are introduced, this statement will be qualified somewhat, but it will remain true in essence).

Gravity

Try to jump as high as you can against the pull of gravity. No matter how hard you try, you always come back down—and rather quickly. You might think that the force of gravity is fairly strong.

If you were one of the lucky few who has had a chance to do this experiment on the moon, you might arrive at a different conclusion. If you conducted this experiment on the surface of a small celestial body, such as an asteroid 1 kilometer in diameter, you would find that by jumping up you would entirely escape the gravitational pull of the asteroid and float off into space. The "escape velocity" at the surface of such an object is about 1 meter per second, or 2 miles per hour. This is how fast you would have to be traveling just after your jump to escape the pull of gravity and head off irretrievably into outer space.

Just as the electromagnetic force between two objects depends on the total amount of electric charge contained by each, the gravitational force between two objects depends on something related to the overall *size* of the two objects. In the thought experiment above, one object's size (ourselves) was fixed, and we saw that the strength of the gravitational force between ourselves and the given planetary body depended on the size of that body. If you performed the experiment carefully with a large number of objects, you would discover that the *mass* of the two objects establishes the strength of the gravitational pull between them (for those not familiar with the difference between "mass" and "weight," it is acceptable for our purposes to use them interchangeably). Thus, the "charge" associated with the gravitational force, completely equivalent to the notion of the electric charge of an object for the electromagnetic force, is just the *mass* of the object (note 2.3).

The gravitational force is *not* strong; in fact, it is quite weak. The gravitational force we experience on the surface of the earth feels fairly strong, but then think of the magnitude of the gravitational charge of the object that exerts the force on us; it's the mass of the entire earth! If we were to remove the earth and replace it with something of a more everyday size, such as another person, the gravitational pull between us and the other person would be unnoticeable—among the sensations associated with someone moving close to you, gravitational attraction is absent. Put another way, the gravitational attraction between two electrons is roughly 10^{42} times weaker than the electromagnetic repulsion between them — a difference of 42 powers of ten (or orders of magnitude). Gravity is intrinsically weak.

Like electromagnetism, the gravitational force is inarguably long ranged. The gravitational force binds the planets into orbit around the sun, clusters the matter of the universe into galaxies, and even generates the overall structure of space and time in the cosmos.

The range of the gravitational force is, in fact, identical to that of the electromagnetic force. They both fall with the inverse square of the separation between the correspondingly charged objects. Why then are the large-scale properties of the universe established by gravitational effects rather than by the effects of the much stronger electromagnetic force? The answer is due to a quirk of nature that, for the time being, is beyond us to explain.

Electric charge comes in two "signs" — positive and negative. Furthermore, the balance between positive (nuclear) and negative (electronic) charges seems to be exact, so that the *net* electric charge of the universe is precisely zero. Thus, strong electromagnetic forces are, by and large, limited to locales where for one reason or another excess positive or negative charge has temporarily accumulated, such as on the surface of the comb after passing through your hair. However, gravitational charge comes in only one sign — there does not appear to be any negative mass in the universe. No objects fall upward when released in the earth's gravity. Thus, even though the gravitational force exerted by any small chunk of matter is quite small, every little chunk of matter in the universe makes a small positive contribution. The resulting effects, added up over all of the matter in the universe, are colossal.

The Strong Nuclear Force

Beyond the realm of what we can experience and explore with our own senses lies the workings of the two nuclear forces. Not only are these two forces inaccessible to our tactile senses, but even their most qualitative description requires some surprising and counterintuitive notions.

The strong force is the force responsible for the binding of "nucleons" — protons and neutrons — into atomic nuclei. Since World War II and the development of nuclear warheads and nuclear energy sources, most people have been aware of the awesome power that is associated with the aptly named *strong force*. Surprisingly, the protons and neutrons that are bound into nuclei are neutral with respect to the charge of the strong force.

We now know that neutrons and protons are composed of more funda-mental objects known as *quarks*—more specifically, "up-" and "down-"type quarks. Quarks are the indivisible carriers of the strong force charge in the same sense that electrons are indivisible carriers of the electromagnetic charge (in fact, quarks are also indivisible carriers of electric charge; the electric charge of the proton is simply the sum of the individual electric charges of its constituent quarks).

An atom is a collection of negatively charged electrons bound to the pos-itively charged nucleus by the electromagnetic force. Atoms are electrically neutral, containing no net electric charge because the negative charges of the electrons exactly cancel out the positive charge of the nucleus. Thus for the most part, two neutral atoms will not interact with each other electro-magnetically: The electric charge of each is zero. However, when two atoms get very close to each other, the fact that the negative charge carried by the electrons is separate and distinct from the positive charge carried by the nu-cleus allows a weak residual force to develop that can bind the atoms into molecules. This residual binding force, while entirely due to the electro-magnetic interaction, is much weaker than the force that would arise if the protons of one atom were attracted by the electrons of the other, without the repelling effects of the other's electron cloud.

An analogy, very apt as we'll see, for the consideration of the strong nu-clear force, is provided by an image of a white tennis shirt on the screen of a color television. To a viewer sitting on a couch, relatively far away from the screen, the white image has no effect on (causes no interaction with) the color-sensing receptors in the eye. When the eye of the viewer gets within a millimeter or so of the image on the screen, however, the image begins to resolve itself into the profusion of minuscule, colored dots from which it is composed. Resolved in this way, these dots interact with the color sensors of the eye but in a much weaker way than they would in the case of, say, a red image, for which every dot would be red. There is an interaction between the picture and the color receptors of the eye, but because of the overall color neutrality of the picture, the interaction is short ranged (you need to get very close to see the dots) and relatively weak.

The nucleons (protons and neutrons) in the nucleus are essentially strong-force atoms—quarks bound together into objects that have no net strong-force charge, just as conventional atoms have no net electromagnetic charge. The force that binds the nucleons together into nuclei is a residual, molecular-type interaction between the nucleons like that for the molecular electromagnetic force, which is uncharacteristically weak relative to the actual force that underlies it.

Yet, it is this relatively weak residual force between the strong-force-neutral nucleons in the nucleus that provides the tremendous power released in nuclear explosions and power generation. However much respect we hold for this conventional (residual) nuclear force between the nucleons in the nucleus, it is only the tip of the iceberg with respect to the strength of the true, underlying strong force. Were we able to harness *this* force, modern nuclear power would seem feeble in comparison. Nuclear power, however, relies on the existence of substances such as uranium-235 and plutonium, whose nucleons are configured in ways that render the nucleus only marginally stable. These substances can be induced to break up (fission) with relatively little prodding, leading to the possibility of a chain reaction. No similar states of matter seem to exist for which the quarks themselves beg to be prodded into a more stable configuration, with the corresponding release of a truly staggering amount of energy. This is probably a good thing.

So the strong force is indeed strong. It is also *in principle* long ranged, like the electromagnetic and gravitational forces. All matter that is measurably stable, however, is assembled from quarks that are, as we've seen for the case of nucleons, strong-force-charge neutral when combined. Thus, for the reasons just discussed, the strong force is *in practice* short ranged. The bundling of quarks into strong-force-neutral protons and neutrons masks the long-range nature of the strong interaction just as the bundling of electrons and nuclei into electrically neutral atoms masks the long-range nature of the electromagnetic force. Were free quarks ever observed, however, the strong force between them would be measured as long ranged—fundamentally, the strong force is long ranged.

As intimated by our television-picture analogy, the charge associated with the strong force is known as *color charge*. Unlike electromagnetism and gravitation for which there are only one type of charge (electric charge and mass, respectively), there are three *separate* types of color charge, arbitrarily referred to as *red*, *green*, and *blue* charges. For example, a red quark (note 2.4) has a red charge of one and green and blue charges of zero, and so forth. By contrast, an electron always has an electric charge of minus one (in units of the proton charge).

While all quarks are, as far as we know, fundamental particles, not all fundamental particles are quarks. A separate category of fundamental particles known as *leptons*, of which the electron is the best-known example, are, unlike quarks, neutral with respect to the color charge of the strong nuclear force. Being indivisible fundamental particles unto themselves, leptons are not composed of color-force-neutral combinations of quarks. Instead, they are color neutral simply because nature, for reasons not yet understood, deemed that they should possess none of the color charge that the strong nuclear force concerns itself with.

The Weak Nuclear Force

At last we come to the most obscure of the forces of nature—the weak nuclear force. The most widely recognized consequence of the weak nuclear force is the process of nuclear beta decay in which unstable atomic nuclei transform themselves via the emission of an electron (and a "neutrino"). For the most part, the weak force is shrouded in obscurity: It doesn't bind atoms or their nuclei together, and it isn't responsible for the gyration of the planets around the sun or any other aspect of our everyday experience.

Paradoxically, to the particle physicist, the weak force is the most interesting of all. Outside of the fact that it is weak (which, as we'll see, is really an illusion), its properties are the richest and most complex of the known forces. The weak force is really the maverick—the rebel—among the forces. A fair number of physical laws held sacrosanct by the other three forces are violated, some of them grossly, by this rogue force. In fact, the violation of a number of these laws is essential for the existence of life.

For example, consider the symmetry between matter and antimatter. Each fundamental particle has an antimatter counterpart that is like the matter particle in every way except that the charges associated with each force (other than that of the gravitational force—there appears to be no such thing as negative mass) are of opposite sign. For example, the antimatter counterpart of the negatively charged electron is the positively charged *positron* (discovered by Carl Anderson of the California Institute of Technology in 1933) and the antimatter counterpart of the positively charged proton is the negatively charged antiproton (discovered by Owen Chamberlain and Emilio Segrè at the University of California at Berkeley in 1955); put these two antimatter constituents together and you get antihydrogen (produced by humans for the first time at the pan–European CERN laboratory in 1995). The antimatter counterpart of a given type of quark is simply the corresponding antiquark. Protons and neutrons are composed of up- and down-type quarks; antiprotons and antineutrons are composed of up- and down-type antiquarks.

Any two objects that are matter/antimatter counterparts of each other have the curious property that when they come in contact with each other they annihilate and convert their combined mass into an equivalent amount of energy, admitted, as we'll see in the next chapter, by Einstein's famous relation, $E = mc^2$ (the released energy is just the annihilated mass times the square of the speed of light). Thus, a universe composed of equal parts matter and antimatter would be hostile and unstable, not the sort of place that chunks of matter the size of planets could exist in relative peace and stability for billions of years.

Luckily for all of us, there *is not* a symmetry between matter and antimatter. There's a tremendous amount of matter in the universe, but naturally occurring antimatter particles are rare (note 2.5). Somehow, matter was preferred over antimatter as the universe evolved. Yet, the three more familiar forces—the electromagnetic, gravitational, and strong nuclear forces—are all unprejudiced in their effects on matter and antimatter. Whatever these forces do unto matter, they do equally unto antimatter. If one of these three forces is responsible, say, for the decay of some unstable particle, then the rate at which that particle decays will be identical for both the particle and its antiparticle partner. All other aspects of the decay—the nature of the decay products and the distribution among these products of the energy left over after the decay—will be identical for particle and antiparticle. Furthermore, in any reaction caused by these three forces, if new particles are produced, they will always be produced with equal amounts and types of matter and antimatter. How can these forces thus account for a universe in which matter dominates so convincingly over antimatter? They cannot.

At Long Island's Brookhaven National Laboratory in 1964, James Cronin and Val Fitch (then at Princeton University) discovered that, for electrically neutral kaons (a relatively common type of subnuclear particle), whose decays were known to be due to the weak force, there is a subtle difference in the decays of neutral kaons and neutral antikaons. This profound discovery revealed a mechanism by which the matter/antimatter *asymmetry*, a critical ingredient for the development of life, could have arisen in the primordial universe. It is only this fourth force—the feeble but irreverent *weak* force—that exhibits this surprising but essential property. For this discovery, Cronin and Fitch were awarded the 1980 Nobel Prize in Physics.

To incorporate and to understand this property within a consistent theoretical framework is one of the great challenges and opportunities facing today's particle physicists. Recent advances in particle accelerators and detectors allow for the development of exacting experiments that explore this phenomenon to a degree unimaginable back in the time of its discovery. Two such modern experiments (the so-called B Meson Factory experiments) are now underway—one at the Stanford Linear Accelerator Center (SLAC) at Stanford University in California and the other in Japan at the KEK National Laboratory.

There is but one type of weak charge, which can be both positive and negative, known as *weak isospin.* As for the other types of charge, weak isospin charge is a property of any given object; electrons, for instance, have a weak isospin of $-\frac{1}{2}$, while protons have a weak isospin of $+\frac{1}{2}$.

The weak nuclear force is, well, weak. Neutrinos, ghostlike fundamental particles that interact only via the weak force, can pass entirely through the body of the earth without being deflected. Unless you have tremendously sensitive apparatus and a good deal of patience, you can only sense the workings of the weak force if, for some reason, none of the other forces are at play in the process under study. However, since there are a number of laws respected by the other three forces that are violated by the weak force, the study of any process violating one of these laws (such as the symmetry between matter and antimatter) is a direct study of the weak force. Once you know where to look, it's actually not that hard to conduct sensitive studies of the weak force. Many have been done, and we now know a great deal about the workings of this force.

Finally, the weak force is short ranged. Two objects interacting via the weak nuclear force need to be within 10^{-18} meters of each other to do so. On this scale, even the inside of the diminutive proton (radius 10^{-15} meters) is a vast, cavernous expanse. The weak nuclear force is truly a subnuclear force.

Final Word on the Four Forces

The cornerstone of the Standard Model of particle physics is the *unification*—the melding into an overarching theory governed by a single, common physical law—of the electromagnetic and weak nuclear forces. It's not surprising that it took physicists until relatively recently (the late 1960s) to achieve this, for on the surface this is an absurd proposition. The electromagnetic and weak forces could hardly be more different. One is responsible for almost everything we experience daily; the other is responsible for *nothing* we experience daily. One is relatively strong, the other seems quite weak. The charges associated with each force are completely independent; one set of fundamental subnuclear particles, the neutrinos, are electrically neutral (have electric charge zero) but do possess weak isospin. The electromagnetic interaction is long ranged, which leads to effects that can be easily observed by the unaided senses (recall the experiment at the beginning of the chapter), while the other operates only over inconceivably short distances. Showing that these are merely two different facets of the same underlying physical law is truly a daunting task.

3

The Great Reawakening

～～～⌇⌇⌇⌇⌇⌇⌇⌇～～～

The Modern Physics Revolution

In his 1894 address at the dedication of the Ryerson Physical Laboratory at the University of Chicago, the American physicist Albert Abraham Michelson set forth what has become one of the most memorable forecasts in the history of physics: "The most important fundamental laws and facts of physical science have all been discovered . . . Our future discoveries must be looked for in the sixth place of decimals."

The notoriety of this premature eulogy lies in the juxtaposition of its broad acceptance in the waning years of the nineteenth century against the staggering developments of the early twentieth century, collectively known as the "modern physics revolution." Within ten years of Michelson's pronouncement, a Swiss patent clerk by the name of Albert Einstein had completely reshaped our notions of space and time; it is ironic that Michelson's own work (for which he was awarded the 1907 Nobel Prize) provided the initial experimental basis for Einstein's theory of relativity. Within thirty years, the development of quantum theory had forced humankind to discard the Newtonian notion of scientific determinism.

The work of these early "modern" physicists so thoroughly awoke physics from its complacent torpor that the twentieth century stands as the greatest century the field has ever known and perhaps ever will (another prognostication!). The beacon of physics was relit so brightly that it illuminated reaches of scientific exploration that lay far outside the imagination of the classical physicists of the nineteenth century.

Einstein's Relativity

The first of Einstein's two great theories of the nature of space and time—special relativity—is particularly beautiful. It rests on two deceptively simple postulates. The first is that the speed of a beam of light as it moves through a vacuum will always be measured at exactly 299,792,458 meters per second, regardless of how much you might speed up in an attempt to catch up with it (this is what the meticulous work of Michelson demonstrated, to no one's greater surprise than his own). The second postulate is that the laws of physics are the same for all observers, independent of how any given observer may be moving through space. Another way that this second postulate can be stated is that there is no experiment one can do that will distinguish one observer's frame of reference as somehow being "preferred" (closer to being at rest with respect to the universe as a whole) over that of any other observer.

From these two seemingly benign statements flow a host of apparently nonsensical, yet experimentally verifiable, conclusions that completely rearrange one's sense of space and time—the very fabric into which the workings of the universe are stitched. Rulers shorten, time slows, the boundary between the notions of space and time erodes, matter and energy become interchangeable facets of a single physical quantity, "mass-energy," and so on.

To a particle physicist, special relativity is second nature. It is simply a tool of the trade. It is as familiar to him or her as, say, the federal tax code is to an accountant. Fortunately, relativity is much less arbitrary and substantially easier to fathom than the tax code.

From our discussion of special relativity, it will be important to remember the relationship between mass and energy. This relationship, as we will see, is as follows: Mass and energy are essentially the same thing.

Before the work of Einstein, the energy E_K of an object not under the influence of any external force was thought to be one-half the product of the particle's mass, m, and the square of its speed, v:

$$E_K = \frac{1}{2}mv^2.$$

The subscript K, for "kinetic," indicates that this energy arises if and only if the object is in motion; that is, if and only if the speed v is not zero.

This quantity plays a central role in our description of the physical world because energy is *conserved* (an extremely important and well-founded experimental fact). No matter what happens, there is always the same amount of energy around before something happens as there is afterward. If an object moving with speed v collides with some other objects, whatever kinetic energy our original object gains or loses (when slowed or sped up by the collision) will be *lost or gained*, in some form or other, by the system of objects with which it collides. Energy conservation will play a critical role in a number of places later on in this book.

Einstein modified this relation considerably, and in so doing, he penned the most celebrated relation in all of physics (but which, by the way, is *not* known as "Einstein's equation" to card-carrying physicists):

$$E_m = mc^2,$$

where c is the well-known, ever-unchanging speed of light—the velocity of the propagation of James Clerk Maxwell's wavelike electromagnetic force-field disturbances (note 3.1).

Ignoring the factor c^2 for the moment, we see that this equation states unambiguously that mass is the same thing as energy. The factor of c^2 indicates just how much energy is associated with each kilogram of mass m. The fact that this proportionality constant, the speed of light, is so large (299,792,458 meters per second) means that the amount of energy associated with everyday amounts of mass is shockingly large. For example, a single gram (one-thirtieth of an ounce) of mass, if somehow converted entirely into wall-plug-style electrical energy, would produce the equivalent of an entire day's output from a large (gigawatt) modern-day power plant. The idea that an *inanimate* object possesses energy—and a lot of it—is counterintuitive. But such counterintuitiveness is the nature and, in large measure, the beauty of Einstein's theory of relativity.

Thus, in addition to the "kinetic" energy associated with motion, we must also consider the energy E_m associated with the *mass* of an object. This "mass-energy" contributes to the object's overall energy regardless of whether it is in motion. Simply, the object's total energy is just the sum of its mass-energy and kinetic energy.

In a collision, both of these contributions to the colliding objects' energies must be taken into account when doing the before and after energy-balance accounting required by energy conservation. If conditions are right—say, if a particle is colliding with another particle that just happens to be its antimatter counterpart, leading to the mutual annihilation of both particles—the result can be a complete conversion of mass-energy into kinetic energy. Alternatively, and again if the conditions are right, the conversion of kinetic energy into mass-energy can also be complete.

In this latter fashion, relatively light and relatively ordinary particles (such as electrons and antielectrons, or positrons) can be hurled against each other with great opposing kinetic energy, only to have that kinetic energy be converted into the mass-energy of an exotic, and heavy, new particle—the kind of new particle that spirits theoreticians to their blackboards and the study of which leads to great advances in scientific understanding. This approach has been a central theme in much of experimental particle physics throughout its fifty-or-so-year history.

This connection between mass and energy has become so familiar to particle physicists that they have taken on the habit of quoting the masses of their favorite particles exclusively in terms of the associated mass-energy. In fact, it's even worse than that. The unit of energy used by particle physicists is not the familiar joule or erg of introductory physics but rather the electron-volt, or eV, which is a unit wholly inspired by the particle physicist's favorite toy—the particle accelerator.

Think of an accelerator as an immense and exceedingly expensive TV set. Electrons from a source inside the set are accelerated toward a target—the screen—by the electric force. So, forgetting about the screen for the moment, a TV set is just an electron accelerator.

The magnitude of the accelerating electric force, times the distance over which the acceleration occurs, is known as the potential difference, or voltage. An electron accelerating through a given voltage obtains a well-defined amount of kinetic energy (energy of motion); if the voltage is exactly 1 volt the electron obtains precisely 1 eV of kinetic energy during the acceleration (for comparison, wall-plug voltage is about 100 volts).

The essential point regarding the electron-volt unit of mass-energy is that an object with a mass-energy of 1 eV, and without any kinetic energy to start with, would release precisely this much kinetic energy if it were annihilated by its antimatter counterpart.

The electron-volt is a *general* expression for a particular amount of energy. The energy of any object can be expressed in electron-volts, whether that energy is possessed by an electron in the TV set, the tennis ball that rockets back and forth over the net during the match you are watching on that TV set, or the caloric content of the chips and soda that you consume as you watch the match.

In these terms, the electron, which is the lightest particle whose mass has been definitively measured, has a mass of about 511,000 eV. Again, this means that if you somehow convert the electron's mass-energy to kinetic energy, the amount of kinetic energy released will be 511,000 eV—the amount of energy an electron would gain after accelerating through a hefty 511,000 volts of electrical potential. Similarly, the mass-energy of a proton is about 938,000,000 eV. The heaviest known fundamental particle, the top quark, has a mass-energy of about 175,000,000,000 eV (note 3.2). This also is about the reach, or kinetic energy per beam particle, of today's most energetic particle accelerators. In contrast, the maximum energy of an electron in a TV set is a few thousand electron-volts, so if we think of particle accelerators as TV sets whose images somehow reflect the fundamental workings of the universe, then these are big TV sets indeed.

Although it's not directly related to the subject matter of this book, we can't really take leave of our discussion of Einstein's relativity without at least a mention of the general theory of relativity. The formulation of this theory took ten years of Einstein's professional life—from 1905 to 1915—which he later admitted to being a difficult and uncertain period. In the end, it was well worth the effort because what came out was a theory whose reshaping of the special-relativistic notion of space and time was as radical and profound a departure from special relativity as the latter was from the classical, prerelativistic, common-sense notions of space and time. The universe of general relativity is one in which the pedestrian, three-dimensional world of our senses is twisted and distorted through higher dimensions, in ways not perceivable by our senses, just as the curvature of a flat piece of paper into a sphere would not be perceivable to a two-dimensional person living on the sheet of paper. The curvature of space-time is related to the distribution of mass-energy in the universe according to a single equation, properly known as "Einstein's equation." Just as for special relativity, the general theory is readily verifiable; it now serves as the basis of the field of cosmology—the theory of the origin and evolution of the universe.

Quantization: The Next Great Leap

The second component of the modern physics revolution (Einstein's relativity being the first) was the development of quantum mechanics. Although no single figure played as central a role in the development of quantum mechanics as Einstein played in the development of relativity, one can point to two physicists whose work and ideas propelled the rapid crystallization of the quantum hypothesis into a fully rigorous and quantitative theory. These two physicists, Erwin Schrödinger (of the Schrödinger equation) and Werner Heisenberg (of the Heisenberg uncertainty principle), were largely responsible for the synthesis of a number of disparate and somewhat qualitative notions into a concise and powerful theory during the few short years of the mid-1920s. To mention only these two does a great disservice to numerous others, Einstein among them, whose willingness to interpret certain physical phenomena in clever and radical ways laid the groundwork for the quantum renaissance of the 1920s.

If a gas composed of a vast number of similar but free atoms, such as that contained in the tube of a neon light, is excited, the gas will begin to glow. This glow is due to the emission of an immense number of brief flashes of light from the individual atoms in the gas. If you shine this light through a prism so that it is broken up into its constituent colors, you will not see the familiar "rainbow" spectrum that passes continuously from a rich purple through green and yellow to red. Instead, the light is composed of ten or twenty discrete colors, or "spectral lines"; the rest of the rainbow is missing. The colors and patterns of observed spectral lines are characteristic of the particular type of atom composing the gas—a principle that is often used to identify the composition of unknown gases.

The reason that a discrete, rather than continuous, spectrum is observed is that individual atoms in the gas cannot emit light of any color; they can only emit certain "permissible" colors. We say that the atomic system is *quantized,* in the sense that we can assign a number (a quantum number) to each of these well-defined colors. This number simply acts to delineate, or quantify which of the various possible colors we see when we pick out one particular spectral line emerging from the prism. This property—that physical systems can only exhibit behavior associated with their assumption of one of a limited number of discrete, permissible states—is the essence of the phenomenon of quantization.

Quantum mechanics states that all physical systems, not just atoms, are quantized. For the effects of this quantization to be noticeable, however, the system must be small. Even for large systems, such as a piece of iron heated to a white-hot glow, which when viewed through a prism appears to emit the full rainbow of possible colors, the possible states are still discrete. It's just that there are so many states that are so closely spaced that it looks like the permissible states of the system constitute a continuum of possibilities—a continuous rainbow of color—even though they really don't. For the excited gas of the neon light, however, each atom emits its light independently of all the other atoms in the neon plasma, so that system of interest is really just the single neon atom, which is small enough that quantum behavior is quite evident.

Quantum theory gives us an explicit notion of the meaning of the word "small," that is, of the physical scale at which quantum mechanical effects become appreciable. This number, known as Planck's constant (h), is, along with the speed of light (c), one of a handful of fundamental quantities that characterize the scale of natural laws. The value of Planck's constant can be looked up in any physics text: $h \approx 6.6 \times 10^{-34}$ joule-seconds, where the joule is an everyday amount of energy—specifically, the amount of energy required to lift 1 kilogram (2.2 pounds) about one-tenth of a meter (4 inches or so) (note 3.3). The minuteness of Planck's constant—about thirty-three orders of magnitude (factors of ten) less than one—tells us that the natural scale of quantum mechanics is small indeed (note 3.4).

Of Balls and Beams: Waves versus Particles

Even more than the phenomenon of quantization, quantum mechanics is concerned with the *wavelike* properties of matter. It was through quantization, and not wavelike behavior, that the physicists of the early twentieth century first began to unlock the secrets of the small-scale behavior of matter. Thus, the designation "quantum mechanics" is a bit of a historical accident; a name that more directly encapsulates the essence of the new physics' departure from classical notions would be "wave mechanics." Somehow, though, "quantum" sets a more revolutionary tone than "wave," so perhaps quantum mechanics is a better choice—at least from a rhetorical point of view.

Particles are hard and discrete, like miniature billiard balls. When two particles collide, they bounce off each other, heading off with directions and speeds different from those before the collision. Even more important, however, particles enjoy the following property: you can tell where they are. If you are asked about the location of a given particle, you can develop a procedure (such as taking a picture of it) to answer the question definitively. So, for a particle, the notion of position is a sensible one. At any given time, the particle is localized at a well-defined position—a position that, at least in principle, can be determined by experimentation.

Waves are everything that particles are not. Bob a stick steadily up and down in a pond and a wave emanates forth, eventually filling the pond with evenly spaced undulations traveling outward from the stick. Exactly where in the pond is the wave? It's not a sensible question to ask. The wave is not localized at a single position; it's everywhere in the pond at once.

In fact, the wave isn't really a "thing" at all, is it? Certainly, the wave carries with it some energy, which can be substantial, as anyone who has spent time at the beach knows. But the wave itself is nothing more than the molecules in the body of water moving up and down in an organized way. The particles (in this case water molecules) seem to be real objects, taking up space and so forth, but the wave is nothing more than a description of the way in which those particles convey energy—from the bobbing stick, the wind, or whatever.

Now, if two sticks are bobbed up and down at different points in the pond, the two waves that are created don't bounce off of each other, as particles would be expected to. Rather, they move through each other, *interfering* with each other as they do, but not changing their individual courses.

Picture two equal-sized waves that are passing through each other. If at some point both waves are peaking, then the peak is doubly high (constructive interference); if a peak from one wave meets a trough from the other, they cancel each other out (destructive interference) and the pond surface is at the same height as it was before either wave arrived. Interference is an essential wave property and is one of the features that distinguishes waves from particles. If two entities interfere with rather than bounce off of each other, then those entities must be waves.

If waves are not characterized by position, then what characterizes them? In a nutshell, four things: wavelength, frequency, amplitude, and phase. There are other ways to characterize a given wave (such as the speed at which the wave travels through the water), but those other properties can always be expressed in terms of the four listed above.

Imagine yourself in a sailboat afloat on the ocean. You set anchor so that the boat is at rest and climb the mast. Looking ahead, you see an endless series of uniform wave crests stretching out toward the horizon. The crests move toward you, wash into the bow, pass along the boat, and then roll away from its stern. You want to radio to your friend on shore, who gets seasick on heavy days, the properties of the waves so that he can decide whether to risk it and join you. He has a lot of experience with being seasick so he wants to know everything of relevance about the waves before he makes his decision.

First, you notice that the boat bobs up and down in a regular cycle as the waves wash by the boat. You measure the number of times the boat bobs up and down each minute — the *frequency* of the wave — and write it down.

Second, you notice that there is a consistent difference in height between the peaks and troughs. This difference, known as the *amplitude*, determines just how violently the ship bobs up and down between successive wave crests, so you measure it carefully, for certainly your not completely seaworthy companion will want to know this.

Third, you notice that the peaks and troughs of the ocean waves are separated by a well-defined distance — the *wavelength*, or distance between successive peaks of the wave. This peak-to-peak distance is the same whether you are looking at the ripple closest to the boat or the one a mile away. You measure the wavelength, and write it down.

Finally, you recognize that if your friend is to know exactly when the boat will be at the top of its bob and when it will be at the bottom, you need to measure something about the *phase* of the wave — when your watch reads exactly noon, are you sitting on top of a wave crest, in a trough, or somewhere in between? How will you measure and convey this piece of information in a respectably quantitative fashion?

The solution to this problem is provided by the following critical insight: Riding waves is very similar, in a certain way, to going around and around in a circle. This insight will be an essential component of our development of the notion of *gauge theory* in chapter 8, so you will want to keep it in mind.

Figure 3.1 shows a dog tethered to a pole in the front yard of a house. The dog awakens from her nap, and looks around. Very quickly, she gets bored and tries to move away from the pole. Pulling taut against the tether, the dog can only go in circles about the pole. After one full circle, or 360 degrees of turning about the pole, she is back where she started.

Now with some work, and for want of anything better to do on a summer afternoon, you could probably train this dog to snap a picture from a camera mounted to her collar every time she walked over an X painted on the ground. Assume that, somehow or other, you had managed to do this and that the dog was taking a nap on such an X. When she wakes up, she snaps

Fig. 3.1. Going around in circles is very similar to floating up and down on an endless series of waves. When the dog returns to the same point (the X-shaped dog bones) on the circle, everything looks the same as it did before she ran around the circle. Likewise, when you float back to the crest of the next wave, everything looks the same as it did when you were on the crest of the previous wave (see fig. 3.2).

a picture; then, as she goes around and around on the tether, she snaps another picture every time she gets back to her former resting place.

What do you see when you develop the film? Thirty-six full-color landscapes of exactly the same thing. The point is this: When the dog goes around a full circle—exactly 360 degrees—she gets back to where she started from. Just looking at the pictures, you have no idea whether she snapped all 36 pictures standing stationary on the X or whether she went around once, or more than once, between each snapshot.

Similarly (see fig. 3.2), after bobbing down one crest and up to the top of the next wave crest, even though you know that you have bobbed up and down one wave, there is nothing you can see that verifies this. You are on top of the wave, just as you were before you started bobbing. If you took a picture every time you were on top of a wave crest, someone looking at the developed roll of film would have no idea whether you snapped the whole roll quickly while on the crest of a single wave or, instead, snapped away at a leisurely pace, waiting for one or two waves to pass by between pictures. Either way, all you would see in the picture is an infinite number of waves receding toward the horizon, the nearest crest of which you happened to be floating on when the picture was taken. Just like going 360 degrees around a circle, you're effectively back where you started: at the crest of one of an infinite sea of waves.

You now notice that at precisely noon the boat is exactly at the bottom of a trough, halfway between successive crests. Making use of this circle analogy, you know how to record your *phase*—where you are, quantitatively, relative to the next wave crest. Since you are exactly at the bottom of

Fig. 3.2. After riding exactly one wave cycle, from the crest of one wave to the crest of the next, there is no change in the way the world appears. Nothing you can see allows you to determine which wave crest you are on, just as, in going around a circle, there is nothing you can see that tells you how many times you've gone around the circle.

the trough of the wave, then you are halfway between successive crests—halfway between 0 degrees (the top of one crest) and 360 degrees (the top of the next crest). Since halfway between 0 and 360 degrees is 180 degrees, you write down your phase at noon as 180 degrees. Halfway around the circle puts you on its opposite side; halfway between crests of the undulating wave puts you on the part of the wave that opposes the crest; that is, it puts you in the trough. This is the direct, and exceedingly useful, analogy between going endlessly around in circles, returning time and time again to the same point, and bobbing endlessly up and down on ocean waves, returning time and time again to the crest of an endless succession of indistinguishable waves.

From this observation of the wave's phase, and given your record of the frequency (rate at which the wave crests wash by), your friend should be able to figure out exactly when it is that he will find himself at a crest, trough, or anywhere in between. You are to be congratulated for being so clever and complete in your measurements, or at least so it would seem. You radio your findings back to shore, and then head below deck for a game of solitaire while awaiting your friend's decision.

At last, the long-awaited answer comes back. The wavelength, frequency, and especially amplitude, all seem amenable to him. However, he cannot believe that you went to all the trouble to measure and report the *phase* to him. How could it possibly make any difference to him whether he's in a trough at 1:01 or on a peak at 2:07? In fact, he is so discouraged by your apparent lack of critical thinking skills that he's decided to stay on shore and reevaluate his commitment to his relationship with you.

Well, that's the way friends are sometimes. But if you're interested in quantum mechanics and, particularly, in its application to the Standard Model of particle physics, then he's taught you a very important lesson. For in quantum mechanics, the overall phase just doesn't matter. Not only does the phase not influence anything we might care about, but there's *no experiment* that you can do to determine the overall phase of a quantum-mechanical system. The quantum-mechanical phase of a physical system is simply unmeasurable (note 3.5). Since, as we'll shortly see, *all* systems have wavelike properties, this is an important point of which to take heed.

Paradoxically, it is precisely this irrelevance that places the notion of quantum-mechanical phase at center stage in the mind of particle physicists. Dogged adherence to this principle—this meaninglessness of phase—coupled with its generalization to ensure consistency with the tenets of relativity and the quantum theory of force fields, leads directly to a profound reinterpretation of the fundamental nature of the behavior of matter. It's not that the phase of quantum mechanical systems becomes relevant but rather that the irrelevance of phase is understood to be, in and of itself, tremendously relevant. The rigorous formulation of this notion, known as *gauge theory*, is a theory of the relevance of irrelevance; within this oxymoronic inspiration lies one of the most profound intellectual leaps in the storied history of particle physics.

Wave-Particle Duality 1: Particles of Light

That light is a wave, and not particle-like, seems obvious now that we've sharpened up our notions of what it takes to be a wave or a particle. If you shine two flashlight beams so that they intersect, they don't slam into each other and go crashing to the ground. Instead, they pass through each other essentially unaffected, as you would expect of waves in good standing. Light can be focused and reflected and can diffract around corners, illuminating regions that are not in the direct path of the original light beam (this is why shadows of objects tend to have fuzzy edges). All of these properties are due to the essential wavelike property of interference.

Waves are, in essence, an organized form of energy transfer through a "medium"; in the case of water waves, the medium is the water at the surface of the pond. What is the corresponding medium for light?

The answer, as mentioned in chapter 2 (and thanks to James Clerk Maxwell, the greatest of the many heroes of the theory of electromagnetism), is that light is a wavelike disturbance in the electric and magnetic force fields, that propagates through space at, not surprisingly, the speed of light, $c = 2.997 \times 10^8$ meters per second, or about 186,000 miles per second. The electromagnetic field exerts forces on objects possessing electric charge, such as the molecules inhabiting the receptors of the human retina; if the force-field disturbance associated with a ray of light reaches your eye, it induces chemical reactions in those receptors. The sensation of color is nothing more than the brain's way of differentiating between different frequencies of oscillation (waving) of the strength of the force field associated with the given ray of light. Brightness is also easily explained; it's simply the amplitude of that oscillation.

Visible light falls roughly within the range of 4×10^{14} to 8×10^{14} oscillations per second, with the red end of the rainbow at lower frequencies and the indigo end at higher frequencies. Directly below the lower-frequency end of the visible spectrum lies the realm of the infrared, while directly above the high-frequency end of the spectrum lies that of the ultraviolet.

A very hot object—such as the filament of an electric stove—begins to glow red as it heats up. Made even hotter, its color changes from red-hot to white-hot. Whatever the color, if you put your hand anywhere near the object (without touching it, mind you), you can sense the heat radiating from it. The light you see, and the infrared light you do not see, are both composed of electromagnetic waves, which transmit some of the thermal energy of the hot object to the surface of your hand. Experiments conducted in the nineteenth century provided accurate measurements of the spectrum—the relative amount of energy contained in each range of color of radiated light—emitted by hot objects. Among other things, these measurements showed that, by and large, the spectrum of emitted radiation depends only on the temperature of the hot object and is more or less independent of what it is that's being heated up.

This spectrum, classical physicists reasoned, is thus a basic attribute of matter and so should be possible to understand. Turning their attention to this problem, though, the physicists of the late nineteenth century hit a snag: the resulting classical *blackbody* theory (note 3.6) of the spectrum of the radiation of light from hot objects was in stark disagreement with the measurements that had been made. In fact, this classical theory predicted that an *infinite* amount of energy would be radiated at high frequencies—a failing that came to be known as the *ultraviolet catastrophe*. The physicists of that day recognized this was absurd, but that's what they got when they applied their classical theory of electromagnetism in an attempt to understand the spectrum of glowing objects.

In 1900, after much thought and reexamination of the assumptions that went into this classical blackbody theory, a soon-to-be-famous German physicist by the name of Max Planck hit on an idea that seemed to resolve the problem. Planck's hypothesis, when used to modify the classical blackbody theory, not only resolved the ultraviolet catastrophe but also produced a prediction for the spectrum of radiation that was in astounding agreement with experimental observations. Planck's hypothesis is simply stated: The energy E transferred by electromagnetic waves (light) of a given frequency (color) f must come in discrete, *quantized* packets of magnitude

$$E = hf,$$

where h is Planck's diminutive constant.

In other words, the quantity hf represents the *minimum* amount of energy that a given color of light can possess as it is radiated. Light of the frequency f must come in packets of energy hf, so the energy of light radiated from the object at a frequency f can be $2hf$, $3hf$, and so on, but nothing in-between. The emission of light is quantized, with the minimum allowed energy hf corresponding to the energy of one *quantum* of light. In fact, we now appreciate this hypothesis as the first conjecture that nature is quantized, although Planck himself had no idea, at the time, just how profound a leap he had made.

To understand how Planck resolved the ultraviolet catastrophe with his hypothesis, consider its implications. The bigger the frequency f, the more energy E it takes for the hot object to release a single quantum of light, which is the minimum amount of energy the object needs to give up to radiate at that frequency. For frequencies that are too high (too far into the ultraviolet end of the spectrum), this minimum energy requirement is just too demanding. Thus, instead of the prediction put forth by the classical theory of an *infinite* amount of energy radiated in ultraviolet light, the theory modified by the hypothesis of light quanta (flashes of light with a well-defined minimum energy content of hf) led to a prediction of no radiation for very high ultraviolet frequencies. Because it takes so much energy to create an ultraviolet light quantum, no energy is radiated in the ultraviolet, and the ultraviolet catastrophe is resolved.

At first, Planck did not consider his conjecture about the quantization of light to be that revolutionary. He assumed there was some detail about the way in which the light was emitted that no one quite understood. Five years later, however, in the same year as his publication of the special theory of relativity, Einstein published a paper on the *photoelectric effect* in which electrons are knocked out of certain materials when those materials are illuminated with visible or ultraviolet light. Based on careful observations of the behavior of the photoelectric effect, performed and published by a number of experimental researchers, Einstein was able to argue convincingly that the photoelectric effect was explicable only if the quantization of light into individual packets of energy $E = hf$ was a fundamental property of light itself, rather than of the emission process, as Planck suspected.

But, in the photoelectric effect, the individual packets of light behaved as if they were particles! Each released electron was knocked out of the material by a single packet, or quantum, of light that, to knock the electron out of the solid, had to collide with the electron in the jarring manner of a particle. This radical idea was eventually accepted by the international physics community; soon afterward the American physicist Arthur Compton coined the term *photon* for these indivisible particles of light. Interestingly, Einstein's 1921 Nobel Prize, which followed three years after Planck's own Nobel, was awarded more for Einstein's explanation of the photoelectric effect than it was for his theories of relativity. Today, Einstein is respected for the former but revered for the latter.

So is light a wave, as Maxwell said, or a particle, as evidenced by the photoelectric effect? We need to put aside our calculations, switch off our experimental apparatus, grab a cup of tea, and reflect for a moment on what we have learned. Light is clearly a wave—the evidence for this comes directly from our everyday experience with flashlights, lenses, and especially these days, lasers. When you look more carefully, though, at phenomena for which Planck's tiny constant sets the scale of activity, there's plenty of experimental evidence that demands we treat a ray of light as if it's composed of a large number of light quanta *particles*.

So, which is it, wave or particle? Which of these two apparently exclusive poles is representative of the fundamental nature of light? The answer, Einstein argued, is *both*. In some contexts, light behaves as if it is composed of little particles moving along together. In other contexts, it behaves like a wave—something our intuition has a difficult time associating with a "real" object that can bounce off other things. Yet the wave *is* the particle, and vice versa. Light has a dual nature, part wave, part particle.

This is the notion of wave-particle duality—a notion central to quantum mechanics in general and particle physics in particular. Indeed, the photon, and particles like it, play a central role in the Standard Model of particle physics and will figure heavily in the discussion to come.

If you have an electric stove, turn it to "high" until it glows red. (If you have a gas stove, as I do, you can hold a needle in the flame until the needle turns red.) Take a good look at the light. The explanation of what you are seeing, as simple a phenomenon as it may seem, launched one of the greatest revolutions in the history of scientific thought.

Wave-Particle Duality 2: Matter Waves

A common conviction of physicists is that, at its most fundamental level, nature prefers uniformity and simplicity over disparateness and complexity. In 1924, a French prince and graduate student by the name of Louis-Victor de Broglie was struck by a sudden insight. He hypothesized that Einstein's conjecture of the dual nature of light represented merely the tip of an iceberg and that the notion of wave-particle duality should extend to all of nature. If light, which common sense suggests should be classified as a wave, can exhibit particle-like properties, then why can't matter particles, such as electrons, protons, atoms, molecules, dust, animals, planets, and so forth, exhibit wavelike properties?

According to Einstein, the connection between light's wavelike property of frequency (f) and its particle-like property of kinetic energy (E) is given by Planck's relation $E = hf$. Simply stated, de Broglie's hypothesis was that this relation should also hold for matter. If we look carefully enough, de Broglie argued, we should see that matter can exhibit wavelike properties, consistent with Planck's relation between a photon's energy of motion and its frequency.

In fact, this is not quite the form in which de Broglie's hypothesis is most commonly stated and readily applied. Rather than being expressed in terms of kinetic energy and frequency, de Broglie's hypothesis is usually expressed in terms of wavelength λ and momentum p. If you're unfamiliar with the distinction between momentum and kinetic energy (energy of motion), don't worry: throughout this book, it will be fine to think of them as one and the same. In any regard, a straightforward mathematical argument that I won't reproduce here shows that, in these terms, de Broglie's hypothesis can be restated as

$$\lambda = \frac{h}{p}.$$

This expression is properly known as the *de Broglie relation*, and states that any material object exhibits wavelike properties, with a wavelength inversely proportional to its momentum: the bigger the momentum/energy, the smaller the associated wavelength. And as usual, the scale of the object's wavelike behavior is set by Planck's constant h, which we know to be very small.

What's small to a human, however, is not necessarily small to an atom: de Broglie's relation tells us that the wavelength of an electron that has been accelerated through 50 volts of electrical potential (i.e., with an energy of 50 eV; see discussion in this chapter titled "Einstein's Relativity") is about 2 angstroms, or about 2×10^{-10} meters. This, it turns out, is roughly the spacing between the atoms of a crystal. It also turns out that the regular, uniform spacing of atoms in a crystal lattice are ideal for studying the effects of interference—the phenomenon that is so uniquely characteristic of *wavelike* behavior.

In 1927, the American physicist Clinton Davisson was studying the way 50 eV electrons bounce, or scatter, off a plug of pure nickel. A minor failure of his apparatus led to a contamination of the nickel sample, and so he had to repurify it by heating the sample in an oven. Although he didn't immediately recognize it, the nickel sample became crystallized in the process. When he again began studying the scattering of the 50 eV electrons, he noticed a peculiar effect—the scattering was much weaker in general, although for particular scattering directions, it was quite pronounced.

Davisson quickly realized that such behavior would be expected if the nickel sample had crystallized and *if the electrons were exhibiting wavelike behavior*—if the electron waves reflected from each of the individual, regularly spaced atoms in the crystal were interfering with one another to form the pronounced pattern that was observed. This would only work if the wavelength associated with the electrons' wavelike behavior was roughly the same as the spacing between the nickel atoms in the crystal lattice. This is precisely the prediction of de Broglie's hypothesis, so both de Broglie (1929) and Davisson (1939) eventually found themselves in the company of the Swedish king, receiving their respective Nobel prizes. One had predicted and the other had shown that particles of matter—electrons—exhibit wavelike properties.

So, waves (such as light) have particle-like properties, and matter particles (such as electrons) have wavelike properties. The concept of wave-particle duality is, as de Broglie conjectured, universal. In general, *any* corporal object has associated with it a wavelength, which can easily be calculated using de Broglie's hypothesis. For instance, I estimate the wavelength of this book, traveling with the momentum required to propel it toward the wastebasket in the corner of your room, to be about 10^{-35} meters.

This is a development with a notable practical application. Have you ever seen an atom or a molecule through a conventional microscope? It's impossible, even with the best microscope, because you can't see any feature of the sample under study with a size smaller than the wavelength of the wave you're using to illuminate the sample. Visible light has a wavelength of between 4×10^{-7} and 7×10^{-7} meters, while atoms and molecules are about 10^{-10} meters across, which is too small to be seen with visible light. However, a relatively languid 50 eV electron has a wavelength of about 2×10^{-10} meters. Increase the energy a little (recall that electron beams in TV sets typically have an energy of a few thousand electron-volts), develop a way to focus electron beams (say, with magnetic forces), and you now have a hope of recording ultraprecise images. Thus, the field of electron microscopy was born.

But why stop there? What happens if, instead of a 1,000 eV electron, you bombard a sample with, say, 100 million eV electrons from a particle accelerator? The corresponding electron wavelength will be much smaller (recall that wavelength is *inversely* proportional to momentum/energy); in fact, it will be about 10^{-15} meters. Correspondingly, Robert Hofstadter's experiments at Stanford University in the late 1950s were the first to "see" that the proton was not pointlike but had, in fact, a measurable radius of about 10^{-15} meters. Buoyed by this success (which was further bolstered by Hofstadter's receipt of the 1961 Nobel Prize in Physics), a much larger accelerator was assembled—the Stanford Linear Accelerator Center (SLAC)—on Stanford University land. The accelerator began operation with electron energies of 10 billion eV (10 GeV) and corresponding electron wavelengths of 10^{-17} meters. This allowed physicists to peer deeply inside the proton, leading to the discovery of its internal constituents, known as quarks, in the late 1960s.

Heisenberg's Uncertainty Principle

Were there to be an award for the physicist who most reshaped the way we think about the world around us, it might well go to the German scientist Werner Heisenberg for his development of the uncertainty principle. The statement that, at some level, the world is unknowable—not just as a practical matter but fundamentally so—struck a fatal blow to the notion of scientific determinism that had taken root 250 years before due to the work of Isaac Newton.

The uncertainty principle follows directly and necessarily from de Broglie's assertion that matter should possess wavelike properties. Consider an object whose momentum/energy is known exactly, not just very precisely but really, truly *exactly*. Then, according to de Broglie, its wavelike nature (or, in the lingo of quantum mechanics, its *wave function*) should be represented by a wave of wavelength $\lambda = h/p$. Again, λ is the wavelength (in meters, yards, or whatever), p is the object's momentum (closely related to its kinetic energy), and h is Planck's constant.

A wave is not localized. There is no point in space to which you can point and exclaim, "See. The wave is right there." Recall the sailor bobbing up and down in the boat—the wave of wavelength λ extended as far in front and behind as the eye could see. The wave was characterized by its wavelength, frequency, amplitude, and (essentially irrelevant) phase but *not* its position because it had no definable position. Heisenberg reasoned that if de Broglie is correct, a particle with a precisely known momentum p must exhibit the properties of a wave with wavelength $\lambda = h/p$, which is completely unlocalized: *if the momentum of an object is exactly known, then absolutely nothing can be known about its position.*

The exact value in meters of the wavelength λ is not material to the discussion. The point is simply that if the particle's wavelike properties correspond to a definite wavelength λ, then the particle behaves like a pure wave, which is completely unlocalized (at any given time, the undulations of a pure wave extend infinitely far forward and backward in space), so nothing whatsoever can be said about its position.

The above is not quite the Heisenberg uncertainty principle but is rather just a special case of it, for what if the momentum is not known precisely but is known to lie instead within some range? In other words, what if the object's momentum is known to some degree, but there's some *uncertainty* about what its momentum really is? The following paragraphs will address this question, and with the help of figures 3.3 and 3.4, lead us to a formal statement of the uncertainty principle.

Say you put together a slingshot that, when drawn back as far as it can go, fires pebbles that have momentum between 0.9 and 1.1 kilogram-meters per second (kg-m/s; the standard yardstick for measuring momentum). So, your slingshot fires objects with momentum of about 1.00 kg-m/s, but for any given object, you will be uncertain of the momentum by about 0.1 kg-m/s. Now, in this case, de Broglie is not so incisive. He is not able to tell us that the wavelike properties (wave function) of any given object from the slingshot will be characterized by a wave of a single wavelength $\lambda = h/p$. Instead, the de Broglie relation tells us that the wave function will be composed of a combination of waves, with wavelengths varying between $h/0.9$ and $h/1.1$ meters. If there's a range of momenta p at play, then there must be a corresponding range of wavelengths λ at play.

To understand what's meant by the expression "combination of different waves," think of two children on opposite sides of a pond making waves of slightly different wavelengths by bobbing sticks up and down in the pond at slightly different rates. From each stick, a wave will emanate outward; when these waves meet, they will combine (really, *interfere*) to produce a more complicated wave form.

Heisenberg knew that when you combine waves with only slightly differing wavelengths, something interesting happens: the combined wave becomes localized. Say that you add waves together, with wavelengths that differ, but lie in the narrow range between $h/0.9$ and $h/1.1$ meters. Even though each individual wave is completely unlocalized, undulating infinitely far backward and forward in space, the *combination* of the waves enjoys a region where all the waves are peaking—adding together (interfering constructively) to generate an even bigger wave form. This region of constructive interference is surrounded by a region where the peaks of some of the waves are cancelled by the troughs of the other waves (destructive interference); in these regions, the size of the combined wave form is zero. But the object cannot exist in a region where its wave function is cancelled out to zero, so the object represented by the combination of waves will not be found in the region where the waves cancel each other out. The presence of the object is restricted to locales for which the combined wave function is nonzero: the object is localized.

The accompanying diagrams of figure 3.3 illustrate this point. Figure 3.3*a* shows an object whose momentum is known perfectly, so is represented by a single wave of wavelength $h/1.00$ meters; the wave undulates smoothly throughout space and exists everywhere in space with equal probability. For figure 3.3*b*, the object is represented by the combination of two waves with slightly different wavelengths: one with $\lambda = h/0.9$ meters and the other with $\lambda = h/1.1$ meters; you can see regions where the combined undulations of the two waves cancel each other out. Figure 3.3*c* shows the combination of several waves with slightly differing wavelength between $h/0.9$ and $h/1.1$ meters; localization is becoming evident. Finally, in figure 3.3*d*, the infinitude of waves of all possible wavelengths between $h/0.9$ and $h/1.1$ meters are combined, representing the case of the slingshot: the momentum of the pebble could be anything between 0.9 and 1.1 kg-m/s. In this case, the localization is complete. You can see that there is a well-defined region in which the particle exists, and that the probability of finding the object quickly goes to zero as you look farther and farther away from the region of localization.

Localization is a defining characteristic of material objects; it makes sense to ask, does anyone know where my glasses are? since eyeglasses, as objects, are localized, and it's conceivable that some observant individual might actually be able to tell you where they are. So if objects are to have wavelike properties, then this is how we need to represent the wave function of objects—as the combination of an infinite number of waves with wavelengths $\lambda = h/p$ corresponding to some range of momenta p. The extent of this range of momenta is just the extent to which you know the object's momentum—it's the uncertainty on the object's momentum. In practice, because Planck's constant is so small, everyday objects will be well localized even if the object appears to be completely at rest to a casual observer. Even if very small, there's always some amount of uncertainty in the motion of everyday objects that allows them to be localized.

The essence of the uncertainty principle is as follows. Figure 3.3d represents the wavelike characteristics of the pebble just as it leaves the slingshot. While the pebble is indeed localized, it is so only within a certain region of space and not at a single point in space. For our pebble with wavelengths between $h/0.9$ and $h/1.1$ meters, the pebble's position at this moment in time is not specified by a single, well-defined point in space but rather by the small region of space defined by the extent of the hump in the object's wave function. The object is in there somewhere, but within those bounds, the object's position is uncertain, not merely as a practical matter but *fundamentally.*

In other words, all you can say about the position of the pebble is that it lies somewhere within the hump shown in figure 3.3d. If you measure the pebble's position by, say, bouncing light off it, you will indeed find it at some well-defined point within the hump, but you cannot predict ahead of time (before the measurement) just where inside the hump you will find the pebble. The pebble's position (before you disturb its wave function by trying to measure it) is uncertain to that degree.

So, even with this complete localization, allowed by the fact that the pebble's wave function consists of the combination of an infinite number of waves with wavelength between $h/0.9$ and $h/1.1$ meters, there's still uncertainty in the position (location) of the pebble. And now the critical point:

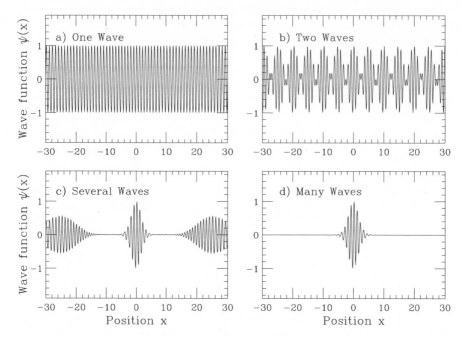

Fig. 3.3. If an object's wave function is represented by a single wave, the wave function extends uniformly throughout space. The object is no more likely to be in one region of the universe as it is in any other region. As we begin to add together waves with differing wavelengths (*b* and *c*), we see that there are regions of space in which the object is more likely to be found than others. If we add together the infinite number of waves corresponding to all wavelengths within a certain range of wavelengths about the original wavelength of *a*, the object becomes completely localized. There is one region of space in which the object is to be found, and the object will *not* be found anywhere else.

the property that determines the extent of the region of localization of the pebble, and thus the uncertainty in the pebble's location, is the *magnitude of the uncertainty in the pebble's momentum*. The *less* uncertain the momentum (the better the momentum is known), the *larger* the region of localization (the wider the hump in figure 3.3*d*) and the *more* uncertain the position.

Figure 3.4 shows just this—the localization achieved when the degree of uncertainty in the momentum is reduced. For this case, waves are combined with a range of wavelengths between $h/0.95$ and $h/1.05$ meters, corresponding to the lesser range of momenta between 0.95 and 1.05 kg-m/s— exactly half that of figure 3.3. The corresponding uncertainty in position in

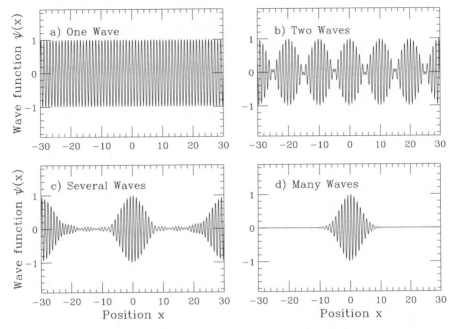

Fig. 3.4. This figure is very similar to figure 3.3 except that the range of wavelengths we add together to form the object's wave function is halved. From *d*, we see that the effect of halving the range of wavelengths—of halving the *uncertainty* in wavelength or, by de Broglie's relation, halving the *uncertainty in momentum*—is to double the range of possible positions. In other words, lessening the uncertainty in momentum increases the uncertainty in position. This is the essence of Heisenberg's uncertainty principle.

the pebble's wave function (shown in fig. 3.4*d*), for this case of lesser momentum uncertainty, is double that of figure 3.3*d*.

The comparison of figures 3.3*d* and 3.4*d* evokes the essence of the uncertainty principle. Since the uncertainty in the momentum grows when the uncertainty in position shrinks and vice versa, their product—the result of multiplying the two uncertainties together—might reasonably be expected to stay the same. This is the quantitative message of the uncertainty principle. If we let Δp represent the uncertainty in the object's momentum (about 0.1 kg-m/s in the case of fig. 3.3, or 0.05 kg-m/s in the case of fig. 3.4) and Δx represent the uncertainty in its position (the widths of the hump in fig. 3.3*d* or 3.4*d*), then Heisenberg's arguments tell us that

$$\Delta p \Delta x = \frac{h}{4\pi} :$$

the product of the uncertainties of an object's momentum and location is the constant $h/4\pi$, where $\pi \cong 3.14159$ is, as usual, the magical ratio between the circumference and diameter of a circle (note 3.7). In particular, if Δp becomes zero (no uncertainty on the momentum, i.e., momentum known exactly), then Δx must become infinite to compensate (if you multiply zero by anything *less* than infinity you get zero, which won't satisfy the above equation), and nothing at all is known about the position x. This is the special case that we discussed in the beginning of this section.

The uncertainty principle is one of the striking revelations of quantum mechanics, and it pervades the theory as a fundamental tenet that cannot be violated. Any result of the full theory of quantum mechanics must be demonstrably consistent with the uncertainty principle.

More interestingly, it is a revelation that seems to address questions of deep philosophical import. With the advent of the uncertainty principle, determinism, the notion that the laws of nature set forth an inextricable course of events from which no deviation is possible, becomes indefensible. According to the uncertainty principle, the exact course of events is fundamentally unknowable. There is always some uncertainty in the physical properties of any given object; not even nature herself knows how this uncertainty will resolve itself the next time the object makes its influence known, say, by the interaction with another object by way of one of the four forces. It's not just a matter of building a better instrument for determining these properties. The exact value is simply unknowable, even in principle. Many have gone on to conjecture that Heisenberg's uncertainty principle is the very source of human free will, but this remains to be demonstrated. Heisenberg's principle is on such solid empirical footing that it is now accepted as one of the fundamental attributes of the natural world.

Another interesting and philosophically important notion that follows from the uncertainty principle is the loss of distinction between the observer and the observed. What is meant by this?

To say that an object whose momentum is known to be within a certain range will have a corresponding uncertainty in its position is *not* to say that the object's position can never be determined with greater precision. Consider an object whose momentum is known so precisely that the uncertainty in its position is large, say, a full meter (again, according to the uncertainty principle, an object with a precisely known momentum will have a poorly known position). You want to measure the position of this object more accurately, so you place a chamber of some special gas in the path of the particle. During the course of the particle's motion through the gas, it collides with a gas molecule, causing the molecule to emit a flash of light. Now, at the exact time that you observe the flash of light, you know that the object is in the vicinity of that particular gas molecule. You now know the position of the object to great precision, to about 10^{-10} meters, which is the size of a gas molecule. However, the uncertainty principle cannot be violated, so your knowledge of the momentum of the particle is now correspondingly much worse.

The point is this: to observe the particle, *it must interact with something from the system of the observer* (in our case, the observer's system is the gas, the apparatus that detects the light flash, the electronics that amplify the signal from the apparatus, and the human or computer that records the amplified signal). That interaction itself necessarily influences the object—alters the wave function that governs the object's physical properties—in such a way that the uncertainty principle is never violated. One can never precisely and simultaneously determine both the position and momentum of an object because in determining one, the process of observation always changes the other to some new, and indeterminate, value. The physical system of the observed object and the system of the observer become thus intertwined in the process. If you want to make an observation of some physical system, you cannot just consider the properties of that system in isolation; to really understand what's going on, you have to consider the properties of the observed system *and* the properties of the observing system, and how they interact. The true system under consideration must always be the combination of the system of the observed with that of the observer. In the (thoroughly verified) theory of quantum mechanics, the distinction between the observer and the observed has to be discarded.

Physical quantities whose uncertainties are linked via the uncertainty principle are known as *conjugates*. As we've just seen, position and momentum are conjugate, but there are also a number of other conjugate quantities. Energy and time are conjugate; for example, the certainty to which you can determine the mass-energy of an object is inversely related, according to the uncertainty principle, to the amount of time you hold it in your sights as you measure its mass-energy. Angular position and angular momentum are also conjugate—the degree to which you can determine the angle through which an object has spun is inversely related to the degree to which you know how fast the object is spinning. In all cases, the product of the uncertainties of conjugate quantities that specify the disposition of an object will be roughly equal to Planck's constant, as it is for the conjugate quantities of position and momentum.

A Master Formula for Quantum Mechanics:
The Schrödinger Equation

So far, we've considered only free objects—objects that execute an unwavering, linear motion through space, absent any influence of the forces of nature (note 3.8). But nature is only interesting to the extent that things influence one another through these forces, so we need to understand how to incorporate forces into our thinking about the wavelike nature of matter. This was first accomplished in 1926 by a forty-year-old Austrian physicist at the University of Berlin named Erwin Schrödinger, in the guise of a celebrated mathematical relation known as the Schrödinger equation.

Schrödinger did not find it most natural to introduce forces into the newly emerging quantum theory in terms of their push and pull on the object being described but rather in terms of the *potential energy* associated with the force doing the pushing or pulling. Let's take a moment to introduce the notion of potential energy.

An object in motion has momentum and kinetic energy associated with its motion. The faster the motion, the greater the energy/momentum associated with that motion.

Consider a car out of gear (in neutral) coasting up a hill, and ignore the friction in the wheel bearings and between the tires and road. As the car climbs the hill, it loses speed. Its motion slows, so it loses kinetic energy. But remember that energy is a *conserved* quantity. The car must have as much energy at the bottom of the hill, when it was moving fast, as it does at the point on the hill when it stops and begins rolling backward down the hill again.

So what happens to the energy that was in the form of kinetic energy (energy of motion) when the car was at the bottom of the hill? The answer is that it gets converted to gravitational potential energy as the car coasts uphill, slowing as it rises against the pull of gravity. At exactly the point at which the car reaches its maximum height, just before it begins to roll back down the hill, its speed is zero, and all of the car's kinetic energy has been converted to gravitational potential energy. The amount of potential energy possessed by the car at that point is exactly equal to the kinetic energy the car possessed before it started coasting up the hill.

The reason this gravitational energy is called "potential" is that it has the potential to do work on the car, thereby restoring the car's original kinetic energy. When the car reaches the bottom of the hill after coasting back down, it has the same speed as it did just before it started up the hill (as long as there's no friction). The gravitational potential energy fully realizes its potential to convert itself to kinetic energy of motion as the car coasts backward down to the bottom of the hill.

If we had slipped blocks behind the wheels of the car just as it reached its maximum height on the hill, we would have locked the potential energy in place. But the energy would still be there, a very concrete aspect of the car's physical state, just waiting for someone to remove the blocks so that it could exercise its potential to expend itself in the task of creating kinetic energy. Potential energy is a form of energy that's stored in an object by virtue of the object being under the influence of some force.

Potential energy depends on the location of the car. In the language of mathematics, we would say that the potential energy *is a function of* the car's location. In this case, the location of the car at any point in its coast is just the distance x along the road that the car has traveled from the bottom of the hill. At the bottom of the hill, the potential energy is zero—no kinetic energy has yet been transmuted to potential energy. As the car rolls up the hill, at each successively higher point the speed, and thus the kinetic energy, is less. This is precisely because the value of the gravitational potential energy is correspondingly greater at each successively higher point—the higher the car, the more potential there is for gravity to do work on the car as it eventually rolls back down to the bottom of the hill. Because overall (kinetic plus potential) energy is conserved, then whatever the car loses in kinetic energy it must gain in potential energy.

If we denote the value of the gravitational potential energy by V, then we would write $V(x)$ to represent this potential energy function. The expression $V(x)$ is nothing more than mathematical shorthand for the statement "the potential energy V has a well-defined value at any given location x, which is just the value $V(x)$ of the function at the point x."

In our example, the potential energy is gravitational potential energy and the object is the car. However, in general, the potential energy could be due to any of the four forces and the object anything that bears some amount of the charge appropriate for that force (recall that mass is the charge associated with the gravitational force).

What follows is the full mathematical expression of the Schrödinger equation as you would see it in a physics textbook. To be candid, this is a restricted form of the Schrödinger equation, appropriate for one-dimensional motion only—the car is restricted to move forward and backward along the road and cannot veer off the road into the second dimension or rise toward the heavens into the third dimension (note 3.9). Nevertheless, what follows is a useful relation, and one that physicists have spent much time working with over the years:

$$-\frac{h^2}{8\pi^2 m}\frac{d^2}{dx^2}\psi(x) + V(x)\psi(x) = E\psi(x).$$

As those who have mastered calculus may recall, this is a differential equation, involving the operation of differentiation in one or more of its terms (in our case, represented by the derivative operator d^2/dx^2 in the first term). Don't worry. You need not understand how to solve this equation to continue on from this point. However, a few words of explanation are in order.

In everyday algebra, an equation represents a precise quantitative relationship between an unknown quantity and one or more known quantities, allowing the value of the unknown quantity to be determined by the application of mathematical rules. For a differential equation, the unknown is not a single quantity but rather an entire function. For the case of the Schrödinger equation, this unknown function, represented by the symbol $\psi(x)$, is the object's wave function.

In a nutshell, the Schrödinger equation is a prescription for determining the wave function $\psi(x)$ of an object. It is a complete prescription, containing everything that can possibly be of use to establish the properties of an object of mass m and energy E under the influence of some force whose potential energy function is $V(x)$. If you think of this differential equation as a game that physicists play, then the way they win the game is by finding all possible wave functions $\psi(x)$ that solve this equation. It's the function $V(x)$ that specifies the physical situation—an electron under the electromagnetic influence of a proton in a hydrogen atom has a certain $V(x)$; a car coasting up and down the hills of a country road has a different $V(x)$ and so on.

What, then, does the wave function $\psi(x)$ represent? Again, the Schrödinger equation represents all we need to know to determine the physical condition of an object, while incorporating all the tenets and constraints of quantum mechanics. The wave function $\psi(x)$, derived from the Schrödinger equation through its rules of solution, thus represents all that can possibly be known about the physical state of the object. It is an economical encoding of the full gamut of physical information that would be accessible to anyone attempting to measure the properties of the object.

Although $\psi(x)$ has no physical meaning, any physical property of the object can be determined once $\psi(x)$ is known. If you want to know the probability of finding the object at any point in space, you simply perform a specific procedure on $\psi(x)$—in this case, just squaring (multiplying it by itself once) the value of $\psi(x)$ at that particular point in space. If you want to know the object's kinetic energy, you perform a different procedure (in this case, involving taking some derivatives, that is, performing a little calculus). If you want to know the object's speed and direction of motion (to the accuracy permitted by the uncertainty principle), there's a procedure for finding that and so forth.

Now, if the particle is perfectly free (not under the influence of any force), then there is no potential energy $V(x)$, or $V(x) = 0$, so you can forget about that term in the Schrödinger equation. It turns out that, in this case, the solutions $\psi(x)$ to the Schrödinger equation are just the pure waves with which we began our discussion of quantum mechanics, with a wavelength given by $\lambda = h/p$, the original conjecture put forth by de Broglie. So, the Schrödinger equation does just what we wanted it to—it allows us to generalize de Broglie's hypothesis to the case where the particle is not free but is rather waving around quantum mechanically under the influence of a force.

Since this theory is called quantum mechanics (although, as mentioned previously, it may well be better described as wave mechanics), we'd better discuss where the notion of quantization fits into the picture.

When we do put in the forces using the term in the Schrödinger equation containing the potential energy function $V(x)$, we find that, for our given $V(x)$, solutions to the Schrödinger equation exist only for certain "allowable" values of the total energy E. This general property is true for any potential energy function $V(x)$ (note 3.10). The particular allowable values of E depend on the particular form of the function $V(x)$, but the fact that only select values of E work is a general property of solutions $\psi(x)$ of the Schrödinger equation.

Thus, the Schrödinger theory is consistent with observed quantum behavior, such as the fact that only certain colors (energies) of light can be emitted by a given atom—those energies correspond to the change of the quantum state of the atom between two of its allowable energies. The energy of the emitted light quantum, or photon, is just the difference between the energies of states before and after the change (energy conservation again!). Since only certain energies are allowable for the states, then only certain energies (colors) are allowable for the emitted light.

By the way, since each type of atom has its own unique potential energy function $V(x)$, the emitted colors are characteristic of the type of atom that's emitting (or absorbing) the light. This, for example, is how astrophysicists can determine the composition of stars by analyzing their light as it reaches earth.

Finally, notice that the Schrödinger equation consists of three terms: two to the left of the equals sign (separated by the "+" sign) and one to the right of the equals sign. The first term is the mathematical representation of the procedure that, once you know $\psi(x)$, tells you how to determine the kinetic energy possessed by the particle at any location x. The second term, to the right of the "+" sign, is the potential energy times the value of the wave function at the location x. The third term, to the right of the "=" sign, is the total energy times the wave function $\psi(x)$.

So, if we look at the factors that multiply the wave function in the Schrödinger equation, we find that to the left of the equals sign we have the sum of the kinetic plus potential energies at the point x, while to the right of the equals sign, we have the total energy. Thus, the Schrödinger equation is just the wave-mechanical statement that the sum of the kinetic and potential energies at any given point is just equal to the total energy—the Schrödinger equation is simply the quantum-mechanical version of the notion of energy conservation. From this quantum-mechanical formulation of energy conservation arises the full set of constraints that prescribe the possible quantum mechanical wave functions for the object. This again illustrates the central importance of the idea of energy conservation (note 3.11).

It is a testimony to the essential role played by quantum mechanics that its development spawned a number of entirely new fields in physics, including atomic and molecular physics, the physics of solids (which includes the physics of semiconductors and microelectronic devices), and particle physics. It was with the development of these fields that the phenomenon of specialization arose, and scientists who contributed broadly across several fields of physics became rarer (but not extinct; the late Italian American physicist Enrico Fermi being a particularly notable exception).

Beyond this point, however, our discussion must reflect this trend toward specialization, so we'll need to leave the exploration of these other rich fields for another time. We can do this without regret because what lies ahead on our chosen path is certainly worth the journey. Indeed, our next destination, quantum field theory, is a further development of basic quantum theory, considered by some to be as profound a leap forward in our understanding of the fundamental workings of nature as the original quantum theory itself.

4

The Marriage of Relativity & Quantum Theory

———~~~~~~———

Relativistic Quantum Field Theory

In 1927, the field of physics was in an interesting state. Based on the bold hypothesis of Louis-Victor de Broglie, Erwin Schrödinger and Werner Heisenberg had put forward a new theory of the behavior of matter that fundamentally reshaped the way in which physicists view the natural world. As the only theory providing a description of the behavior of matter on atomic scales, quantum mechanics seemed destined to become a cornerstone of modern physical science. At the same time, however, the physics community was in no position to be smug about its accomplishments, for the quantum mechanics of Schrödinger and Heisenberg was inconsistent with the other great success of the modern physics revolution—Einstein's relativity.

Schrödinger and Heisenberg were able to explain why the electron in a hydrogen atom remains in perpetual orbit rather than spiraling quickly in to fix itself inertly against the proton. Their theory also provided a precise accounting for the observed energy levels, or allowable states, of the hydrogen atom. Yet, for their *nonrelativistic* quantum theory, space and time played distinct roles, quite at odds with Einstein's notion of a single, integrated underlying fabric of space-time. Despite the great achievements of Schrödinger and Heisenberg, it was clear that much work remained to be done.

This challenge ignited the imaginations of a number of brilliant young physicists, who found themselves in close contact under the guidance of Neils Bohr and his institute for physics in Copenhagen. Barely a year passed before a reformulation of quantum mechanics consistent with the tenets of relativity was put forward by the Englishman Paul Dirac. The ensuing development of relativistic quantum mechanics and the quantum theory of fields took some time, but with the work of Richard Feynman and others, relativistic quantum field theory reached its current state of perfection in the late 1940s.

Force Fields

To a physicist, a *field* is some aspect of the properties of a region of space that can be quantitatively assessed at every point in that region. The room in which you sit has a temperature that varies from here to there; the set of numbers that represent the readings of a thermometer at each point in the room is referred to as the "temperature field" of the room.

An electrically charged object at some point in the room will attract or repel a second electrically charged object placed in the room. The strength and direction of that attractive or repulsive force depends on where in the room the second object is placed. The electric force field or, more simply, the *electric field* of the first object is just the set of numbers that represent the strength and direction of the force that the second object would experience at any given point in the room.

Isaac Newton's theory of gravitation and James Clerk Maxwell's theory of electromagnetism are classical, or non-quantum-mechanical theories of the gravitational and electromagnetic fields. For classical field theories, the field associated with the force in question can be represented by a function of space and time (note 4.1). Mass (gravitational charge) generates a gravitational field in a prescribed way, electric charge generates electromagnetic fields in a somewhat different but equally definitive way, and so forth.

Although any given charged object is localized, existing at a well-defined point in space, the field generated by the charge (such as that of the electrically charged object we placed in the room just above) extends throughout space. Thus, in this classical picture, fields are responsible for a phenomenon of action-at-a-distance, the ability of an object carrying the charge associated with a certain force to influence another object that also carries the appropriate type of charge but is not in direct contact with the first object. The earth orbits the sun, not because the sun reaches out a fiery arm and turns the earth about it but because the considerable mass of the sun generates a gravitational field throughout the solar system, through which field the massive earth moves, and whose resulting continual tug keeps the earth from flying off into interstellar space.

Knowing the field, it's easy to work out the strength of the force that a charged object would feel if placed in the field without knowing anything about the charge or charges that generated the field in the first place. The force that the object would feel is simply given by the value of the field at that point in space and time, multiplied by the magnitude of charge possessed by the object.

For those who like math, the following example illustrates this point. If an object possessing an electric charge of magnitude Q is placed at rest a distance r from a stationary electric charge of magnitude q, the expression for the electric force exerted on the charge Q by the charge q is quite simple, and was determined by a series of careful experiments performed by the French physicist Charles Coulomb in 1785. The electrostatic, or "Coulombic" force, F_Q exerted on the object carrying charge Q obeys the inverse square law

$$F_Q = k\frac{Qq}{r^2};$$

doubling the distance r between the two charges reduces the strength of the force by $(½)^2 = ¼$ and so forth. The electrostatic force constant k is a number, determined by experiment, that expresses the overall strength of the electrical force between two charged objects; k is a fundamental constant that is completely situation independent; no matter what the charged objects are, or what values their charges and separation take, k always has the same value (note 4.2).

To introduce the electric field, we rewrite this in terms of the equivalent expression

$$F_Q = (Q) \times \left(k\frac{q}{r^2}\right). \tag{1}$$

Then, by fiat, we declare the magnitude E of the electric field generated by the charge of magnitude q, at a distance r from that charge, to be the second of these two factors:

$$E = k\frac{q}{r^2}. \tag{2}$$

With this definition, we see that the force felt by the object of charge Q is given by

$$F_Q = Q \times E; \qquad (3)$$

it's simply, as stated above, the product of the strength E of the electric field at the object's location, times the magnitude of the object's charge. Knowing the electric field E at every point in space, we know the force. We don't have to worry about what generated that field—a single stationary charge q, as in this example, or a multitude of q's placed at different positions about the object with charge Q. The electric force on the charge Q will always be related to the resulting electric field according to the simple expression (3).

But why bother. Isn't this just a vacuous mathematical manipulation? Who cares whether we express the force between two charged objects directly using expression (1) or break it up into two steps and make use of expressions (2) and (3)? In fact, the latter seems less economical, arbitrarily requiring the introduction of an unnecessary physical entity known as electric field.

The answer lies within Maxwell's fully unified theory of electromagnetism. In this theory, time-varying electric and magnetic fields can support each other, leading to electromagnetic-field disturbances that propagate endlessly through free space, *in the complete absence of any nearby electrically charged objects.* So, the fields do seem to possess an independent physical meaning. Furthermore, Maxwell's theory not only predicts the possibility of such electromagnetic field disturbances, it also relates the speed at which this disturbance travels through space to the electrical force constant k and the corresponding magnetic force constant μ_0 (note 4.3). Plugging in the numbers for k and μ_0, Maxwell found that such a disturbance would propagate through space with a speed of about 299,800,000 meters per second, or about 186,000 miles per second—the speed of light.

Thus, Maxwell succeeded in explaining the origin of light, nature's premier mode of transporting energy and information between physical (and biological) systems. Without the introduction of the notion of a field, this work would not have been possible.

As we reshape our notion of force fields from the point of view of quantum field theory, we should keep one thing in mind. Quantum field theory is not really, in and of itself, a single theory of nature. Instead, it is something of even greater stature. Quantum field theory is really a framework on which all current theories of the fundamental interactions of matter are based. These theories, which bear such names as the electroweak theory, quantum electrodynamics, quantum chromodynamics, supersymmetry, quantum gravity, and the like, whether supported by experimental evidence or merely conjectural, are all different deployments of the principles of quantum field theory and its generalizations. We'll soon see why it is that physicists have so much confidence in this field-theory framework.

The Quantization of Fields

For conventional quantum mechanics, objects move in the presence of forces whose field strengths vary smoothly from point to point in space. These fields are incorporated into the Schrödinger equation (the quantum mechanical wave equation that the all-encompassing wave function must satisfy) in terms of their associated potential function $V(x)$. The potential function represents the amount of work the object must do against the force in moving to the position x, for any position x. The requirement that the wave function satisfy the Schrödinger equation associated with the particular potential function $V(x)$ leads to the quantization of the motion of the object; when probed, the object will be found in one of a finite number of possible states, each of which corresponds to a specific value of the object's total energy.

Quantum field theory takes this notion one step further, asserting that the fields themselves are quantized. This view of the nature of fields is radically different from that of conventional quantum mechanics.

From Maxwell's classical perspective, light is a self-supporting disturbance in the electromagnetic field that carries energy from the original source of the disturbance (such as a wiggling electric charge in a faraway star) to the observer. But according to the quantum-mechanical notion of wave-particle duality, in some contexts, a ray of light behaves as if it were composed of an immense number of tiny particles, known as *photons*. These tiny, indivisible packets of light are prime candidates for the quanta of the electromagnetic field. The electromagnetic field, rather than being a quality of space that varies smoothly from point to point, instead becomes something entirely different: an assemblage of photons, each with a specific energy given by

$$E = hf,$$

where h is Planck's constant and f is the frequency (color) of the photon. Since photons, as particles of light, travel at the speed of light, any attempt to develop a quantum-mechanical description of their behavior must be consistent with Einstein's relativity. Since the Schrödinger equation is not, we will need to introduce at least one new quantum-mechanical wave equation in this chapter.

Particle Exchange: A Fresh Look at the Nature of Forces

If the fields are quantized into discrete bundles that carry energy and behave like particles, how do we understand the mechanism by which they exert their characteristic forces? To answer this, let's strap on a pair of skates and head down to the local rink.

Picture two ice skaters facing each other, one of whom holds a large ball in her hands. As she throws the ball toward the other skater, she recoils against the action: as the ball flies away, she begins to slide backward. The harder she throws the ball, the quicker she recoils, but once the ball is thrown, her speed is set, and it doesn't change again.

In catching the ball, the second skater absorbs its momentum and energy, and goes sliding back in the other direction. The speed at which he recoils after catching the ball again depends on how hard the ball is thrown, but once the ball is caught, his speed is set too, and it doesn't change again.

So, the result of the exchange of the ball is simply that both skaters slide off in opposite directions, *just as if they had exerted some sort of force on each other.* From the point of view of quantum field theory, this is exactly how forces are conveyed. Replace the two skaters with two electrons, and the ball with a photon. Seen in this way, the repellent force between the two electrons is the result of an exchange of one or more photons—the quantum of the electromagnetic field—between the two electrons.

This point of view seems completely at odds with the action-at-a-distance principle of classical field theories. However, once you know how to make use of it, you can show that for two charged objects sitting at rest quantum field theory predicts that the force (and field) will have just the properties that one would expect from the classical approach, as if they were indeed influencing one another according to action-at-a-distance. So quantum field theory doesn't invalidate the classical theory of fields; it merely substitutes a more refined notion of how those fields are generated. Ultimately, this allows for the extension of the force-field interpretation of the interactions of matter into regimes for which the classical approach is inadequate—regimes for which quantum mechanics and relativity must be taken into account.

Where do the photons come from that are exchanged between the two electrons as they repel each other? They come from nowhere. They are created (thrown) and absorbed or annihilated (caught) for the sole purpose of transmitting the force. Before and after their brief passage between the electrons, they do not exist.

But this is fine because according to the tenets of relativity, it is *not* true that matter can neither be created nor destroyed. Instead, what is true is that energy can neither be created nor destroyed (as long as mass is properly accounted for as energy according to the expression $E = mc^2$); energy is conserved. If at some instant in time there are just two objects (two electrons) and at a slightly later instant there are three objects (two electrons plus one photon), then everything is fine, as long as the total energy of the system is unchanged.

Instead of having the two electrons repel each other at rest, what if you throw them at each other? If you work out the energy balance, making sure that energy is conserved at every step of the way as the electrons repel each other, you find that some of the exchanged photon's energy will be accounted for as mass. The exact amount of mass required of the photon depends on how hard the electrons are thrown at each other and how close they come before they repel, but in this case, the photon always has to have some amount of mass.

But photons don't really weigh anything (how much does a light beam weigh?), so how could this particle-exchange-force idea possibly be right? We are saved by Heisenberg's uncertainty principle. The uncertainty (ΔE) in the energy of an object is related to its lifetime (Δt) according to

$$\Delta E = \frac{1}{\Delta t}\frac{h}{4\pi},$$

with h, as usual, as Planck's constant.

The photon that is exchanged lives for only a short time Δt (while it flies at the speed of light between the two electrons). As a result, there is a substantial uncertainty in its energy, and, because mass and energy are closely related, its mass. Thus, in the odd world of relativistic quantum mechanics, a given particle, known experimentally to have certain mass (e.g., zero for the photon), can have an entirely different mass if it lives only for a brief period as it is exchanged between two interacting objects. Technically, we say that such a short-lived particle, required by energy conservation to have a mass different than its known value, is *off mass-shell* (note 4.4). Such particles are also referred to as *virtual* particles.

The Four Forces Again

From the point of view of quantum field theory, the electromagnetic force is a consequence of the exchange of virtual photons—the quanta of the electromagnetic field—between electrically charged objects. It's easy to extend this notion to the description of the other three forces. Each force is generated by the exchange of the quantum of the associated force field between objects that carry the charge appropriate for that force (electric charge for the electromagnetic force, mass for the gravitational force, color charge for the strong nuclear force, weak isospin for the weak nuclear force).

To describe any given force, we need to state what the associated force-field quanta are and describe their properties (e.g., mass, charge, and "spin," the last of which will be described below). Developing a quantum field theory for a given force requires the identification and classification of the field quanta that get exchanged when the force is at play.

In the case of the electromagnetic force, there is a single field quantum (the photon), but in general, there need not be just one. For example, the weak nuclear force has associated with it not one but three separate field quanta. Two of these, the "W bosons," are like each other in every way, except they have opposite electric charge; the third, the "Z^0 boson" (or simply Z boson) is heavier and electrically neutral. One can ask why it is that two of the three quanta of the weak nuclear force carry the charge associated with the electromagnetic force. For now, we'll say that this is merely one of the properties of the weak force-field quanta that make the weak force what it is and leave it at that. In the bigger picture, this suggests that the electromagnetic and weak nuclear forces are intimately related to one another.

In reality, there's a bit more to the specification of the quantum field theory associated with a given force than the identification and description of its associated field quanta; there's another ingredient that needs to be included. This ingredient is the nature of the "coupling" between the field quanta and the charged objects that toss these quanta back and forth—the details governing the process of the creation of a given field quantum by the tossing object and the absorption of the quantum by the receiving object.

One aspect of this coupling is its overall strength—the stronger the coupling, the more quanta are exchanged in a given interaction, or the greater the likelihood that quanta will be exchanged at all (for our skaters, a stronger force would correspond to a greater likelihood, in any given instant, that a ball would be exchanged between them). The strength of a force's coupling to an object is related to the magnitude of the associated charge carried by the object. For example, the more mass (gravitational charge) carried by an object, the more gravitational field quanta it will toss out or absorb every second and the greater the resulting pull of gravity on the object will be.

A second aspect of the description of this coupling between a field quantum and appropriately charged objects is referred to as its *space-time property*. The weak nuclear force is particularly interesting because of its tendency to irreverently violate basic symmetry principles, such as the equitable treatment of matter and antimatter. It is the space-time property of the coupling that determines, among other things, whether this (and other) symmetry principles are respected or violated and, if so, to what extent.

In practice, this is not as much of a complication as it may appear because there are only a handful of different space-time properties consistent with the tenets of Einstein's relativity that are available to the coupling between field quanta and charged objects. In addition, once the field quantum is identified, its characteristics (specifically, its spin) restrict the possible space-time properties of the coupling even further. So, once the field quanta are identified and their properties determined, it's usually fairly clear what the appropriate space-time property of the coupling must be, and then the theory is essentially complete.

The development of quantum field theory was a great leap forward in our understanding of nature. It provides a description of the behavior of forces with phenomenal quantitative power, while at the same time simplifying and extending our description of those forces.

Maxwell's classical theory of electromagnetism, triumph that it was, is stated in terms of four rather intimidating interrelated (differential) equations, the development of which spanned a century of painstaking experimentation and brilliant theoretical strides. With quantum field theory, one need only identify and describe the electromagnetic field quantum (the photon) and its straightforward coupling to electrically charged objects, and the job is done. Moreover, the resulting quantum field theory of electromagnetism is more broadly applicable than the classical theory, reproducing the results of the classical theory for the everyday applications that drove the development of classical electromagnetism, while providing the appropriate modification of that description in the case that high energies (relativity) or short distances (quantum mechanics) are at play. Even more, the particular properties of the strong and weak nuclear forces (especially the fact that these forces are short ranged, operating only over nuclear scales of 10^{-15} meters or less) make it impossible to deduce the corresponding Maxwell-like classical field equations for these interactions. Their description is *only* possible within the framework of quantum field theory.

Feynman Diagrams

One of the primary proponents of the development of quantum field theory and its use in the description of the electromagnetic force was the American physicist Richard Feynman (note 4.5). Feynman's contributions play a central role in the formal, quantitative structure of quantum field theory, most of which lies beyond the scope of this book. However, one of these contributions, the representation of quantum field theory processes in terms of "Feynman diagrams," provides a straightforward way to visualize and describe the possible interactions between objects subject to the influence of a given force. Feynman diagrams will be of great use to us throughout the remainder of this book.

To understand the meaning of a given Feynman diagram at the level appropriate for our discussion, there is one mathematical hurdle to overcome: the description of the motion of a particle in terms of a graph of position versus time. Such a graph is known as a space-time plot, or space-time diagram.

Take a look at figure 4.1, the coordinate axes of a graph with the horizontal (x) axis representing the position of the object and the vertical (t) axis representing the time at which the object has the given position x in space. By convention, distance increases from left to right, and time increases (elapses) from bottom to top, as you might expect. If you like, you can think of the x-axis as a meterstick (or yardstick) with numbers increasing to the right, while the t-axis represents the readings on a ticking stopwatch.

Figure 4.2 shows how an object at rest would be represented on such a graph. If it's at rest, with a speed of zero, then it's not going anywhere. Its position, represented by its location along the x-axis, is fixed in time. You might

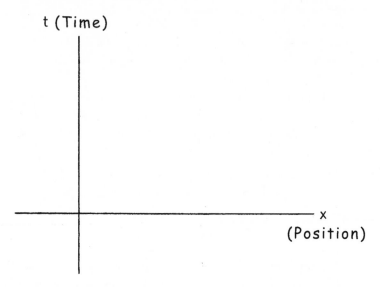

Fig. 4.1. For the axes of the position versus time (space-time) plot, distance *x* increases to the right, and time *t* increases toward the top of the page.

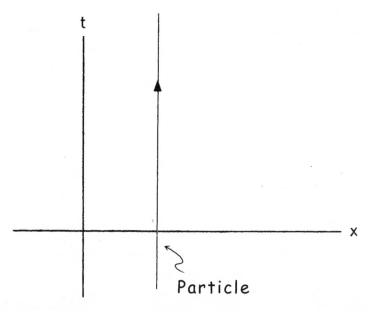

Fig. 4.2. For an object at rest, the position *x* is fixed, so only the time changes. Note that the arrows point from lesser to greater times: the particle is moving *forward* in time, as you would expect!

expect such a particle to be represented by a dot on this graph, but re-member, time waits for nobody or, for no object, so for this particle at rest, time marches inexorably forward. Even though this particle is happily en-joying its repose, going nowhere in space (i.e., possessing an unchanging position along the space, or x-axis), it is traveling through time, so it is rep-resented by a line parallel to the t-axis—a line that has associated with it a constant value of the spatial position x but an ever-increasing value of the associated time t. Furthermore, this passage through time has a well-defined sense—physical objects, we know from experience, must always travel for-ward in time. This sense of elapsing time is represented by the arrow, which points from lesser to greater time.

Now, take a glance at figure 4.3a. The dotted lines show us that when the particle is at the (space-time) point A, it has time T_1 and position X_1. At some later time T_2 (remember that time elapses from bottom to top), the particle is at the space-time point B and has a greater reading, X_2, for its spa-tial position; as time elapses, the particle moves toward larger and larger readings on the meterstick. Thus, this graph represents a particle moving through space, from left to right, with some speed. The exact value of that speed, in meters per second, is not important for us to specify. The arrow on the line again reminds us that the particle is moving forward through time. Similarly, the line in figure 4.3b represents a particle moving with some speed or other, but in this case, it is going in the opposite direction, from right to left.

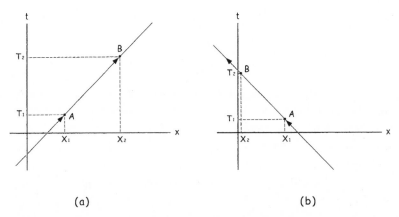

(a)   (b)

Fig. 4.3. In a, the object is further to the right at the later time T_2, so it's mov-ing to the right. Similarly, in b, the object is moving to the left.

Now we are ready to draw our first Feynman diagram. Since these diagrams are Feynman's way of depicting the interaction of two objects by way of a given force, to be definite, we'll pick the electromagnetic force, and for the objects, we'll pick two electrons, which carry the appropriate charge (electric charge) for influencing each other by way of the electromagnetic force. Since this is supposed to be a Feynman diagram, and we know how much Feynman loved his quantum field theory, this diagram had best represent the conveyance of the electromagnetic force from the point of view of field theory.

Take a look at figure 4.4a. Using the interpretive skills developed in the last few paragraphs, what we see represented there is the following: At early times (toward the bottom of the plot), two electrons approach each other, one from the left moving toward the right, and the other from the right moving toward the left. At point A, the electron on the left emits (tosses out) a photon (labeled γ), the quantum of the electromagnetic field. This photon carries energy and momentum away from the electron that emits it, causing that electron to recoil and to move back toward the left. The photon moves to the right until it is absorbed by the second electron at point B, at which point the second electron recoils to the right. This process of photon exchange occurs over a very short distance, typically, the length of the photon's path will have atomic dimension (10^{-10} meters) or less, so the details of the photon exchange process are lost to the observer. What the observer would see is two electrons, originally moving toward each other, repelling each other and scattering back in the directions they each came from. What this diagram represents is one way, in fact, the simplest way, that two electrons can influence each other through the electromagnetic force from the point of view of quantum field theory.

Figure 4.4b is similar to 4.4a except, in this case, the photon is emitted by the right moving electron at point B, and absorbed by the left-moving electron at point A. Again, we have to step back and ask what the observer would see: two electrons, originally moving toward each other, repelling each other and scattering back in the directions they came from. In other words, exactly what the observer would see in figure 4.4a. The processes are identical in the sense that there's no way for the observer to tell whether process 4.4a or 4.4b caused the electrons to scatter away from each other. To make that distinction, the observer would have to cut into the diagram and detect the photon, but then the observer's apparatus would absorb the photon, and it wouldn't make it over to the other electron to play out the

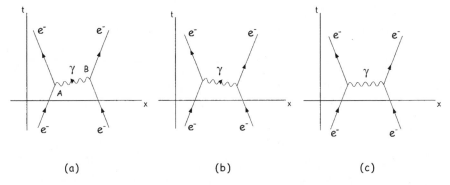

Fig. 4.4. Two separate but indistinguishable ways in which two electrons can influence each other through the electromagnetic interaction (a and b). The Feynman diagram for these processes is shown in c, which represents both the possibilities of a and b. In other words, a graduate student charged with calculating the probability that the process represented by the Feynman diagram c will take place must add together the probabilities of processes a and b.

transmission of the force. The process we were interested in studying would be kept from happening.

Since we do want the two electrons to scatter, we cannot disturb the scattering process to observe the photon and determine whether the electron on the left or the electron on the right threw out the virtual photon. So, there's no point in distinguishing between the two possibilities. The scattering process is the combination of these two different indistinguishable and identical pieces (fig. 4.4a and 4.4b). Thus, we might as well represent the process as shown in figure 4.4c, which we know will always really mean the combination of figure 4.4a and 4.4b. Figure 4.4c is the true Feynman diagram for this process.

Our Feynman diagram only represents motion in one dimension. We know, however, that things, in general and electrons, in particular, move through three-dimensional space, not just back and forth on a line. Even if the electrons are really approaching each other from opposite ends of a line segment, after they influence (scatter from) each other, they will usually be found going off in some completely different direction, which requires the other two dimensions of space to describe. So, the Feynman diagram (fig. 4.4c), you would say, cannot really represent, in full generality, the process of the scattering of two electrons through the exchange of a single photon.

This is not the case. No matter how many dimensions you need to describe the motion of the electrons before and after the scatter, it's still fundamentally the same process—two electrons influencing each other through the exchange of a single photon. The purpose of Feynman diagrams is to categorize and describe the different possible types of interactions between a given set of interacting particles through a given force and not to represent the exact motion (speed and direction) of the particles after the interaction. After all, this is quantum mechanics, and the uncertainty principle guarantees that the result cannot be the same every time the scattering is performed. What characterizes the type of interaction taking place is simply the list of particles taking part in the interaction (two electrons and a photon in our case) and the pattern of interconnection of these particles in the Feynman diagram (the photon connecting the two electron trajectories). Mathematicians would like us to say that the geometry of the diagram doesn't matter, just the topology.

Feynman diagrams are so useful because they contain all the information that one can possibly know about the process that is represented (here, the electromagnetic interaction between two electrons conveyed, or mediated, by the exchange of a single photon). A person who knows quantum field theory can look at the Feynman diagram associated with a given process and with enough paper and pencil lead (or pen ink if they're really experienced) turn this diagram into an explicit, quantitative calculation of everything that can possibly be predicted about the interaction.

This calculation will reveal two things. The first, known as the *total cross section*, is the probability that the two particles moving toward each other will actually exchange the photon and interact rather than just pass by each other unheeded. The second, known as the *differential cross section*, is just the relative probability of any given final result of the scatter, assuming that the scatter really did occur. Choose the energy and direction you would like the particles to have after the interaction has taken place. The differential cross section provides the probability that your chosen energy-direction combination will be achieved, relative to the probability of any other possible energy-direction combination.

Think of the popular boardwalk game in which you pay a few dollars for a chance to throw balls at a target that, when hit, will cause a particularly annoying individual to fall into a vat of cold water. The total cross section for this process is related to the probability that any given throw will dunk this fellow. The differential cross section is related to the probability that, given that the target has been hit and this grating fiend is going to take a drink, the ball you threw will bounce off the target with a given speed and direction—such as briskly back into your face, providing the irrepressible cad with even more ammunition for his taunts once he climbs back out of the tank.

What more could you ask to know about a process like this in which two electrons influence each other for an exceedingly brief period and then go on their scattered ways as if they had never met each other (note 4.6)? The description in terms of the total and differential cross sections is complete—it gives us all the possible a priori (before-the-fact) predictive power admitted by the probabilistic nature of quantum mechanics.

Vertices and the Minimal Interaction

In this chapter, we have come to know the electromagnetic force as the force between electrically charged objects caused by the exchange of one or more photons between the charged objects. As we'll see shortly, the use of Feynman diagrams allows us to expand our notion of what the electromagnetic force really is and, in fact, will even incline us to drop the use of the expression "electromagnetic force" in favor of the more general notion of the "electromagnetic interaction" (although we'll continue, somewhat loosely, to use the two terms interchangeably throughout the remainder of the book).

Take another look at figure 4.4c, the Feynman diagram depicting the mutual repulsion of two electrons through the exchange of a single virtual photon. Is this the only possible type of interaction between these two electrons? Is there another possible way for these two electrons to influence each other?

There is, and figure 4.5 is an example of just such a process—the interaction of two electrons through the exchange of two photons. From what we know about quantum field theory, there's no reason why this shouldn't happen—and it does! As Feynman has said, anything that can happen does happen. Since good old-fashioned quantum mechanics lies at the heart of this picture, it's not that the force is exerted through the exchange of one photon some of the time and two photons other times. Rather, whenever the electrons repel each other and go flying back, there's a certain probability that it was due to the exchange of one virtual photon and a certain other probability that it was due to the exchange of two virtual photons. We

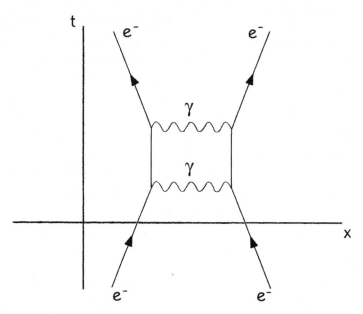

Fig. 4.5. The Feynman diagram representing the interaction of two electrons through the exchange of two photons.

can never know (without disturbing the interaction) which it was, so it was both, every time. This progression doesn't stop with the exchange of one or two photons—any number of photons can be exchanged, so the true electrostatic repulsion of the two electrons comes about by a combination of all of these (and other) possibilities.

In general, there's always an infinite number of different ways that any given process (such as the scattering of two electrons by way of the electromagnetic force) can take place, each one represented by its own unique Feynman diagram. The electrons can exchange any number of virtual photons—one, two, six, forty, whatever. If this is the case, then how can quantum field theory be of any use at all? If there's an infinite number of difficult calculations to do every time you want to make a prediction, how can you ever hope to finish the calculation?

What saves us is the coupling between the photon and the electron. Look again at figure 4.5. Every time an electron and photon meet in the Feynman diagram, we have to remember to take into account the coupling strength between the electron and the photon. This coupling is less than one; for the electromagnetic force, it's about $\frac{1}{100}$ (note 4.7). For every intersection between the electron and photon we add to get the new diagram, we pay a price of about 100 in the probability that that diagram contributes to the overall process. Comparing figures 4.4c and 4.5, we see that figure 4.5 has two extra electron-photon intersections, so we would expect it to be roughly $(\frac{1}{100}) \times (\frac{1}{100}) = (\frac{1}{10,000})$ as likely as figure 4.4c. It doesn't make a whole lot of difference whether you include the more complicated diagrams involving two, three, and more exchanged photons. Only if you're interested in getting a really precise prediction would you be compelled to include these more complicated higher-order diagrams.

Thinking about Feynman diagrams in this way brings about a sea change in our thinking about the way objects interact. We can start thinking of these intersections, or *vertices*, between electrons and photons, rather than the *exchange* of photons, as the defining element of the electromagnetic force. Every photon exchange involves two such electron-photon vertices, so perhaps the vertices are indeed more fundamentally characteristic of the electromagnetic interaction than is the process of photon exchange.

We can make use of our new fundamental element—the electron-photon vertex—to construct a strikingly different Feynman diagram (fig. 4.6). It begins, at early times toward the bottom of the diagram, with a real (nonvirtual) photon, from a laser beam perhaps, coming from the left, approaching an electron coming from the right. At point A, the photon is absorbed by the electron at an electron-photon vertex—the vertex that, in our revised point of view, is the fundamental and characterizing element of the electromagnetic force. (Note that the vertex at A is identical to either of the vertices in fig. 4.4c—a wiggly line representing a photon terminates on a solid line representing an electron.) Then, at point B, the diagram employs another vertex at which the electron spits out a photon—a different photon, which, in general, has a different energy than the initial one.

Technically, to conserve mass-energy at all times, the mass of the electron has to be slightly different than its true mass during the time interval between the absorption of the first photon and the emission of the second. But, just as for the virtual photon of figure 4.4c, that's fine because of the uncertainty principle (recall this discussion toward the end of "Particle Exchange: A Fresh Look at the Nature of Forces" in this chapter). In this case, instead of having a virtual photon, we have a virtual electron.

By viewing the electromagnetic force in terms of electron-photon vertices rather than photon exchange, we are inclined to expand our list of processes governed by that force. The photon exchange of figure 4.4c is just

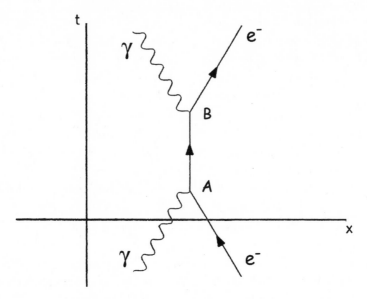

Fig. 4.6. The Feynman diagram representing the most basic way that an electron and photon can interact by way of the electromagnetic interaction. Such interactions are known as Compton scatters, after the American Nobel Laureate Arthur Compton, who identified the process.

one possible mode of interaction afforded by the electromagnetic force. If we consider the more basic *vertices* as the fundamental description of electromagnetism and play around a bit, we get the diagram of figure 4.6, a whole new process in which an electron beam can scatter photons from a light beam by absorbing the photons and reemitting them with different directions and energies.

It's one thing to conjecture about the consequences of describing electromagnetism in terms of electron-photon vertices; demonstrating that this hypothesis is correct is another thing altogether. In this case, experimentation actually preceded conjecture. In 1923, well before the development of quantum field theory, the American physicist Arthur Compton discovered that gamma rays (energetic photons) sometimes lost energy as they bounced off the stationary electrons in a carbon block. Compton was able to show, with further experimentation, that this *Compton scattering* process was consistent with the absorption and reemission, at a different energy, of the gamma-ray photon, the process depicted, in modern terms, by figure 4.6. This work garnered Compton the 1927 Nobel Prize in Physics.

It *is* the more basic and general electron-photon vertex that we should look to as the fundamental component of the electromagnetic force, not photon exchange, which is only one of several modes of interaction that can be constructed from electron-photon vertices. This most basic, irreducible action of electromagnetism is known as the *minimal interaction vertex*. The job of anyone developing a quantum field theory of any given fundamental force is to make a list of all minimal interaction vertices associated with the force and to delineate the strength and space-time property of each interaction vertex.

For electromagnetism, there is only one such minimal interaction vertex. This is the electron-photon vertex, or, more accurately, the fermion-photon vertex, the electron being but one member of a family of particles known as fermions that interact with photons by virtue of their electric charge (we'll learn what "fermions" are shortly; the family of fermions includes both leptons and quarks, not all of which possess electric charge). This vertex has a coupling strength equal to about $\frac{1}{100}$ and, technically speaking, the space-time property of a vector. That's it. That's basically all there is to our quantum field theory of electromagnetism, known as *quantum electrodynamics*, for which Feynman and two others (Julian Schwinger of the United States and Shin'ichiro Tomonaga of Japan) won the 1965 Nobel Prize. Well, not quite all, because as construed above, the theory doesn't quite work. Fixing it up requires another leap into the world of the counterintuitive; it was this leap that most impressed the 1965 Nobel Committee. We'll discuss this problem and the technique of renormalization that solved it toward the end of the chapter.

Note that no one ever talks about the electromagnetic force between electrons and light, and for good reason—there isn't any. Light doesn't get bent by charged objects. Take a house key laden with a painfully large amount of static charge (easily produced on a dry day on a plastic playground slide) and hold it up to a light, whether from the sun, a light bulb, or a laser beam. The charge on the key will not bend the light passing near it. Electrically charged objects do not exert a force on light, but they do occasionally interact with it by absorbing and reemitting it in a different color (with a different energy). However, two statically charged objects do exert a force on each other. In both cases, light and charge or charge and charge, the workings of the interaction can be described in terms of a single fundamental component, the electron-photon vertex. Only in one of the two cases is a force exerted, but in both cases, there is an interaction. Thus, in quantum field theory, we speak most correctly in terms of theories of fundamental interactions; forces (as we usually construe them in terms of the action-at-a-distance interplay between two appropriately charged objects) are just one facet of the more general notion of an interaction.

In the mid-1800s, Faraday's introduction of force fields liberated our thinking in a way that led eventually to the explanation of the phenomenon of light. In the mid-1900s, the *quantization* of these very same fields led to a further liberalization of our thinking, allowing the concept of force to be generalized to the notion of interaction, thereby permitting the incorporation of a host of new types of behavior into our understanding of the fundamental workings of nature.

Relativistic Quantum Mechanics, Antimatter, and Spin

At this point, we digress to consider some of the "relativistic" aspects of relativistic quantum field theory (which we have shortened to quantum field theory, or even field theory, in the previous discussion). We'll begin the digression with the introduction of *angular* momentum.

Let's take another look at Planck's constant h. Planck's constant sets the scale for the most basic quantum mechanical behavior, including the wavelength of material objects (de Broglie's relation: $\lambda = h/p$, where p is the object's momentum) and the uncertainty principle ($\Delta p \Delta x \geq h/(4\pi)$). The unit of Planck's constant—the measuring stick used to establish its (minute) size—is joule-seconds, where the joule is the unit of energy. The joule-second, it turns out, is the unit of something called angular momentum.

To get a feel for angular momentum, consider a carousel at an amusement park spinning about its axis (note 4.8). During the ride, the carousel has a lot of energy of motion. If you were to try to stop the carousel by planting your feet and grabbing one of the horses, the horse would certainly win the contest. Yet, this energy of motion (kinetic energy) is not in the form that we are used to. This kinetic energy is not associated with the motion of the carousel through space—the carousel is fixed at a point and, despite its great energy, poses no threat to anyone with enough wits not to walk into it. The energy derives not from the motion of the carousel through the amusement park but rather from the spinning of the carousel about its axis.

Analogous to the carousel, particles such as protons and electrons can also spin about their axes. Unlike the carousel, fundamental particles cannot be made to spin faster or slower or have their spinning motion stopped. The intensity of spinning, the angular momentum, is a fixed property of the particle. Just like a particle's mass, the angular momentum associated with this spinning motion, known simply as spin, is a characteristic of the particle at hand. The direction of the axis about which the particle is spinning *can* be changed, but the rate of spinning about that axis, no matter which way it points, cannot be changed.

We've seen that quantum mechanics is required to understand and describe the behavior of individual particles. Thus, we might expect that quantum mechanics and, in particular, its attendant scale factor h, might have a role to play in the spin of fundamental particles, especially since the unit of h, as we've seen, is just that of angular momentum.

The amount of angular momentum possessed by an electron or a proton is $\frac{1}{2}\hbar$, where for expediency we have introduced the reduced Planck's constant $\hbar = h/2\pi$ (to be read as "h-bar"). An individual photon, on the other hand, possesses an angular momentum of $\hbar$. We often forget about the $\hbar$ and just say that electrons and protons are spin-$\frac{1}{2}$ and that photons are spin-1.

The set of particles with half-integer spin (spin equal to an odd number of half multiples of $\hbar$, such as $\frac{1}{2}\hbar$, $\frac{3}{2}\hbar$, $\frac{5}{2}\hbar$, and so forth) are known as fermions. It's a testament to the incredible breadth of Enrico Fermi's work that, in addition to Nobel-caliber accomplishments in nuclear physics, particle physics, and the physics of solid materials, he was also the first to work out the quantum-mechanical laws governing the interplay of objects with half-integer spin.

The corresponding set of particles, with spin a pure integer multiple of $\hbar$ ($1\hbar$, $2\hbar$, $3\hbar$, and so on), are known as bosons. The work of Satyendra Nath Bose on the quantum-mechanical properties of systems composed of identical bosons impressed Einstein, who was subsequently able to secure Bose a professorship at India's Dacca University with a recommendation written on the back of a postcard, despite the fact that Bose never bothered to earn his Ph.D.

Each of these classes of particles—fermions and bosons—plays a distinct role in the world of particle physics. Spin-$\frac{1}{2}$ fermions (such as electrons and quarks) are the components of what is conventionally thought of as matter, while the fundamental bosons (such as the photon) are the conveyers of force.

Schrödinger's and Heisenberg's theory of quantum mechanics was a stunning success but only as long as none of the particles in the system under study were traveling so fast that Einstein's relativity needed to be taken into account, in which case their nonrelativistic theory fell to pieces. In tackling this problem in the late 1920s, the set of young physicists who initiated the development of relativistic quantum mechanics established themselves as the brightest of their generation (note 4.9). As we'll see, the imposition of relativity on the new quantum theory did not come easily, but through these physicists' brilliant reinterpretation of what had seemed like fatal flaws in their work, a much deeper and richer understanding of nature arose than anyone had dared anticipate.

Recall the discussion of chapter 3. The Schrödinger equation is the quantum-mechanical statement of energy conservation: The energy due to motion (kinetic energy) and the energy due to fighting against the field of whatever force you are considering (potential energy) is just equal to the total energy, a number that is fixed for all time (since energy is a conserved quantity) as long as the system under consideration is not disturbed from the outside. Relativity, though, tells us that the mass of the particle that we are trying to describe quantum mechanically also needs to be included in the total energy. This requires a modification of the wave equation.

To keep things from being too complicated, the proponents of the relativistic theory decided to begin with the description of *free particles* — particles moving freely through space, not under the influence of some external force. So, the term in the Schrödinger equation associated with fighting back and forth against the force, the potential energy or $V(x)$ term, could be forgotten about. This is really no loss at all, for in relativistic quantum *field* theory, which grew out of this formulation of relativistic quantum mechanics, the forces are reintroduced through the minimal interaction vertices and the exchange of field quanta. So, free-particle relativistic wave equations are all that we'll need to consider.

What comes out of this consideration is the quantum mechanical version of *relativistic* energy conservation for a free particle. The sum of the kinetic energy and the energy associated with the particle's mass is just the total energy, which again must be conserved. (If we create a virtual particle at a minimal interaction vertex, it's now OK because the mass-energy of this newly created particle is correctly taken account of in the energy balance of this *relativistic* formulation of quantum mechanics. The modeling of a force through the creation and absorption of virtual particles is an intrinsically relativistic point of view.) The resulting (differential) equation is known as the Klein-Gordon equation, and there's nothing to stop us from solving it (instead of the nonrelativistic Schrödinger equation) to find the wave function of a particle moving with relativistic speed.

There is, however, something to make us unhappy with some of the wave functions we wind up with. Half of the various possible solutions of the Klein-Gordon equation have *negative* energy. In addition, in quantum mechanics, when you square a particle's wave function (multiply it by itself at every point in space), you get a function that gives the probability of finding that particle at any given point in space. If you square the wave function of a negative-energy solution to the Klein-Gordon equation, the probability is negative, a result that makes no physical sense (note 4.10). Negative energy solutions with negative probabilities—this does not sound healthy.

Many would have seen this as a fatal flaw but not Paul Dirac. Dirac was convinced that a meaningful relativistic wave equation ought to exist. He analyzed the failings of the Klein-Gordon equation to find the cause of the negative probability solutions, enabling him to propose a wave equation with the same physical content (kinetic energy plus energy due to mass equals total energy for a free particle) but which avoided the plague of negative probability. This task was not particularly straightforward because what Dirac ended up with was a "matrix equation"—four separate but intertwined differential equations that needed to be simultaneously satisfied by its wave function solution. Messy though this approach may seem, it immediately provided two wonderful dividends: while the negative energy solutions remained, the negative probabilities were gone, and the new equation, in its complexity, precisely incorporated the behavior of spin-½ particles.

With these successes, Dirac was convinced that the final hurdle—the explanation of the negative energy solutions—must be surmountable. His insight was that the negative energy solutions could be interpreted as positive energy solutions for *antimatter* particles that would behave in every way like their particle counterparts but would have opposite charge. This suggestion was unspeakably bold. No one had seen or even thought of antimatter at that time; it wasn't even yet the stuff of science-fiction novels. Dirac proposed it to satisfy his conviction that his candidate for a relativistic wave equation describing spin-½ particles had to be correct. Antimatter was simply the ingredient that he needed to make it all work. So he proposed the existence of a whole new set of particles with very specific properties that nobody had ever dreamed of before.

In 1932, Carl Anderson of the California Institute of Technology, in a balloon-borne experiment in the upper atmosphere, observed a particle with all the properties of an electron except that it bent the wrong way in the experiment's magnetic field. Anderson immediately knew what he had

seen: the positively charged partner of the negatively charged electron—
Dirac's antimatter *positron*. It was not long after this that both Dirac and
Anderson were invited to Stockholm (Dirac shared the 1933 Nobel with
Schrödinger, while Anderson received his in 1936).

The place in quantum theory that Dirac carved out for antimatter, a necessary component for the self-consistency of the relativistic quantum theory,
remains to this day, although the precise interpretation of the role of antimatter in the relativistic theory (i.e., the precise way in which it solves the
negative energy problem) has undergone some evolution. The modern interpretation is due to none other than Feynman and is the interpretation
most appropriate for application to quantum field theory. This interpretation is an essential step in our discussion. Feynman's take on the issue,
which dates from 1948, is a curious one. It has become such an accepted
part of the particle physicist's thought process that it's easy for professional
physicists to lose touch with just how unusual it is.

The wave function of a free particle that solves the (nonrelativistic)
Schrödinger equation can be split into two factors—one dictating how the
wave function varies from point to point in space and the other dictating
how the wave function varies from instant to instant in time. The factor $\psi(t)$
describing the time (t) dependence is rather simple:

$$\psi(t) = e^{-i\alpha Et},$$

where E is, as usual, the particle's total energy (kinetic energy plus the energy due to the particle's mass) and α is just a fixed number whose value
doesn't really concern us. The number $e = 2.718281828$ is the base of natural logarithms, and i is the "imaginary" value of $\sqrt{-1}$. We don't need to
know what it means to have an imaginary number as an exponent (note
4.11); what we *do* need to know is that the quantity $-\alpha Et$ in the superscript
of this equation involves the product of the particle's energy E and the time
t at which you want to know the value of the wave function ψ. If t is positive, and the energy E is positive, then $-\alpha Et$ is negative (less than zero),
simply because of the minus sign.

For wave functions that solve relativistic wave equations, such as Dirac's
wave equation for spin-½ particles, half of the solutions are of this form,
while the other half are of the form

$$\psi(t) = e^{+i\alpha Et} = e^{-i\alpha(-E)t}.$$

(Remember, when you multiply two minus signs together you get a plus sign, so that these two expressions for $\psi(t)$ are indeed the same; the minus sign in front of the α multiplied by the minus sign in front of the E in the right-most expression give the plus sign in the middle expression.) So, looking at the right-most expression, and comparing it to the previous equation, we see that this half of the solutions have negative energy. As previously mentioned, this was the cause of considerable consternation until Dirac was able to identify them as solutions describing the behavior of antiparticles.

Feynman, however, looked at this function and saw something that was apparently as ridiculous as it was obvious: If we remove the minus sign from in front of the E and instead place it so that it is in front of the t, then for these nettlesome solutions we can write

$$\psi(t) = e^{+i\alpha Et} = e^{-i\alpha E(-t)}.$$

Comparing this to our original equation, we see that we are, in some sense, saved. The energy E is again positive, as our common sense requires. However, the price we have to pay for this repair job seems quite steep: That the wave function now depends on $-t$ rather than t means that this wave function describes a particle *moving backward in time*. For some reason, this didn't faze Feynman. As boldly as Dirac predicted antimatter five years before it was discovered, Feynman made the statement that as far as the quantum theory was concerned antiparticles are simply particles traveling backward in time.

As outrageous as this sounds, when you apply it to what we know about quantum field theory, everything falls neatly into place. Consider again the minimal interaction vertex of quantum electrodynamics, the quantum field theory of the electromagnetic force, shown in figure 4.7a. An electron moving through space to the right emits a photon (eventually to be absorbed by another charged particle, we presume) and, in doing so, recoils to the left. What if, at the vertex, the photon connects the incoming positive-energy-electron solution to an outgoing *positron* (antielectron) solution rather than an outgoing electron solution? As shown in figure 4.7b, this particle, which we can think of as an electron traveling backward in time, must recede *back* in time from the electron-photon vertex as indicated by the arrow. Now we see why we bothered to carry those heretofore meaningless arrows around in our Feynman diagrams. When you change the direction of the arrow, you switch between descriptions of particles going forward in time and antiparticles—particles going backward in time.

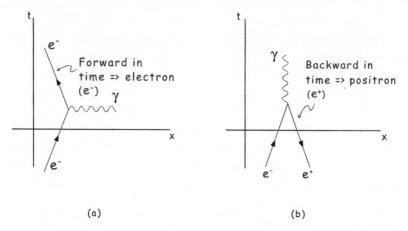

Fig. 4.7. The electron-photon minimal interaction vertex (*a*). If we instead have the outgoing electron going away from the vertex backward in time, the same minimal interaction vertex now represents the annihilation of an electron with its antimatter partner (a positron), forming a photon (*b*). Note that, regardless of whether the particle travels forward or backward in time in our quantum field theory calculation, particles we observe in the laboratory always travel forward in time, so what we observe is, as stated, an electron and positron fusing together (annihilating) to form a photon.

Now, no one has ever observed a particle traveling backward in time. What an observer of this process would see is not a particle receding in time from the interaction vertex but rather an antiparticle *positron* proceeding *forward* in time toward an interaction at the vertex.

And what does this positron appear to do in that interaction? Look again at figure 4.7*b*. The interaction happens at a specific time, which is just the height of the vertex along the vertical (time) axis. Before this time, there are an electron and its antimatter counterpart, a positron, approaching each other. After this time, there is just a photon. The electron and positron have destroyed each other! This diagram represents the annihilation of matter and antimatter, the conversion of mass (remember that the electron and positron have identical, nonzero mass) into a state of pure energy. It is the relativistic quantum field theoretical rendering of the most famous of Einstein's assertions about relativity: $E = mc^2$. Even more notably, the process represented by this diagram represents another entire class of interactions, succinctly described by the single common denominator of quantum electrodynamics: the vertex of the electron-photon minimal interaction.

Feynman's receding time interpretation of the negative energy solutions is wonderfully general. Not only did it allow the Dirac equation to become the basis of the description of spin-$\frac{1}{2}$ particles in quantum field theory, it also resurrected other attempts to make quantum theory consistent with special relativity. For instance, Feynman's approach simultaneously solves both the negative energy and negative probability problems of the Klein-Gordon equation, quandaries that had set Dirac on a course that eventually led to the development of the Dirac equation. Successful as the Dirac equation may be, it only provides a description of spin-$\frac{1}{2}$ particles. To develop a full, general relativistic quantum field theory of particle interactions, we need relativistic quantum mechanical descriptions of other types of particles. For example, for the electromagnetic interaction, the spin-1 photon must be described if we want to put electrons and photons together into minimal interaction vertices. With Feynman's approach, it was soon recognized that the Klein-Gordon equation provides the basis for the description of spin-0 particles. From the comparison of the Klein-Gordon (spin-0) and Dirac (spin-$\frac{1}{2}$) equations, it was possible to deduce the proper form for the description of spin-1 particles (known as the *Proca equation*), allowing for the description of photons and other spin-1 particles.

Finally, one can ask whether Feynman's receding-time approach is merely a formal development, allowing for the description of antimatter within the domain of quantum field theory, or whether it is a physical discovery, reflecting some deeper truth about the differing relationship of matter and antimatter to the underlying fabric of space-time. It seems that, as of yet, there is no definitive conclusion; perhaps the answer will become clear in the deeper context of some future step forward in our understanding of nature.

The Living Vacuum

The development of relativistic quantum field theory was a great leap forward. Field theory simultaneously simplified and expanded our understanding of the fundamental mechanism that underlies the way in which objects in the universe influence one another. Inspired as it may seem, however, the theory as presented so far simply doesn't work. The problem lies in a set of interaction processes represented by a progression of Feynman diagrams that we have so far ignored.

Consider again the mutual repulsion of two electrons through the electromagnetic force. In its most basic form, this interaction is represented by the exchange of a single virtual photon, as shown in the space-time plots of figures 4.4a and 4.4b, which together are represented by the single Feynman diagram of figure 4.4c. This interaction can also proceed by the exchange of any greater number of virtual photons (e.g., the process involving the exchange of two virtual photons is shown in fig. 4.5). But every time you add another minimal interaction vertex involving a photon and an electron to the Feynman diagram, the probability that the interaction takes place according to the new diagram is about 1% of that represented by the original diagram. Since every time you add another virtual photon you have to add two new vertices to the diagram, the mistake you make by ignoring the possibility that more than one virtual photon can be exchanged is rather small.

Now that we've understood how it is that antiparticles fit into the picture, however, we are prepared to discuss an entirely different way in which the basic diagram (fig. 4.4c) describing the repulsion of two electrons can be modified. This modification is represented in figures 4.8a–4.8c.

In these figures, we see that the picture of electron-electron repulsion presented by quantum field theory admits the following possibility: the photon exchanged between the two electrons can, through a standard minimal interaction vertex, instantaneously turn into an electron-positron pair, only to revert to a photon again at a second vertex. (Note that in figs. 4.8a and 4.8b, one of the particles in the loop in the middle of the diagram travels forward in time and, as a result, is an electron and the other travels backward in time and, as a result, is a positron.) This second photon is then ab-

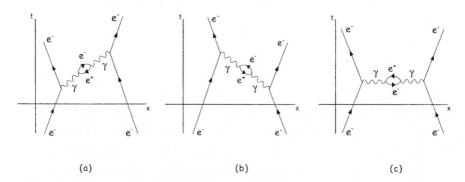

Fig. 4.8. Two separate but indistinguishable ways in which two electrons can influence each other by exchanging a photon that temporarily fluctuates into an electron-positron pair (a and b); c is the corresponding Feynman diagram.

sorbed by the second electron, completing the transmission of the force. As for figure 4.4a and 4.4b, in figure 4.8a, the photon is emitted by the electron coming in from the left and absorbed by the electron coming in from the right, while in figure 4.8b, the roles are reversed. Also as for figure 4.4, the two possibilities (figures 4.8a and 4.8b) are indistinguishable to any observer, so they are packaged into the single Feynman diagram of figure 4.8c.

Again, we might question whether the process represented by the diagram of figure 4.8 is consistent with the notion of mass-energy conservation since the mass of a system consisting of an electron and a positron (both of which have a mass-energy of about 511,000 eV) is decidedly *not* the same as the mass of the photon (zero) that produced them. Again, we are rescued by Heisenberg's uncertainty principle. Since the electron-positron pair exists only for a brief period, its mass-energy is uncertain, so mass-energy can still be conserved, even though the nominal mass of the electron-positron pair is different than the nominal mass of the photon. Just as the short-lived and unobserved photon exchanged between the electrons in figure 4.4 is known as a virtual photon, electron-positron pairs such as that of figure 4.8 produced by an instantaneous fluctuation from and back to a photon via two minimal interaction vertices are known as virtual electron-positron pairs.

Comparing the Feynman diagrams of figures 4.4c and 4.8c, we see that the latter diagram includes two extra vertices, so, again, we would expect the probability of the more complicated, higher-order process of figure 4.8 to be $(\frac{1}{100}) \times (\frac{1}{100})$, or about one ten-thousandth, less likely than that of figure 4.4. However, counting vertices is not the only thing that goes into the calculation of the interaction probability of a given Feynman diagram; it's merely a crude rule of thumb. There is a big difference between the higher-order diagrams of figures 4.5 and 4.8c that leads to a much different result for the interaction probability.

The essence of the difference is this: When the virtual photon fluctuates into the virtual electron-positron pair, its energy must be shared, part going to the electron and part to the positron. But there are a large number—in fact, an infinite number—of different ways the energy can divide itself between the positron and electron. The energy of one of the two can be pretty much anything, as long as the other has the correct amount of energy to compensate and to make the total add up to the virtual photon's energy (note 4.12). When all these different possibilities are taken into account, the interaction probability of figure 4.8c, instead of being 10,000 times smaller than that of figure 4.4c, is larger—much larger. In fact, because there's an infinite number of ways for the process to take place, the calculated interaction probability is infinite.

This is an unacceptable state of affairs. When this more complicated diagram is included (which it must be—recall Feynman's rule that anything that can happen must happen), we find that our quantum field theory cannot predict the relative probability of electrons scattering off each other into the various directions that you might mount a detector because the probability of scattering into any given direction is predicted to be infinite. Since the most that a probability can be is one, something is wrong.

So, after all its great insights and advancements, some of them strikingly confirmed by experiment, are we to conclude that quantum field theory is useless, having been pecked to death by these ephemeral matter-antimatter fluctuations?

Luckily, no. But it took some serious soul searching by Feynman and others to recognize the source of the problem and to dream up a fix for it. The approach that was eventually successful in overcoming this hurdle is known as *renormalization*.

Feynman and others argued that quantum field theory is so profoundly successful on a number of fronts that it must be correct. So, when two electrons repel each other, it must be true that the description of this process includes the diagrams of both figures 4.4c and 4.8c (as well as those of fig. 4.5, although as mentioned, these don't much matter). The way to incorporate these diagrams without having the theory fall apart, Feynman argued, is to step back for a moment and think hard about exactly how the theoretical predictions of quantum field theory relate to physically observable quantities that can be measured in the lab.

Think again about the process represented by figures 4.4 and 4.8. What is observed in either case is the scattering of one electron off another. What goes on between the two electrons during the scatter is, according to the most basic tenets of quantum mechanics, off-limits to the prying eyes of experimentation. The intermediate particles cannot be directly observed without disturbing the whole process so profoundly that it can no longer be interpreted as simply one electron scattering off another.

Let's say that the electron coming in from the left in figure 4.8 is the "projectile" in the scattering experiment, while the one coming in from the right is the "target" (feel free to switch projectile and target if you wish; it will make no difference to the argument). Now consider figure 4.9, which is identical to figure 4.8c except that there is now a dashed circle encompassing the target electron's electron-photon vertex and the virtual electron-positron pair. There's nothing physical about the circle; it's just there to evoke an interesting interpretation of the process of figure 4.8c. Figure 4.8c inclines us to think of the process as one in which the projectile emits a virtual photon that fluctuates momentarily into a virtual electron-positron pair and then is absorbed (again as a virtual photon) by the target electron.

Figure 4.9, however, suggests that we think of the process instead as the exchange of a *single* virtual photon between a projectile and target electron in which the target electron is not just the bare electron represented by the single deflecting line but rather the electron "dressed up" by everything inside the circle—the bare electron itself plus the fleeting "vacuum fluctuation" engendered by the virtual electron-positron pair and the other virtual photon. From this point of view, it's not that the exchange mediating the repelling interaction between projectile and target can occur in two different ways (figs. 4.4c and 4.8c) but that the target electron can have two different forms, the bare form of figure 4.4c or the dressed form of figure 4.9.

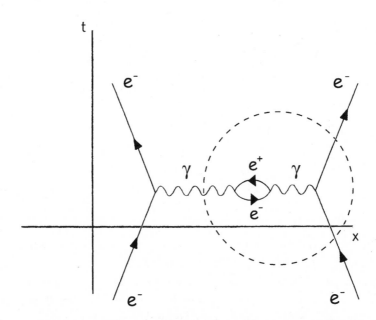

Fig. 4.9. The Feynman diagram of figure 4.8c again. In this case, the dashed circle is included to suggest the interpretation that the electron on the right is composed of the "bare" electron that comes in from the right and exits to the right plus the encircled electron-positron vacuum fluctuation.

Since this is quantum mechanics and you can't ever tell what's going on between the two electrons, who is to say which is the correct point of view? They're both equally valid, so let's pick the interpretation that serves us best in trying to connect our theoretical picture with the realities that confront us in the lab. We'll see that this is the second interpretation—that of figure 4.9 rather than of figure 4.8c.

We now introduce the essential realization that underlies the procedure of renormalization: Whenever one makes a measurement involving an electron as a target, one is simultaneously measuring all possible manifestations of the target electron, the bare target electron of figure 4.4 as well as the target electron dressed by the fluctuation of the virtual photon into an electron-positron pair in the vacuum immediately surrounding the target electron. The bare electron of figure 4.4 is not a physically observable particle; what is observable is the electron that is the combination of the bare electron of figure 4.4 and the dressed electron of figure 4.9. In fact, it's not just a matter of including the processes of figures 4.4 and 4.9. Now that we know the rules for cobbling together electron-photon minimal interaction vertices, we can surround the target electron by an arbitrarily complex morass of virtual electron-positron pairs connected together by equally virtual photons. Figure 4.10 shows one of the infinite number of possible diagrams that we could draw. Indeed, the vacuum surrounding the target electron is veritably seething with these vacuum fluctuations, and if you want to talk about the *physical* electron, the one you will sense when you do experiments on electrons, you had better include in your considerations the full set of these fluctuations.

Thus, when you measure anything at all about the electron—say, the strength of the electron's charge, which you must do by an experiment such as repelling another charged object from it—you are not measuring the strength of the charge of the bare electron but rather the strength of the charge of the bare electron plus whatever effects are added in by this fool's gallery of virtual vacuum fluctuations. What we can then do is adjust, or renormalize, the charge of the bare electron so that, when you add in the effects of the seething vacuum surrounding the electron (represented by diagrams such as fig. 4.10), the result of the corresponding field theory calculation yields precisely the measured value of the electron charge.

Since the charge you measure by scattering off an electron is really that of the bare electron and its attendant cloud of vacuum fluctuations and since you can never measure the bare charge directly (the dubious under-

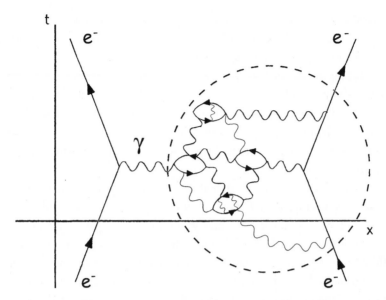

Fig. 4.10. An even more complicated process by which two electrons can scatter off one another. The vacuum is indeed a busy place, and to understand what really goes on when objects influence each other, we are compelled to take the "business" of the vacuum very seriously.

world of virtual particles will always be there), then you are free to make the bare charge whatever you need to in order to make the effective charge of the electron plus vacuum entourage agree with observation. You need not worry about what the value of the bare electron's charge becomes in this process. The charge of the bare electron cannot be measured. In fact, the value that you need to choose for the bare electron charge to compensate for the infinite probabilities imposed by the diagram of processes like those of figures 4.9 and 4.10 is, predictably enough, infinity.

So the charge of an electron, theorists tell us, is infinite. But they also tell us that no experiment you ever mount will measure that infinity. Who cares if this seems ridiculous? This choice, outlandish as it may seem, restores the ability of quantum field theory to make reasonable predictions for the outcome of any experiment that can be done to measure the electron's charge.

If, for a proposed quantum field theory of any of the forces of nature, the adjustment of a *finite* number of parameters (such as the bare electron charge and mass) renders the calculation of the outcome of *all* observable processes finite, then the theory is workable and is a possible fundamental theory of that force. Such theories are called renormalizable. If this is not the case, then the theory is classified as unrenormalizable and is not a candidate for the fundamental description of the force.

Looking toward the latter part of this book, there is a certain subset of the possible implementations of quantum field theory, known as *gauge theories*, whose underlying structure tend to make them renormalizable. For this and other equally fundamental reasons, which will be discussed in chapter 8, gauge theories enjoy an elevated position in the society of particle physics theories.

In this chapter, the implementation of quantum field theory under discussion is specifically that of quantum electrodynamics, the fundamental theory of the electromagnetic interaction. In this implementation, it is indeed true that calculations of all observable processes are rendered finite by the adjustment of a few underlying parameters. Quantum electrodynamics is renormalizable and is our current candidate for the relativistic quantum theory of the electromagnetic interaction. In due time, we shall see that quantum electrodynamics is itself a gauge theory, in fact, the simplest possible gauge theory that can be constructed.

Infinitely charged particles that can travel forward or backward in time and a vacuum that's as alive and shimmering as the air above a suburban parking lot on a hot summer's day—these are just a few aspects of the profoundly bizarre world-view that quantum field theory would ask us to take to heart. Doing so, though, we are led to consider some interesting consequences.

The first of these is a process that flies in the face of everyday common sense, while simultaneously providing hard evidence for the existence of the sea of virtual electron-positron pairs that populate the vacuum. By arranging the participants in the right way, we can join together two minimal interaction vertices to represent the process of figure 4.11: a real photon, coming in from the left at the speed of light after being emitted from a distant source, suddenly finds itself absorbed by an electron-positron vacuum fluctuation (point A in fig. 4.11). Instead of fluctuating back into a photon, one of the two charged particles (in this case the positron) finds itself absorbing a second photon that happens to be nearby (point B in fig. 4.11). After the dust clears, we're left with a *real* electron-positron pair that go happily off on their own, both with the possibility of being sensed by a well-placed particle detector (note 4.13).

This diagram, a necessary consequence of the tenets of relativistic quan-

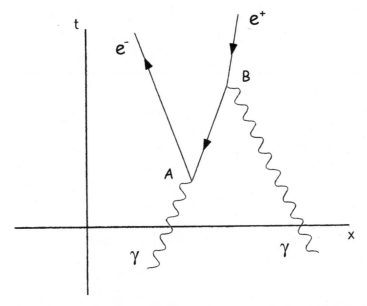

Fig. 4.11. The Feynman diagram for the most basic process by which two photons can fuse to form an electron-positron pair that flies off in independent directions: matter is "conjured from light."

tum field theory, makes the rather profound assertion that it's possible to conjure matter from light. We all know that when two flashlight beams are shined against each other, nothing happens. They pass right through each other without effect. However, quantum field theory predicts that if one beam is made energetic enough (by, say, bouncing it off an energetic electron beam), then every once in a while two of the photons from the flashlight beams will interact, with the very concrete result that two matter particles (one electron and one positron) are produced from the collision of the two light beams.

In a 1995 experiment at the Stanford Linear Accelerator Center (SLAC), directed by Adrian Melissinos of the University of Rochester, the leading edge of a powerful laser beam that had been Compton scattered (bounced off an energetic electron beam) to high energy by the SLAC electron beam was directed back against itself. Every once in awhile, one of the high-energy scattered photons combined with a photon in the unscattered laser beam in the manner of figure 4.11 to form an electron-positron pair that was subsequently identified in a detector surrounding the collision point of the two beams (note 4.14).

Although not for the first time, but perhaps more poignantly than ever before, this experiment showed that the living vacuum—this seething maggot's nest of virtual particle-antiparticle pairs—is very much alive and real. For there, making their presence known via highly characteristic flashes of light in the detector of the Melissinos experiment, were just those pairs of electrons and positrons, jarred into a concrete existence by photons from the two colliding light beams of SLAC experiment E144.

Another surprising and counterintuitive consequence of our quantum-field-theoretical reformulation of electromagnetism was pointed out in 1948 by the Dutch physicist Hendrik Casimir. Casimir recognized that, if all space were in fact alive with virtual electron-positron pairs, then there ought to be a net interaction between the virtual particles in this not-so-vacuous vacuum and the free electric charges in an electrically neutral conductor (note 4.15).

Casimir predicted that two parallel metal plates, electrically neutral and thus uninterested in each other according to classical electromagnetic theory, should be attracted to each other by the effect of the electron-positron fluctuations in the interceding vacuum. This Casimir effect was finally measured in 1996 by Steven Lamoreaux of the University of Washington, confirming the strength of the attraction predicted by Casimir to within the 5% accuracy of the experiment, and providing yet another striking confirmation of the unusual world of quantum field theory.

Another surprising consequence of quantum field theory, with rather profound implications for the whole of particle physics, is the dependence of the charge of the electron (or of any charged object) on the energy of a probe that senses the charge. Consider the scattering of one electron off another through electromagnetic repulsion; suppose that one of the electrons is a probe (such as the electrons in an electron microscope), while the other electron is the target (such as the sample in the electron microscope). According to de Broglie's relation $\lambda = h/p$, as the momentum of the probe electron increases, its wavelength decreases accordingly. Thus, as the momentum of the probe increases, its ability to see the target electron for what it really is—a bare electron stripped of its cloud of virtual hangers-on— becomes better and better (but never so good that it sees the target electron purely as the bare, underlying particle, with its infinite charge). If the concept of renormalization is correct, then as we go to higher and higher probe momentum, the effects of the virtual cloud should diminish, and we should observe that the value of the electron charge increases and becomes closer to its true, bare value (of infinity). This happens very gradually as the energy of the probe is increased. With the advent of modern particle accelerators, however, which achieve electron energies equivalent to the acceleration of the electron through 100 billion volts, the effect should be observable. Current measurements show that, at these high energies, the electron's charge is about 2 percent greater than it is when measured with a low-energy table-top experiment, which is in perfect accord with the predictions of quantum electrodynamics.

This property—the dependence of the charge on the momentum, or scale, of the interacting particles—is known as *running*, and in this sense, all charges run. Intriguingly if you look at the momentum dependence of the electromagnetic, weak, and strong force charges of the fundamental particles, they all appear to be running toward roughly the same value. We measure the variation of the charge from everyday scales out to interaction energies of about 100 billion electron-volts (100 GeV); if we take the observed values and momentum dependence of the charges and use this information to extrapolate to higher momentum, the values all seem to coalesce at an interaction energy of about 10^{16} GeV (note 4.16). (For now, the measurement of the running of gravitational charge is beyond the limits of our experimental capabilities.)

To physicists, this strongly suggests that the various forces of nature are in fact merely different facets of the same underlying "grand unified" interaction, a single natural phenomenon that governs the way objects in the universe influence each other, and would thus be fully responsible for every phenomenon that the universe puts forth. Although substantial strides have been made toward the formulation of this ultimate paradigm, it remains an unattained goal. Particle physicists believe, however, that the next generation of experiments, due to be completed during the period between 2005 and 2020, have the potential to move us substantially toward that goal. Indeed, the discussion of the design of these near-future experiments, some of which are only in their earliest planning stages, is driven in large measure by the search for this "theory of everything."

Stunning Precision

Our archetypical quantum field theory is quantum electrodynamics, the quantum field theory of the electromagnetic interaction. It is to this theory that we turn for the most exacting test of the tenets of quantum field theory.

Throughout this chapter, we have made frequent reference to the electrical repulsion of like-charged objects, in particular electrons. We have barely mentioned magnetism, but certainly our theory of electromagnetism should provide as thorough a description of magnetism as it does of the electrical force.

It is a property of electric charge (the sole type of charge associated with the phenomenon of electromagnetism) that, when it is in motion, it creates *magnetic* fields. When a charged objects orbits in a circle about some point in space, it generates a magnetic field, which is to say, it acts like a magnet. A charged object also generates a magnetic field if it spins about its axis. All electrons spin, possessing an angular momentum of magnitude $\frac{1}{2}\hbar$ associated with their continual and unyielding rotation about their axes. So each electron acts, among other things, as a tiny magnet. In fact, everyday household magnets, such as the one holding your unpaid bills to your refrigerator, are magnetic by virtue of the collective magnetic effect of all the spinning electrons (magnetizing a piece of iron amounts to getting a fraction of the electrons' axes of rotation lined up and pointing in the same direction).

The ratio between the angular momentum ($\frac{1}{2}\hbar$ for the electron) associated with a charged particle's spin and the strength of the correspondingly generated magnetic field is known as the particle's *gyromagnetic ratio*, which is usually represented by the Greek letter μ. Now, recall that the Dirac equation provides the appropriate quantum-mechanical description of spin-$\frac{1}{2}$ particles, such as electrons. One of the nice things that follows from this description is a *prediction* for the value of these particles' gyromagnetic ratio:

$$\mu_{\text{pred}} = \frac{2q}{mc},$$

where q is the particle's electric charge, m is its mass, and c is the speed of light. Since all these quantities are known with tremendous precision for the electron, this expression provides an exacting prediction for the electron's gyromagnetic ratio.

However, it's quite possible to measure the strength of the electron's magnetic field, thereby determining its gyromagnetic ratio experimentally. If you place a magnet (such as a spinning electron) close to another magnet (such as the precisely calibrated magnetic coil in the experiment we're about to discuss), the magnetic part of the electromagnetic force will try to align the first magnet in the direction of the field of the second magnet, just as a compass needle (which is nothing more than a little magnet) tries to align itself with the earth's magnetic field. However, if the first magnet (the electron) is placed in the field of the second magnet so it is not quite aligned, it will begin to rotate (precess) about the direction of the second magnet's magnetic field. It's exactly like a top, if you can remember the last time you played with one. If you get a top spinning and set it down on the table perfectly vertically (aligned with the earth's *gravitational* field, which points up and down), it just stands there. However, if you set it down so it's a bit tilted, it slowly rotates (precesses) about the vertical.

For the case of the two magnets (the electron and the experiment's magnetic coil), the frequency of precession (number of seconds per rotation) depends in a well understood way on the strength of the two magnetic fields. So, if you measure this frequency and you know the strength of the coil's magnetic fields, you can calculate the strength of the electron's magnetic field. This is a beautiful technique because it's relatively easy to measure a frequency with precision. You just need to trap the electron in the region of the coil's field and watch it go around and around for as long as you can bear. You then divide the number of rotations by the time elapsed; that's your measured frequency. The longer you watch it, the more accurate your measurement.

These measurements confirmed, more or less, Dirac's prediction of the size of the electron's gyromagnetic ratio. Modern experimental technique, however, allows electrons to be trapped for a long time, so recent experiments are extraordinarily precise. With all of this precision, a small deviation from Dirac's prediction emerged. If we let g_{meas} represent the ratio of the observed gyromagnetic ratio of the electron to that predicted by Dirac, that is,

$$g_{meas} = \frac{\mu_{obs}}{\mu_{Dirac}},$$

then the result of the most accurate modern experiment is

$$g_{meas} = 1.00115965219$$

to within an uncertainty of about one in the very last digit.

The fact that g is not precisely one means that Dirac's prediction is not quite right. It's not wrong by much, just a little over one-tenth of a percent. But the experiments are so accurate that even such a small discrepancy is

extremely significant, insofar as it is no less than eight orders of magnitude larger than the uncertainty on its measurement.

But, and this is the critical point, Dirac's relativistic wave equation is not our quantum theory of the electromagnetic interaction, it is merely one (albeit very important) ingredient in that theory; specifically, the ingredient that allows us to determine the wave functions of the spin-½ particles (electrons) that enter into the Feynman diagrams of the full theory. So quantum field theory tells us that Dirac's prediction for the electron's gyromagnetic ratio is just a start, and to make an accurate prediction we need to use the full theory. We need to calculate all of the Feynman diagrams associated with the interaction of the quantum of the magnetic field (the photon) with an electron.

Now, as we've seen, when we include the quagmire of virtual particles that crawl continually in and out of the vacuum, there are many diagrams we need to calculate. As usual, the relevance of each possible diagram decreases with increasing number of vertices (as long as we use the appropriately renormalized electron properties), so we can ignore the really complicated ones. However, the experimental result for g_{meas} is so accurate that we need to include some of these more complicated diagrams if the accuracy of our theoretical prediction is to match that of the experiment: We need to include all Feynman diagrams with up to seven minimal interaction vertices. These diagrams are shown in figure 4.12.

The calculation of these diagrams took several years, but in the end, Dirac's original prediction did need to be modified. The value g_{pred} of the predicted modification was

$$g_{pred} = 1.0011596522.$$

Comparing this predicted modification to the measured difference

$$g_{meas} = 1.00115965219$$

Fig. 4.12. The set of Feynman diagrams that were calculated to provide a prediction of the electron's gyromagnetic ratio with an accuracy consistent with that of experiments. The calculation of these diagrams took several years. Reprinted from Donald H. Perkins, *Introduction to High Energy Physics*, 2nd ed. (Reading, MA, 1982), fig. 8.2.

from above shows an agreement to mind-boggling precision. Quantum field theory predicts the value of the electron's gyromagnetic ratio to better than one part in 10^{10}, or about three parts in 100 billion! This establishes quantum field theory in general (and quantum electrodynamics in particular) as the most quantitatively successful theoretical framework ever devised. Quantum field theory—this collection of innovative, counterintuitive, irreverent, and sometimes almost perverse, but inarguably profound, ideas—has received the loftiest possible imprimatur that could ever be afforded to a physical theory: precise, quantitative confirmation to a fraction of a fraction of a fraction (if that) of a gnat's eyelash. It must be right.

5

Patterns in Nature

—∿∿∿∿∿∿∿—

The Fundamental Building Blocks

If you corner a particle physicist and ask her to explain the Standard Model of particle physics to you in 30 minutes or less, about the best she can do is to describe the so-called building blocks of matter, the relatively small set of fundamental constituents that, when cobbled together in various combinations, fully account for the contents of the material universe. The discovery of these indivisible building blocks was a great stride forward, but saying that a list of them represents the full content of the Standard Model is somewhat akin to describing a musical work by citing its progression of chords—it provides a fair picture of the mechanical structure of the work, but doesn't get to the essence of its beauty. Nevertheless, one does need to delineate the progression of chords to understand the work, and likewise, we need to introduce the list of fundamental (indivisible) particles to appreciate fully the conception of nature into which they are incorporated.

Three Quarks for Muster Mark: The Eightfold Way

The rapid development of particle physics experimental technique (in particular, the development of the particle accelerator and the "bubble chamber," which provides a visible record of the paths of individual subatomic particles emanating from a high-energy collision) led to a great expansion of the list of elementary particles in the 1950s and early 1960s. At first, particle physicists were euphoric about the rapid pace of discovery, but as the list grew longer and longer, with no apparent underlying order or structure, the euphoria gradually gave way to frustration and the sense that this burgeoning array of new particles carried with them some important message about the structure of matter that we were simply unable to comprehend. The American physicist Willis Lamb, in his acceptance speech for the 1955 Nobel Prize (for the discovery of subtleties in the spectrum of light emitted from hydrogen atoms that could only be explained via the application of relativistic quantum field theory, and thus provided one of that theory's earliest confirmations) put it this way: "the finder of a new elementary particle used to be rewarded by a Nobel Prize, but such a discovery now ought to be punishable by a $10,000 fine."

In retrospect, the particles that those physicists discovered were not elementary but were instead combinations of a much smaller number of truly (at least as far as we know at this point) indivisible particles known as quarks. The problem is that the aptly named strong force binds these quarks together so tightly inside these prior-day elementary particles that it is impossible to pry an individual quark away and observe it in isolation. The direct experimental evidence for the presence of quarks, which emerges only when the energy of the probe being used to look for them is large, is subtle. It required a good deal of head scratching in the late 1960s to recognize that such probes were recoiling off something small and hard inside of the naively designated "elementary" particles of the 1950s.

But we get a bit ahead of ourselves here. In the early 1960s, a physicist at the California Institute of Technology by the name of Murray Gell-Mann interpreted the patterns observed in the emerging array of elementary particles as being due to a symmetry possessed by a much smaller set of internal degrees of freedom—quarks—without ever splitting those particles apart with high-energy particle beams. This insight established a well-defined structure within the disparate array of elementary particles, an achievement for which Gell-Mann was awarded the 1969 Nobel Prize in Physics.

The development that set the apple cart right again was Gell-Mann's static model of the strong nuclear force, often referred to as the eightfold way. Quantum electrodynamics is a theory of the electromagnetic interaction that purports to describe everything that electromagnetism has to offer, from high-energy matter-antimatter annihilation to the most basic binding properties of atoms. The eightfold way, on the other hand, provided a framework by which the array of fundamental particles, previously a jumbled mass of vaguely related particle states, could be understood in terms of a basic underlying principle. But it did not provide a theory of the interactions of these particles with each other. Gell-Mann's model is a static one in that it can describe and codify the array of particles as they exist "statically" on their own. It does not attempt to describe the "dynamical" rules that such particles adhere to when they bounce off one another. The more complete dynamical theory would have to wait another ten years or so, until the development of quantum chromodynamics, the current candidate quantum field theory of the strong interaction.

Gell-Mann's eightfold way was perhaps the first conscious application of the results of the pure mathematical field of group theory and, in particular, the theory of "Lie groups," to a problem in physics. Lie groups, named after the nineteenth-century Norwegian mathematician Sophus Lie (pronounced Lee), lie at the very heart of the connection between physical symmetries—the patterns that are observed to order the behavior of the natural world—and the detailed content of the theories that describe that behavior (we'll learn a lot more about Lie groups beginning in chapter 6). Nowhere does this connection play more of a role in governing the nature of physical theory than in modern particle physics.

That Gell-Mann was not much aware of the preexisting body of work on Lie groups, and ended up rediscovering most of the necessary mathematical results on his own, does not detract from the beauty of this marriage between the abstract pursuits of the pure, ivory tower mathematicians and the world of theoretical physics. Today, the connection between abstract mathematics and formal theoretical physics is even stronger, and one of the latter's greatest practitioners (Ed Witten of the Institute for Advanced Study at Princeton) was the recipient of the 1990 Fields Medal, the moral equivalent of the Nobel Prize in the world of pure mathematics.

Underlying Gell-Mann's eightfold way, playing out the intangible symmetries of the Lie group, was a cast of three characters, whimsically referred to as quarks in reference to an obscure line from James Joyce's *Finnegans Wake*: "three quarks for Muster Mark" (note 5.1). These three quarks, when assembled in various combinations, would economically account for the myriad of elementary particles that had been discovered over the course of the prior decade. Furthermore, argued Gell-Mann, if the strong force obeys this Lie-group symmetry, then the particles formed by binding the quarks together with the glue of the strong force have to fall into patterns that are characteristic of the Lie group (in the language of the mathematicians, the patterns of the "irreducible representations" of the Lie group). This is precisely what was observed: the formerly daunting array of elementary particles fell neatly within the patterns of Gell-Mann's eightfold way. There were one or two missing particles, associated with unobserved combinations of quarks allowed by the eightfold way, but they were soon discovered, with precisely the predicted properties.

Two of these three quarks can be combined together in various ways to produce ordinary nuclear matter—neutrons and protons. These two quarks were given the names "up" (u) and "down" (d). In the eightfold way, the up quark has an electric charge of $+\frac{2}{3}$, (two-thirds of the charge of the proton), while the down has a charge of $-\frac{1}{3}$ ($-\frac{1}{3}$ the charge of the proton, or one-third of the charge of the electron). So, for example, if you combine two up quarks and a down quark, you have a charge of $\frac{2}{3} + \frac{2}{3} - \frac{1}{3} = +1$—the charge of the proton. In the eightfold way, a proton is just that, a combination of two u's and one d. Similarly, a neutron is one u and two d's, a combination that has no net electric charge.

A third quark was required by the eightfold way to allow for the construction of a set of unstable particles that had been observed only in cosmic radiation and in bubble chambers that intercepted high-energy particle beams at particle accelerators. Relative to ordinary nuclear matter, the properties of these particles was rather strange, so this quark was given the name *strange* (s). To piece everything together properly, the s quark needed to have a charge of $-\frac{1}{3}$, just like the d. But the strange quark differentiates itself from the down quark in other ways; in particular, strange quarks are quite a bit heavier than the down quarks (with a mass of about $\frac{1}{3}$ of that of a proton, while a down quark's mass is only a small fraction of the proton's mass).

To Gell-Mann, the question of whether quarks really exist or are, instead, just a mathematical construct, was not of central importance; it seems that, at the time, Gell-Mann leaned toward the latter point of view. The important thing was that the theory allowed the previously senseless array of elementary particles, the particle zoo, to be neatly arranged in sensible, well-ordered patterns that were understood in terms of a basic and overarching principle: a Lie group symmetry associated with a small set of underlying, indivisible states—the three quarks. Even better, the theory had predicted new particle states that were found in later experiments. The theory worked.

In the late 1960s, a team of physicists from the Stanford Linear Accelerator Center (SLAC) and the Massachusetts Institute of Technology (MIT) conducted a series of experiments in which high-energy electrons were scattered from protons in a target composed of hydrogen. Unexpectedly, they observed a substantial number of scatters for which the electron bounced off of the proton at a sharp angle. The rate of such scatters was much larger than would have been expected if the proton were a uniform blob of charge 1 fermi (10^{-15} meters) across (the size of the proton had been measured on the Stanford campus by Robert Hofstadter and colleagues ten years earlier, an experiment that garnered Hofstadter the 1961 Nobel Prize in Physics).

The results of these experiments were rather surprising. It would be like throwing a ball bearing into a vat of tapioca pudding and having it bounce back at you. It would only be possible if, instead of being a uniform gelatinous blob, the pudding were a suspension of small, hard objects. Every once in a while, when you happened to hit one of these small objects just right, the ball bearing would take a hard bounce and come flying back at you.

In the SLAC/MIT experiment, the electron scattered off the charged proton electromagnetically through the exchange of a virtual photon. Recall the de Broglie relation: If the momentum of the virtual photon is large, then its wavelength is small. Accordingly, it will scatter off a small region somewhere amid the proton's volume of electric charge. In the SLAC/MIT experiment, the wavelength of the photon for the highest-energy scatters was about 10^{-16} meters, or about one-tenth the proton radius. If the proton is a uniform blob of charge, then these electrons will scatter off only a small fraction of the stuff of the proton, and this small fraction won't have much mass. But when something moving very fast bounces off something light, it won't get deflected very much. The fact that a lot of the SLAC/MIT electrons bounced off the proton at sharp angles suggested, instead, that the stuff of the proton was concentrated in some sort of subparticle, which itself is much smaller in radius than both the proton and the wavelength of the virtual photon that it absorbs in the scatter.

The data thus suggested that the proton was mostly empty space, inside of which whirled tiny subparticles whose combined properties produced the bulk properties of the proton. The proton, it seems, is a sort of subnuclear atom.

It was quickly realized that these subnuclear particles could well be the quarks of Gell-Mann's eightfold way and further scattering experiments conducted from this point of view confirmed that they had precisely the properties predicted by that model. For this demonstration of the physical existence of quarks, Jerome Friedman and Henry Kendall of MIT, and

Richard Taylor of SLAC, received the 1990 Nobel Prize in Physics (note 5.2).

Since then, accelerator energies have increased substantially, allowing us to probe atomic nuclei (or, more accurately, to probe the quarks within the nuclei) with wavelengths as small as 10^{-18} meters—one-thousandth of the proton radius—and the high-angle scatters of the incoming probe are still observed. Thus, with photon wavelengths of 10^{-18} meters, we are not breaking up quarks into smaller objects the way we were able to break the protons up into quarks with our 10^{-16} meter wavelength photons. So, to the best of current experimental knowledge, quarks are *the* fundamental, indivisible constituents of nuclear matter. They will be one of the categories of particles that will appear on our list of truly fundamental particles.

No one claims that quarks have been shown definitively to be the fundamental building blocks; it's just that, with any experiment that we can do today, we can't "see" anything smaller than roughly 10^{-18} meters across, and quarks are apparently smaller than that. For now, we can safely label them as fundamental.

Even with the immense economy of description provided by quarks and the eightfold way, the list of the fundamental constituents of matter is still fairly long; there are twelve particles, six of which are quarks. Some theoretical physicists hypothesize a smaller set of even more fundamental particles (commonly referred to as *preons*) from which these twelve particles are constructed, much as the daunting array of particles that so dismayed Willis Lamb are economically constructed from arrangements of the three quarks (u, d, s) of the eightfold way. No experimental evidence exists for such sub-subnuclear particles. But then again, there's really no reason to expect them to be as large(!) as 10^{-18} meters.

Leptons: The (Not-So) Light Ones

Quarks are not the only class of fundamental constituents of matter. We now understand that quarks make up the nuclei of atoms, but we've known for even longer that atoms are not composed of nuclei alone; we also need electrons. Electrons are the most well-known member of a second class of particles, known as leptons, from the ancient Greek word for *light* ("light" as in "not heavy"). As we've discussed, it's the strong nuclear interaction that binds protons and neutrons (well, really quarks) together into nuclei. Electrons are not bound into nuclei. They simply don't carry the charge associated with the strong interaction, so they can't join the protons and neutrons in the tightly bound nucleus. In fact, this is the defining characteristic of leptons: leptons do not carry the strong-force, or color, charge. As a result, they don't interact with other matter as readily (as strongly) as objects constituted from quarks.

In a 1937 experiment with cosmic radiation, Cal Tech physicists Seth Neddermeyer and Carl Anderson (the same Anderson as that of the 1933 positron discovery) observed a second type of particle that did not participate in the strong interaction. This particle was similar in character to the electron but with a much greater mass. While the electron mass-energy is about 0.511 MeV (an "MeV" is a million electron-volts, so this is 511,000 eV) (note 5.3), making it light relative to the proton's mass-energy of nearly 1 billion electron-volts, the mass-energy of this new particle, dubbed the *muon*, was about 100 MeV. So this lepton is not so light; its lack of a strong-force charge is what places it in a class with the electron. Finally, in 1975, a team led by Martin Perl of SLAC (co-recipient of the 1995 Nobel Prize in Physics) discovered yet a third lepton with electron-like properties—the τ lepton, sporting a mass-energy of about 1,780 MeV—slightly less than two billion electron-volts. Light indeed!

Our story of the somewhat inappropriately named leptons does not end here. We can step back and follow an independent thread, dating back to a rather bold hypothesis regarding the radioactive decay of atomic nuclei put forth in 1930 by Wolfgang Pauli. The process of nuclear beta decay changes the nucleus of one type of atom into the nucleus of another (with atomic number different by one from that of the original nucleus) in conjunction with the release of an electron or positron. In fact, the inability of physicists of that day to describe beta decay in terms of the three known forces of nature (gravitational, electromagnetic, and nuclear) prompted them to introduce the fourth force, the "weak nuclear force," the only purpose of which (at that time) was to permit beta decay.

Some careful experiments in the 1920s had shown that energy was apparently not conserved in beta decay. The energy of the decay products, including the mass-energy of the new nucleus and produced electron or positron, was measured to be less than that of the mass-energy of the original nucleus. Even though the putative weak force behind this process was new and poorly understood, Pauli believed that it could not violate the sacred principle of energy conservation, so he hypothesized that there was a third object taking part in beta decay. To have escaped detection in all of the beta decay experiments, this object would have to interact very reluctantly with matter. It would have to be able to pass straight through the detectors in these experiments without leaving a trace.

This object would have to have properties different from those of any previously known form of matter, and so Pauli felt compelled to hypothesize the existence of a wholly new and different type of particle. This new particle would have to be a lepton, with no strong-force charge, for otherwise it would interact readily with the matter in the detectors and have been discovered long before. In addition, this particle could have no electric charge, for even the electromagnetic interaction would be enough to have exposed this particle in the earlier experiments. The great Italian American physicist Enrico Fermi, who eventually made essential use of this particle in the first rigorous theory of the weak interaction (receiving the 1938 Nobel but for more or less unrelated work in nuclear physics), dubbed this ghostlike particle the *neutrino*, which is an Italian-language-inspired way of saying "little neutral one": an electrically neutral (uncharged) lepton.

Partaking only in the weak and gravitational forces, the latter of which is extremely weak even when compared with the weak force, the bashful neutrino truly despises making its presence felt. A beam of neutrinos can travel through the entire breadth of the earth with little attrition. In fact, trillions of neutrinos from the nuclear reactions that fuel the sun pass through your body every second, very few of which (thankfully) pause to introduce themselves. Nevertheless, callous human ingenuity eventually was able to ferret out the hermit neutrino, and in 1958, Fred Reines (who, with Martin Perl of the τ lepton, received the 1995 Nobel Prize in Physics), and Clyde Cowan of the Los Alamos National Laboratory developed a detector that was able to observe the interactions of neutrinos coming from the core of a powerful nuclear reactor.

These days, a surprising amount of physics is done with neutrino beams, including, among other things, the study of the properties of quarks. In fact I owe my 1988 release from the indentured servitude of graduate study to such an experiment.

The neutrino story, even our highly abridged version of it, requires further discussion. In 1962, a small team of physicists from Columbia University in New York led by Leon Lederman, Mel Schwartz, and Jack Steinberger (the three shared the 1988 Nobel Prize in Physics), working at the Brookhaven accelerator laboratory on Long Island, found there were two distinct types of neutrinos. One of them, produced in association with an electron at the point of origination of the neutrino beam, always turned back into an electron in the course of its interaction with the matter of the detector at the end of the beam line. The other, produced in association with a muon (one of the two heavier lepton cousins of the electron) always switched back to a muon in its interaction with the detector.

It might be mentioned here that the great contribution of Columbia University, the home institution of Lederman, Schwartz, and Steinberger, to the development of particle physics is underrepresented in this book. Another such institution (perhaps somewhat reflecting my alumnus bias) is the University of Chicago, Enrico Fermi's home institute after taking his leave of fascist Italy. For better or worse, it was at Chicago that the human race created the first artificial sustained nuclear reaction, unlocking, for the first time, the staggering power of the strong nuclear force. The site of this infamous accomplishment, across the street from the present-day Enrico Fermi Institute, is marked by a small, but particularly haunting, work by the English sculptor Henry Moore.

In any regard, although it was not recognized at the time, the discovery that there are two distinct types of neutrinos was arguably the first step in the incubation of our modern-day view—the Standard Model—of the fundamental interactions of matter. The essence of the Columbia/Brookhaven two-neutrino discovery, from the modern point of view, is that the fundamental constituents—the building blocks—of matter come in repeating patterns, known as *generations*.

In 1962, there were two known charged leptons—the electron and the muon. The two-neutrino experiment demonstrated that each *charged* lepton has a *neutral* lepton partner. The electron has its electron neutrino, which only consorts with electrons, and the muon has its muon neutrino, which will only break quantum mechanical bread with the muon. There are two pairs, or generations, of matter particles: one consisting of the electron and its electron neutrino and the other of the muon and its muon neutrino. Thus, we see a pattern emerge: an electrically charged lepton (electron) is paired with an electrically neutral lepton (electron neutrino), forming an electron-like generation. This generational pattern repeats itself, showing up again in the muonlike generation (charged muon plus neutral muon neutrino).

To get somewhat ahead of ourselves, it is precisely this pattern (which, in addition, is mirrored in the arrangement of the quarks) that provides the essential clue to the profound and powerful connection between the concrete world of natural law and the abstract world of mathematical construct. And, surprisingly enough, it will be through the weak nuclear interaction, the most obscure and uncooperative of the four known modes of causation, that this connection will be made.

When the fundamental constituents are listed on paper, it's conventional to arrange the members of the same generation within parentheses (as shown in fig. 5.1). The muon is represented by the Greek letter mu (μ). The neutrino is represented by the Greek letter nu (v), with the subscript indicating the type (electron or muon) of neutrino. The neutrinos get top billing because their electric charge is larger than that of their associated charged lepton (in the system of whole numbers, 0 is larger than -1).

But there are three, not two, known charged leptons; recall Perl's τ. By the time of Perl's 1975 discovery, the generational structure had been fully established (see "The November Revolution" section that follows), so it was immediately suspected that Perl's lepton was a member of a new generation, a previously undiscovered third generation of matter.

It is interesting that the existence of the third generation was not completely unexpected. This point merits a brief digression.

Antimatter is thankfully quite rare. Otherwise, the matter that might form useful systems such as stars, planets, and the bodies of living creatures, would be subject to annihilation by antimatter. Without some difference in behavior between matter and antimatter, the egalitarian matter/antimatter soup present just after the big bang could not have evolved into the comfortable matter-dominated universe of today.

In 1973, the Japanese theorists Makoto Kobayashi and Toshihide Maskawa, two years before Perl's discovery of the τ lepton, recognized that if there were at least three generations of matter, then it would be possible for the theory of the weak interaction to exhibit a preference for matter over antimatter. In other words, if there are three or more generations of matter,

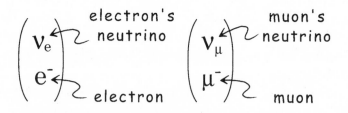

Fig. 5.1. The first two lepton generations.

then it might be possible to explain why it is that almost all of the antimatter present in the primordial universe disappeared, leaving the essential matter abundance of today.

Following this line of reasoning, Kobayashi and Maskawa predicted the existence of the third generation of matter (note 5.4). With this prediction, they also provided one of the first hints that particle physicists might be able to aid cosmologists in their ambitious quest to understand the origin and evolution of the universe, a connection that has grown over the years into a fascinating, successful, and rather well-funded scientific pursuit.

If matter comes in generations, which include one neutral lepton for each charged lepton, and the τ lepton is the lepton of the third generation of matter, then what about its associated τ neutrino? This, too, is an interesting story.

As late as the turn of the millennium, a vexing footnote to the array of constituents of the Standard Model read as such: we know that the τ neutrino must exist, but they're hard to produce in great quantity. The τ neutrino had not yet been seen, but we knew that they were out there somewhere.

Finally, in August 2000, the τ neutrino was unambiguously identified in an experiment at the Fermi National Accelerator Laboratory (Fermilab, currently the world's forefront particle accelerator) in suburban Chicago. In this experiment, a neutrino beam was passed through a vat of photographic emulsion, quite literally taking a three-dimensional snapshot of any neutrino interaction in the emulsion. On review of the emulsion after a prolonged exposure to the beam, a number of the interactions embedded in the emulsion clearly showed the spontaneous emergence of a charged τ lepton—the unambiguous signature of the interaction of a τ neutrino in the body of the emulsion.

Hats off to this persevering team of physicists, for they were able to isolate what, in all likelihood, was the last undiscovered fundamental constituent of matter. Why do I say this?

In 1988, two accelerator laboratories (SLAC in California and CERN, the bastion of pan-European particle physics on the French-Swiss border outside Geneva) began almost simultaneously to produce copious numbers of Z^0 bosons, the neutral member of the set of 3 particles responsible for the weak nuclear interaction. The Z^0 decays rapidly and numbers among its decay products all the different quarks and leptons, as long as the mass-energy of the quark or lepton is less than half of the Z^0 mass-energy of about 90 billion electron-volts.

Charged leptons can be pretty heavy (the τ has a mass-energy of about 2 billion electron-volts); the same is true for quarks. So, fourth-generation-charged leptons or quarks, if they exist, might be too heavy to be produced by the decaying Z^0, thus their existence cannot be ruled out by studying what the Z^0 decays to. However, neutrinos are known to be much lighter, no heavier than about 10 million electron-volts. So, if there were a fourth-generation neutrino, it's a virtual certainty that it would have been discernible through the measurements of the Z^0 decay properties. One of the most prominent activities of the particle physics program over the decade of the 1990s was the measurement of the properties of the Z^0 to high precision, and there are clearly only three neutrinos available for the Z^0 decay. It looks as if three, the minimum number of generations necessary for matter/antimatter symmetry violation, is precisely the number of generations that exist. With the year 2000 discovery of the τ neutrino, all of the members of these three generations (including the quarks) have now been observed.

How much do neutrinos weigh? At this point, our understanding of neutrino masses is in a funny state. We know from studying the decays of the charged leptons, which always involve the emission of a corresponding neutrino, that none of the three neutrinos is heavier than about 10 million electron-volts. This data, though, is also consistent with the hypothesis that all the neutrinos are massless.

However, we now believe that neutrinos do indeed have mass. If neutrinos are not massless like photons but, instead, have differing masses like their charged lepton counterparts, then it is possible for the different neutrino types to oscillate, to change their identity from one type to another as they travel through space.

During the 1990s, initially sparse data hinting at such oscillations, using neutrinos from the sun as a naturally produced neutrino beam, were bolstered with the addition of data from modern neutrino experiments. The leader in this pursuit is the Super-Kamiokande experiment in Japan, which makes use of neutrinos produced by cosmic rays as they bombard the earth's atmosphere. Additionally, an accelerator-based neutrino experiment (the LSND, or Liquid Scintillator Neutrino Detector, experiment at Los Alamos) also reported a likely observation of neutrino oscillations: the appearance of electrons in interactions of their nominally pure muon neutrino beam. These experiments don't appear to be mutually consistent, but it seems fairly certain that at least one of the experiments is observing neutrino oscillations, certain enough, at least, to attract the attention of the 2002 Nobel Committee. Ray Davis Jr. of the University of Pennsylvania and Masatoshi Koshiba of the University of Tokyo—proponents of the experiments conducted with solar neutrinos—shared the 2002 Nobel Prize in Physics. So neutrinos do seem to have mass, although it's unclear at this point what specifically those masses are.

At this point, neutrino physics is a growth field, for this all must be sorted out. Again, this aspect of particle physics reaches beyond the nominal boundaries of the field because if the elusive but abundant neutrino does have mass, then it is certainly an important player in the overall—the cosmological—structure of the universe.

The November Revolution

If there's any single discovery that established the Standard Model as just that—*the* Standard Model—it was the discovery of the "charmed" quark. It is also a classic story of competition between two competing and complementary approaches—those of electron-beam and proton-beam accelerators.

Historically, the field of particle physics has benefited vastly from the interplay of these two approaches—the precise, controlled measurements generally permitted by lower-energy electron machines complement the greater reach of the higher-energy, but intrinsically sloppier, proton machine experiments. In the case of the charmed quark, however, it was not a question of complementarity but, rather, one of direct competition.

In late fall of 1974, experimenters at the SLAC electron-positron collider (SPEAR), independently of and simultaneously with an experiment analyzing the products of catastrophic collisions of the Brookhaven proton beam with beryllium nuclei, obtained clear evidence of a new particle, with a mass-energy of about 3,100 MeV (remember, again, that the "M" in "MeV" stands for million). Although quite different in nature, the data of either experiment provided convincing evidence for the new particle. This new particle received two names: it was called the J particle by the East Coast experimenters (from Brookhaven and MIT, although an entirely different MIT group than that of the discovery of quarks several years earlier at SLAC), and the ψ by the West Coast team (SLAC and the University of California, Berkeley). Today, this particle is known as the J/ψ.

The Brookhaven-based group on the East Coast of the United States had been sitting on its evidence for the new particle, with a journal paper awaiting final approval from the collaboration's leader, MIT professor Samuel C. C. Ting. Ting is nothing if not a very careful experimenter, and even in the face of solid evidence, he had his team ensure that every feature of the experiment's data was understood and explained. The nature of electron machines being as it is, though, the evidence for the discovery of the J/ψ at SLAC on the West Coast was more than just solid, it was numbingly obvious.

In Ting's experiment with the Brookhaven proton beam, the short-lived J/ψ particles were observed by measuring the energy of the electron and positron they occasionally decayed into. At an energy corresponding to the mass-energy of the new particle, there was an excess of about 250 electron-positron pairs over what would have been expected were there no new particle; this excess was culled from the products of billions of proton-beryllium collisions accumulated over a number of months.

In the SLAC electron-positron experiment, led by Burton Richter, the J/ψ was produced through the annihilation of electron-positron pairs from countercirculating beams, the reverse of the Brookhaven experiment's use of the decay to electron-positron pairs. In electron-positron accelerators, the energy is precisely controlled and is quite uniform from particle to particle in the beam. If the sum of the energy of a colliding electron and positron from the countercirculating beams is not precisely equal to that of the J/ψ, a J/ψ cannot be produced in the collision. However, if the sum of the energy is precisely that of the J/ψ, the particle will be produced and produced in great quantity.

In early November 1974, the SLAC/Berkeley team was exploring the energy region around 3 billion electron-volts with the SPEAR electron-positron beams. The electron-positron collision rate didn't quite agree with expectations. As a result, it was decided to explore the energy region in fine steps, taking a little data at a lot of different, closely spaced energies. When the sum of the electron and positron beam energies reached exactly 3.105 billion electron-volts (the mass-energy of the J/ψ) it's not too much of an exaggeration to state that all hell broke loose. The collision rate jumped by a factor of 100. If the combined beam energy was increased or decreased by as little as 0.010 billion electron-volts, the collision rate went right back to normal. If the beam energy was tuned back to precisely 3.105 billion electron-volts, the collision rate went right through the roof again. The conclusion was unavoidable; a new particle, the J/ψ, had been discovered. Several thousand were produced at SPEAR in a single weekend.

With this great collision rate, it was possible to check and cross-check the SLAC data in a matter of days rather than months, and a paper announcing the result was drafted over the weekend of November 9–10, 1974. As the story is commonly told, Ting got wind of the excitement at SLAC and called Richter to inform him of the evidence in the Brookhaven data. Richter then apprised Ting of the details of the SLAC result, and it became clear that they were seeing independent and confirming evidence of the same exotic new particle. The two papers, one from the MIT/Brookhaven group and the other from the SLAC/Berkeley group, were submitted on November 12 and 13, respectively, and appeared back-to-back in the December 2, 1974, edition of the *Physical Review Letters*. For their efforts, Ting and Richter shared the 1976 Nobel Prize in Physics.

In all academic pursuits, there are times when a result or an idea has its authors bursting at the seams to make it public and to mark it once and for all as theirs. It is also often the case that the result or idea has arisen from a certain ripeness—an overall evolution within the particular pursuit toward the result—and, in such cases, there is always the risk that others are rapidly closing in on the same goal. In such cases, the pressure to announce, to publish, to claim the prize at the expense of care and certainty is enormous. Ting, with the profound result within the Brookhaven data clamoring for the open air, opted for the high road and withheld the promulgation of the discovery of the J until certainty was at hand. However, the SLAC experimentalists did not try to scoop the longer-standing Brookhaven data but, instead, worked cooperatively toward a joint announcement. Both groups deserve, in addition to the highest scientific accolades, credit for their restraint.

It could be argued that this restraint cost Ting and Richter half a Nobel Prize. But, in the end, there are no semi-laureates. You're either a Nobel Laureate or you're not. Most of us are not; Ting and Richter are, and deservedly so.

But what exactly is the J/ψ? Why was its unearthing in 1974 deemed a revolution, rather than just an interesting discovery? The answer lies in its relation to a theory connecting the electromagnetic and weak nuclear interactions that emerged in the late 1960s primarily due to the work of Sheldon Glashow (Harvard University as well as various points west), Steven Weinberg (MIT and the University of California, Berkeley), and Abdus

Salam (Pakistan; via Imperial College, London). This theory is today's Standard Model of Particle Physics; for their work in developing the Standard Model, these three theorists shared the 1979 Nobel Prize in Physics.

The Standard Model derives its inspiration from, and is heavily reliant on, the generational pattern exhibited by the quark and lepton constituents of matter. We again need to get a bit ahead of ourselves and discuss some aspects of this model to understand how the J/ψ acted to establish this pattern as a central component of the fundamental properties of the natural world.

Before the development of the Standard Model, it was thought that the weak interaction was mediated by a pair of electrically charged particles known as W bosons (again, a boson is any particle with spin, or intrinsic angular momentum, that is an integer multiple of the reduced Planck's constant $\hbar = h/2\pi$). The W^+ boson possesses a positive charge of the same magnitude as that of the proton, while the W^- boson (the antimatter partner of the W^+) is negatively charged, with the same electric charge as the electron.

In neutrino experiments, the neutrino is detected by the sudden appearance of a charged lepton. Figure 5.2, representing the interaction of a neutrino with a down quark (charge $-\frac{1}{3}$) from a neutron or proton in a nucleus, shows why this is the case.

In figure 5.2a, the neutrino (say, a muon neutrino) emits a W^+ boson; for the overall electric charge not to change, the neutrino must turn into something negatively charged. Since this is a muon neutrino (ν_μ), the negatively charged particle it turns into is a muon (μ^-). The W^+ is then absorbed by the down quark, changing its charge and turning it into an up quark (since up quarks have a charge of $+\frac{2}{3}$, one unit of charge greater than that of the down quark, everything works out). What one sees in the detector after the interaction is the muon (the telltale sign of an incoming *muon* neutrino) and a splash of energy associated with the up quark, which is ripped from the nucleus by the interaction. Figure 5.2b shows the same process except the interaction is instead mediated by the passage of a W^- boson from the stationary down quark to the incoming neutrino. The Feynman diagram for this process is shown in figure 5.2c, which represents the quantum-mechanical combination of both of these (experimentally indistinguishable) processes.

Looking at figure 5.2, we see that the weak interaction associates leptons of the same generation. To interact weakly, a lepton must exchange a W bo-

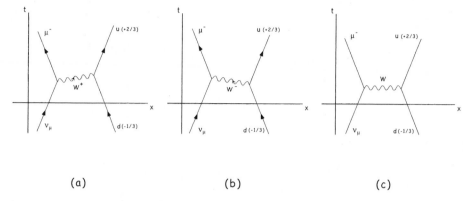

(a)   (b)   (c)

Fig. 5.2. *a* and *b* are the space-time diagrams associated with the scattering of a down quark in an atomic nucleus by a muon neutrino in a neutrino beam through the exchange of a W boson. *c* is the corresponding Feynman diagram for the process. *a* and *b* show why it is that to absorb a W^+ or emit a W^- boson, the down quark must turn into a quark with one additional unit of electric charge (remember the principle of charge conservation). The up quark, with two-thirds of a proton's charge, is an obvious candidate for the quark *q* that exits the interaction.

son; in doing so, it must change its charge. To figure out exactly how to change its charge, the lepton looks within its own generation for the particle with the appropriate charge, and that's what it turns into. An electron neutrino turns into an electron, and a muon neutrino turns into a muon.

On the other side of the Feynman diagram of figure 5.2, you might expect the same thing to be true for the quarks. Perhaps the quarks are also arranged in generations, with the W boson connecting the negatively charged down quark with the up quark—the positively charged quark that lies within its quark generation. But there was a problem: the experimental evidence available before 1974 was incompatible with this point of view.

The eightfold way tells us that there are three quarks—*u*, *d*, and *s*. How do you arrange an odd number of quarks into generations, each with two quarks in it? In reactions such as that of figure 5.2, the up (*u*) quark connects with the down (*d*) quark, leaving the strange (*s*) quark completely out of the picture. Arranging three quarks into generations of two quarks each, someone has to get left out, and that someone is the *s* quark.

Well, more or less. Even this statement needs to be hedged because experimental particle physicists noticed occasional evidence for "strangeness changing" weak interactions, such as that shown in figure 5.3—reactions that are in every way identical to that of figure 5.2, except with the initial quark being a strange (*s*) quark rather than a *d* quark.

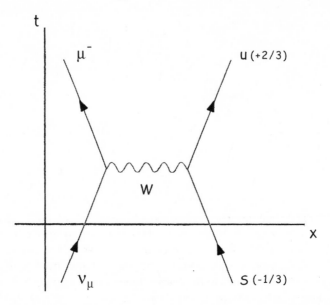

Fig. 5.3. The Feynman diagram representing the process of scattering a strange quark by a neutrino through the exchange of a *W* boson. Again, in the process the strange quark must turn into a quark with a charge of $+\frac{2}{3}$.

Studies of the relative rate of occurrence of the processes of figures 5.2 and 5.3 showed that, in a given neutrino interaction, the likelihood of having the process of figure 5.3 take place was only about a twentieth of that of the process of figure 5.2. So, in early 1974, the problem was not just that it was impossible to arrange the three known quarks into neat generations of two quarks each. In addition, the observation of the reaction of figure 5.3, albeit at a twentieth of the level of that of figure 5.2, showed that one couldn't simply pair the *u* and *d* quarks into a generation, leaving the *s* quark hovering alone on the sidelines. The social life of the *s* quark, while limited, was not nonexistent.

So, it seemed as if the whole idea of generations—pairs of particles differing by one unit of electric charge and related by their behavior in weak interactions—fell apart when you started talking about quarks. This was not good news for the Standard Model of Glashow, Salam, and Weinberg, which is based on this generational pattern.

After the revolution of November 1974, it became clear how to remedy this situation. This remedy requires another entirely new type of exchange force: that of the *neutral* weak interaction.

One of the most compelling aspects of the Standard Model is that it presents a picture in which the electromagnetic and weak nuclear forces are unified—seen to be different facets of a single underlying electroweak force. The electromagnetic field quantum, the photon, is electrically neutral. In developing the unified electroweak theory of their new model, Glashow, Salam, and Weinberg found it necessary to introduce an additional weak interaction field quantum that is complementary to the electromagnetic photon, and thus is also electrically neutral. This third (in addition to the charged W^+ and W^- bosons), electrically neutral weak field quantum is known as the Z^0 boson, the superscript "0" referring to its lack of electric charge.

By exchanging a neutral weak interaction field quantum, neutrinos (say, in a neutrino beam) can interact with matter (say, in the detector of a neutrino experiment) without turning into their corresponding charged lepton. Such a process is shown in the Feynman diagram of figure 5.4—since the muon neutrino emits (or absorbs) an electrically neutral Z^0 boson, it doesn't have to turn into a muon—it can just remain a muon neutrino (in contrast to fig. 5.2 for W boson exchange). Similarly, the d quark that's struck by the

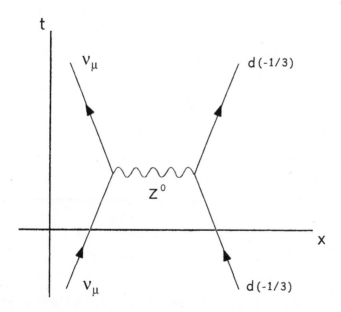

Fig. 5.4. The Feynman diagram representing the process of scattering a down quark by a neutrino through the exchange of a Z^0 boson. Since the Z^0 boson carries no electric charge, the d quark can retain its identity after the scatter.

exchanged Z^0 can remain a d quark after it's been struck. Such processes were observed for the first time in mid-1973 by the Gargamelle neutrino experiment at CERN.

Now, take a look at figure 5.5, a Feynman diagram for which a muon neutrino again exchanges a Z^0 boson with a nucleus in the target, remaining a muon neutrino after the scattering. Unlike the process of figure 5.4, though, in this process the d quark is changed into an s quark when it absorbs the Z^0. Since both the d and s quarks have the same electric charge $(-\frac{1}{3})$, there's nothing obviously wrong with this diagram. The electric charge of the neutrino and quark together is the same before and after the exchange of the Z^0. If the exchange of the charged W^+ and W^- bosons in figures 5.2 and 5.3 can connect the u quark with both the d and s quark, then perhaps the neutral Z^0 boson can connect the d and the s quark directly together, leading to the hypothetical process shown in figure 5.5.

According to the Standard Model, this process should be able to take place sometimes, given the quarks (u, d, and s) known in the early 1970s to be available for participation in the weak interaction. The problem is that

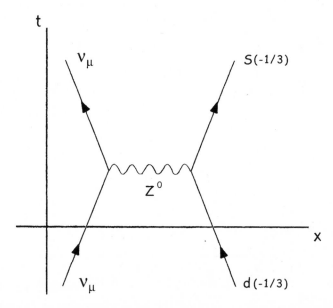

Fig. 5.5. The Feynman diagram representing the process of scattering a down quark by a neutrino through the exchange of a Z^0 boson, but with the d quark turning into an s quark in the process. Since, like the d quark, the s quark has a charge of $-\frac{1}{3}$, there's nothing wrong with this process from the point of view of charge conservation. However, this process does not take place!

processes like that of figure 5.5 are not observed in nature. The incorrect prediction of the existence of processes like those of figure 5.5 was a real stumbling block for the nascent Standard Model.

In 1970, though, Glashow, along with John Iliopoulos, from Greece, and Luciano Maiani, from Italy, realized that they could dispense with this problem if there were a fourth type of quark, whimsically designated the "charmed," or c quark. The u quark connects with both the d and s quark but prefers to connect with the d quark by a margin of twenty to one. If this fourth quark could replace the u quark in figures 5.2 and 5.3, except with an exactly complementary twenty-to-one preference for the s quark over the u quark (rather than the other way around), everything would be fine again.

The idea lurking between the lines here is that this fourth quark allows the critical generational pattern to be established for the quarks, but it does so in a funny way. The generational partner of the u quark is not the d quark but is instead $\frac{19}{20}$ d and $\frac{1}{20}$ s. We say this because the diagram of figure 5.2 is about twenty times as likely to happen as that of figure 5.3. So let's call the generational partner of the u quark the d' quark: d' and not just d since it's mostly d but has a little s in it. Similarly, the generational partner of the c quark is the complementary mixture of $\frac{1}{20}$ d and $\frac{19}{20}$ s, so we'll call the generational partner of the c quark the s' quark: mostly s, but with a little bit of d.

Now, a process such as that shown in figure 5.5 can actually happen in two ways: one for which the d and s both take part as a d' (likely for d, unlikely for s) and the other for which the d and s both take part as an s' (unlikely for d, likely for s). What Glashow, Iliopoulos, and Maiani realized was that these two possibilities would always exactly cancel each other out.

So, if such a fourth quark exists, filling out the generational pattern of the quarks in a way that mirrors that of the leptons, then processes like that of figure 5.5 in which the quark type is changed by the exchange of a Z^0 boson are forbidden from taking place. This would be a good thing, since that's exactly what the experimentalists were seeing (or, actually, not seeing): Z^0 exchange associated with a change of quark type does not take place in nature. This prescription for the introduction of a fourth type of quark, which fills out the generational pattern of the quarks and thereby acts to nullify the possibility of reactions like that of figure 5.5, is known as the GIM mechanism, where GIM is an acronym of its creators' names.

But why introduce the GIM mechanism, with all its complications, to motivate the need for the charmed quark? If you accept the claim, unmotivated as of yet, that the Standard Model is predicated on the existence of

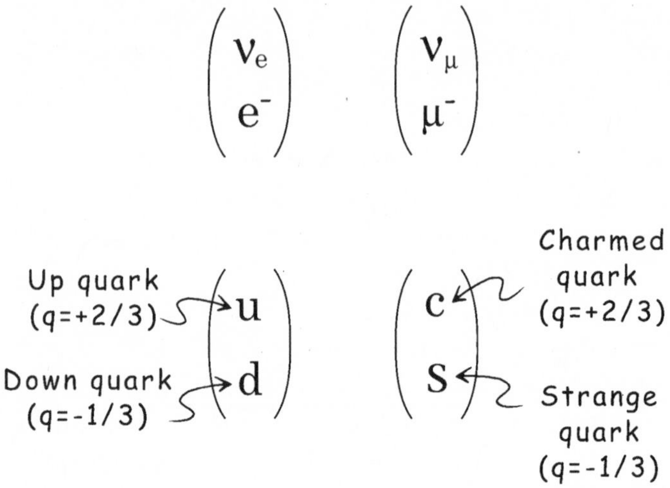

$$\begin{pmatrix} \nu_e \\ e^- \end{pmatrix} \qquad \begin{pmatrix} \nu_\mu \\ \mu^- \end{pmatrix}$$

Up quark
(q=+2/3) → u

Down quark → d
(q=-1/3)

Charmed
quark
c (q=+2/3)

s ← Strange
quark
(q=-1/3)

Fig. 5.6. The first two generations of matter, including the quark generations that mirror the lepton generations.

generational pairs of quarks and leptons, then it seems as if the strange quark needs a generational partner. This should serve as reason enough to motivate the search for the charmed quark, and indeed, that's correct: the charmed quark completes the second quark generation, in a way that mirrors the arrangement of the first two lepton generations, as shown in figure 5.6. But that's just not the way this most-celebrated chapter in the history of particle physics unfolded, and to understand the nuance of the discovery of the J/ψ, one needs to appreciate the GIM mechanism.

In any regard, the 1973 discovery of the Z^0 exchange process begged the following question: Where, if anywhere, is the charmed quark? In 1974, everything hinged on this question: the explanation for the lack of the "strangeness changing" neutral weak (Z^0) interactions of figure 5.5 as well as the establishment of the generational pattern for the quarks, mirroring that of the leptons.

This is where the J/ψ fits in. The electrically neutral J/ψ is not the charmed quark itself, which must have an electric charge exactly two-thirds that of the proton to fit into the generational pattern of figure 5.6. Rather, the J/ψ is a composite particle, a short-lived subnuclear atom composed of a charge $+\frac{2}{3}$ charmed quark and a charge $-\frac{2}{3}$ antimatter anticharmed quark that orbit around each other for about 10^{-20} seconds—the amount of time it takes the charmed quark and anticharmed antiquark to realize that they're matter/antimatter counterparts and annihilate each other, which is why the J/ψ decays.

The revelation of the discovery of the J/ψ, the so-called November Revolution, was that the charmed quark does exist, that quarks do have a generational structure that mirrors that of the leptons, and that the Standard Model may well be a valid description of nature. It was at this point that the Standard Model achieved the central status in the field of particle physics that is reflected by its name.

The charmed quark, and the array of current-day particles that contain it, escaped prior detection due to the charmed quark's large mass-energy (half the J/ψ mass-energy, or about 1.5 billion electron-volts), which placed it beyond the reach of earlier accelerators. In the 1960s, when none of these heavier particle states had been observed, the known elementary particles were fully accounted for with the three quarks (u, d, and s) of Gell-Mann's eightfold way. Rather than rendering Gell-Mann's static model invalid, the inclusion of the charmed quark acted instead to extend the notions of Gell-Mann into a new realm, that of charmed particles. On the other hand, by the time of the discovery of the J/ψ, the new dynamical theory of quantum chromodynamics was beginning to emerge, a loftier and more powerful theory of the strong nuclear interaction than that of Gell-Mann's eightfold way.

Why We Need Top Quarks

The most profound result of the November Revolution was to establish the dance card for the individual quark types by filling in the generational partner of the s quark with the newly discovered c quark. At the time, this established a beautiful symmetry between the known lepton types and the known quark types. The leptons seemed to come in two generations (the electron and muon generations), each of which had an upper member (a neutral neutrino) and a lower member (a negatively charged electron or muon). With the addition of the charmed quark, the quark types were similarly seen to fall into two generations, each with an upper member possessing $+\frac{2}{3}$ of a proton's electric charge (u and c, respectively) and a lower member possessing $-\frac{1}{3}$ of a proton's charge (d and s, respectively).

The connection that establishes two fundamental constituents (two quarks or two leptons) as quantum-mechanical dance partners (members of the same generation) is their association through the weak interaction process of W boson exchange (as shown in the Feynman diagram of fig.

5.2c). On either side of the exchanged W in this diagram, the two participating constituents are two members of the same generation.

As we've seen, there's a caveat to this distinction. Leptons, which we can think of as being puritanical, always respect their generational commitment and will never offend our graces by waltzing with a partner from a different generation. If a muon neutrino emits or absorbs an (electrically charged) W boson, it is always the muon and never the electron that will consort with the muon neutrino to compensate the loss or gain of the W boson's electric charge.

Quarks, however (perhaps because they are a discovery of the late 1960s), like to swing a little. It's not the case, for example, that the s quark connects solely with the c quark during W boson exchange, for ample evidence exists that the Feynman diagram of figure 5.3 does take place. For quarks, then, we are forced to be more open minded in how we establish generational partnerships. It's not that d dances exclusively with u and s dances exclusively with c; instead, what we observe is that d connects preferentially with u and s preferentially with c, but once in a while (roughly 5 percent of the time), d will pick up with c rather than u, or s with u rather than c, a phenomenon known as generational mixing, or, simply, mixing. Such mixing is a good thing, for within the context of the Standard Model, it is exactly this phenomenon that is thought to be the source of the weak nuclear force's divergent treatment of matter and antimatter (again, as long as there are three or more generations to mix in this way). So in this roundabout way, it would seem, mixing is essential to the existence of life (note 5.5).

This beautiful symmetry between the known generations of quarks and leptons did not last long. It was only nine months later, in August 1975, that Perl's discovery of the τ lepton was submitted to the *Physical Review* for publication. The τ, like the electron and muon, needs generational partners to keep the pattern intact. As we've seen, the τ neutrino, the generational partner of the τ, was finally detected in July 2000. But what about the quarks of the third generation?

It wasn't too long after the discovery of the τ that this question was answered. By 1977, Fermilab (in suburban Chicago) had overtaken Brookhaven as the site of the premier American proton accelerator. In an experiment led by Leon Lederman of Columbia University, which was essentially a higher-energy version of the Brookhaven experiment that codiscovered the J/ψ, a particle was observed with properties similar to the J/ψ, only much more massive. Further study revealed that this particle, dubbed the Υ (or upsilon), consisted of a quark of charge $-\frac{1}{3}$ and an antiquark of charge $+\frac{1}{3}$, in orbit about each other. This clearly was a candidate for the lower member of the third quark generation and was accorded the name of bottom, or *b*, quark. This is a truly heavy quark, with a mass-energy of about 4.5 billion electron-volts, about three times as heavy as the relatively heavy charmed quark and about five times as heavy as a proton, which is composed of the even lighter up and down quarks.

Rounding out the dance card this time, with the bottom quark's generational partner, required more than just slamming the tremendously powerful Fermilab (or CERN) proton beam into a stationary block of material. To develop enough energy to produce the sixth and final quark, it was necessary to collide the high-energy proton beam with an equally high-energy countercirculating *antiproton* beam — like the electron-positron collider at SLAC, but with protons and antiprotons instead. This required some effort, but even when these machines were completed at CERN and Fermilab in the early 1980s, the bottom quark's partner eluded detection.

The race was on, and with eventual increases in the proton/antiproton beam energies (to just about a trillion electron-volts each), and some elegant developments in particle detector technology, the sixth quark, dubbed "top" or *t*, was discovered at last at Fermilab in March 1995. Taking place in the era of truly modern particle physics experimentation, the groups that mounted this detection effort (the CDF and D0 collaborations) each consisted of many hundreds of physicists and supporting technologists from all over the world.

If the bottom quark is surprisingly heavy, at roughly 5 billion electron-volts, then the top quark is colossally so, for it weighs in with a mass-energy of no less than 175 billion electron-volts, a single, fundamental particle with a mass almost 200 times that of the proton. This fact is not lost on the theoretical physicists of the world, who have given substantial thought to why this particle's mass lies in a league of its own, and whether perhaps this experimental fact is trying to tell us something important about the way the universe operates. In a nutshell, however, we have no idea why "top" so tips the scales.

The quality of fundamental particles that we have been referring to as "type" in fact has a more formal, if somewhat whimsical and arbitrary, name: "flavor." Thus, there are six flavors of quarks. In some sort of order, they are *d*, *u*, *s*, *c*, *b*, and *t*. Sometimes *b* and *t* go by the names of *beauty* and *truth*, respectively, rather than the more staid nomenclature of *bottom* and *top*.

Similarly, there are six lepton flavors; the complete list of quark and lepton flavors will be presented in an orderly fashion in section "Reprise: The Fundamental Building Blocks." Measuring the properties of the different quark and lepton flavors and understanding their interrelations (such as the parameters that establish the degree of mixing between the generations) is a pursuit referred to as *flavor physics*.

Finally, although so much stock is put into the symmetry between the quark and lepton generations, this property of the fundamental building blocks is not understood in the context of the Standard Model. Within either of the categories—quark or lepton—the origin of the generational pattern is understood and provided one of the central insights that led to the development of the Standard Model. But there is nothing within the Standard Model that suggests why it is that there are equal numbers of quark and lepton generations. To be sure, this symmetry does seem necessary within the Standard Model; without it some of the predictions of the model make no sense. But this is not to say that the reason for the symmetry between the quark and lepton generations is understood.

So, it's simply not yet understood why, if there are three generations of leptons, there should necessarily be precisely three generations of quarks. It could just be an accident of nature, but past experience suggests that a symmetry pattern as simple and clear as this provides an essential, if yet undeciphered, clue about the workings of nature.

Particles of Force

The fundamental particles introduced so far, the six quark and six lepton flavors, are what we currently believe to be the basic constituents of matter. Atoms are made up of electrons in orbit around a nucleus composed of protons and neutrons. The protons and neutrons are themselves composed of up and down flavored quarks. Thus, ordinary matter is composed of material that comes strictly from the first generation, and from this generation, only the electron neutrino is missing as a component of ordinary matter— because it's so weakly interacting that it can never be bound to anything. Nonetheless, given that a flux of trillions of electron neutrinos from the processes that fuel the sun are passing through our bodies every second, perhaps we can even say that the electron neutrino is ordinary matter.

In high-energy cosmic ray events and in high-energy physics laboratories, heavier forms of matter can be formed using components from the second and third generations (the array of matter particles that one can construct from quarks will be examined in the section "The Particle Zoo" at the end of this chapter). In general, these exotic matter states are either, in the case of leptons, the fundamental particles themselves or are assemblages of either two or three quarks, usually only one of which is a member of the second or third generation. In one case, an experiment at Fermilab did succeed in producing atoms consisting of a muon in orbit about a charged pion (up quark and down antiquark combination). Such states are invariably fleeting, decaying away on time scales for which a microsecond is a veritable eternity, but they do leave experimental evidence that uniquely identifies them when they either pass through or decay within modern particle physics detectors.

More specifically, quarks and leptons constitute the set of fundamental (indivisible) *fermions*. Again, a fermion is any particle with an amount of spin angular momentum that is a half-integer (some integer plus one-half) multiple of $\hbar = h/2\pi$, where h is Planck's constant. All of the fundamental fermions, both quarks and leptons, are spin-½, that is, spin on their axes with an angular momentum of $\hbar/2$. Matter, we say, is built up from the fundamental fermions and, in most cases, made of the fermions from the first generation.

But if you build a single object, such as a proton, a pion, an atom, or a ball bearing, out of smaller components, you need somehow to glue those components together so that the object moves around like a single, rigid body; this is where the forces come in. From the vantage point of quantum field theory, forces are mediated by the exchange of field quanta, such as the photon of electromagnetism, or the W^+, W^-, and Z^0 bosons of the weak nuclear force.

All of the mediating particles, the field quanta, are bosons and possess an integer multiple of $\hbar$ of spin angular momentum. Thus, matter consists of fermions (quarks and leptons) bound together by one or more of the forces. The forces, in turn, are characterized by the identity of the exchanged field quanta, or, more precisely, by the properties of the exchanged boson field quanta, and the nature of interaction vertices that connect the exchanged boson field quanta to the fermions that exchange them.

The field quantum of electromagnetism is the photon, usually denoted by the Greek letter gamma (γ) (note 5.6). Photons are massless and have spin angular momentum exactly equal to $1 \times \hbar$; they are "spin-1."

As we have seen, there are three field quanta of the weak nuclear interaction—the W^+, W^-, and Z^0 bosons. All three of these were first observed directly at the CERN SPS proton/antiproton collider in 1983, leading to 1984 Nobel Prizes for Carlo Rubbia (the leader of the experimental collaboration that detected the particles) and Simon van der Meer (for his innovative contributions to the development of the proton/antiproton collider). The superscripts "+," "−," and "0" refer to their electric charge, relative to that of the proton. W and Z bosons are, like the photon, spin-1. Unlike the photon, however, they are quite heavy, possessing roughly 100 billion electron-volts of mass-energy (100 proton masses) each. This, as we'll see in later chapters, is an important point.

The field quantum of the strong nuclear force is known as the gluon, whose tongue-in-cheek name has to do with the fact that gluons mediate the incredibly strong forces that bind quarks together into nuclei. The existence of gluons was first confirmed in 1979 at the PETRA electron-positron collider in Hamburg, Germany. PETRA was the first electron-positron collider with high enough energy to allow for the detection of occasional extra (third) jets of particles emerging from the annihilation of the electron and positron to quark-antiquark pairs. Two of these jets (collimated bursts of subatomic particles) were associated with the quark and antiquark. The third was due to a gluon that radiated from one of the quarks, just as a moving electron in a radio transmitter sends out radio wave photons into the surrounding space (the parallel between the electromagnetic process of photon radiation and the strong-force process of gluon radiation is nearly perfect, once you reach energies high enough so that the latter can happen).

There are really eight quanta of the strong nuclear force field. Gluon is a generic term that pertains to all eight of the field quanta without distinguishing between them. One can never tell which of the eight possible quanta partakes in a given interaction; as far as measurable quantities go (such as mass, electric charge, etc.) they all look exactly the same. The differences between the eight field quanta collectively known as the gluon are subtle. Generically, though, gluons are massless, have spin-1, and are usually denoted "g."

Finally, the field quantum of the gravitational field is known as the *graviton*, whose existence is purely speculative; the graviton has never been observed. Even worse, we don't even know how to formulate a quantum field theory of gravity, so there's not even a theory that makes use of the graviton to calculate gravitational forces. Despite this, we do know that, should anyone ever come up with a workable quantum theory of gravity (and there are quite a number of people trying), its graviton will be massless, with spin-2 (having an intrinsic angular momentum of two times $\hbar$). The symbol for the graviton is g, about the same as that of the gluon. Not to worry, for the gravitational force and its putative field quantum are really no concern of ours.

Of all these force-mediating particles, only those of the weak nuclear force—the W and Z bosons—have any mass. This sets the weak nuclear force quite apart from the other three forces. The mass-energy of the W and Z, about 100 billion electron-volts, corresponds according to the de Broglie relation to a wavelength of about 10^{-19} meters. This is the energy at rest; when you get one of these things moving, thereby adding in additional kinetic energy, the wavelength only gets shorter. So, this is the distance over which the weak nuclear force holds sway. If you're not within 10^{-18} meters or so of an object, you're not going to interact weakly with it.

To reiterate this essential point: The weak force is exceedingly short ranged, a fact entirely due to the mass of the associated W and Z field quanta. However, a massless field quantum can have an arbitrarily low energy/momentum and thus an arbitrarily long wavelength and a correspondingly large sphere of influence.

Because mass plays such a central role in distinguishing the behaviors of the weak and electromagnetic interactions, a central hurdle that had to be overcome in the unification of these two forces is the interpretation and incorporation of the W and Z boson masses. This challenge was met, but at a certain cost: that of the introduction of another particle, known as the "Higgs boson," which plays a role unlike that of any other particle that we know of.

The Higgs, as it is often called, has yet to be observed, and the pursuit of its discovery is currently the single most concerted effort in experimental particle physics. Substantial experimental evidence suggests that, if the Higgs exists at all, its discovery is near in which case the Higgs should be observed before 2010. On the other hand, what the evidence may be suggesting is that there simply is no such thing, a scenario that may well prove more exciting than the former alternative's confirmation of the Higgs' existence. Only time will tell.

Finally, while photons carry none of the charge (electric charge) associated with the (electromagnetic) force mediated by photon exchange, the same cannot be said of the other field quanta. The W and Z bosons, in addition to electric charge, also carry the weak isospin charge associated with the weak force. Gluons also carry their force's charge, the color charge of the strong nuclear force. Each individual gluon carries both a color (positive) and an anticolor (negative) charge. There are enough different types of color charge (so-called red, blue, and green charge) that one can form eight different combinations of a color and anticolor charge (note 5.7). It's the different color charge combinations that distinguish between the eight field quanta of the strong force that are collectively known as the gluon.

This fundamental difference between the photon and the other field quanta has notable implications—both in terms of the relation of this difference to the underlying mathematical description of the mediated force and to the striking consequences associated with having the field quanta carry the very charge with which they interact. In chapter 8, our discussion of bosons that carry the charge of the interaction they mediate will underscore the remarkable dependence of the everyday workings of nature on the obscure mathematical properties of Lie groups.

Reprise: The Fundamental Building Blocks

So far we've met, depending on how you count, seventeen fundamental particles—twelve fermions and five bosons. We have strong suspicions that this list is incomplete (the hypothetical graviton and Higgs boson are the least of our worries in this regard), but for now, this is what we've got. A compilation of these basic building blocks is presented in tables 5.1 and 5.2.

The fundamental fermions comprise three up-type quarks of electric charge $+\frac{2}{3}$, three down-type quarks of electric charge $-\frac{1}{3}$, three electrically neutral neutrinos, and three negatively charged leptons. These twelve particles are arrayed in three generations (table 5.1). The five bosons, all quanta of the various force fields, include the photon, the two W bosons, the Z boson, and the gluon (of which there are really eight). These are listed in table 5.2. The epoch of the discovery of these fundamental constituents extends slightly beyond one century, from Sir J. J. Thomson's 1897 discovery of the electron to the July 2000 announcement of the τ neutrino discovery. The discovery and identification of this set of constituents is one of the major triumphs of twentieth-century science.

There's one thing that's been left out of the tables 5.1 and 5.2—antimatter. This is because the rule for determining the antimatter constituents is so simple that it doesn't bear repeating the tables just to list them. The rule is this: Each matter fermion has an antimatter counterpart that is oppositely charged but otherwise identical to the associated matter fermion. (The charge associated with each of the forces, with the exception of gravity, is opposite for the antiparticle.)

To denote an antiparticle, we just place a bar over the particle's symbol. For example, the antimatter counterpart of the u quark—the up antiquark—is the $\bar{u}$ antiquark, while the electron generation's antineutrino is denoted $\bar{\nu}_e$. The exceptions to this labeling rule are the charged leptons— the antimatter counterpart of the electron (e^-) is the positron (e^+), of the

Table 5.1   Spin-½ Fermions

Electric charge	Generation 1 (mass)	Generation 2 (mass)	Generation 3 (mass)	
$+\frac{2}{3}$ $-\frac{1}{3}$	$\begin{pmatrix} u\ (3\ \text{MeV}) \\ d\ (7\ \text{MeV}) \end{pmatrix}$	$\begin{pmatrix} c\ (1.2\ \text{GeV}) \\ s\ (120\ \text{MeV}) \end{pmatrix}$	$\begin{pmatrix} t\ (175\ \text{GeV}) \\ b\ (4.2\ \text{GeV}) \end{pmatrix}$	Quarks (all interactions)
0 -1	$\begin{pmatrix} \nu_e\ (\leq 3\ \text{eV}) \\ e^-\ (0.511\ \text{MeV}) \end{pmatrix}$	$\begin{pmatrix} \nu_\mu\ (\leq 0.2\ \text{MeV}) \\ \mu^-\ (106\ \text{MeV}) \end{pmatrix}$	$\begin{pmatrix} \nu_\tau\ (\leq 18\ \text{MeV}) \\ \tau^-\ (1.78\ \text{GeV}) \end{pmatrix}$	Leptons (no strong interactions)

Plus antiparticles

Table 5.2   Integer Spin Bosons

Interaction	Boson(s)	Mass-energy[a]	Spin[b]
Electricity and magnetism	Photon (γ)	0	1
Weak nuclear	W^+	80.42	1
	W^-	80.42	1
	Z^0	91.19	1
Strong nuclear	Gluon (g:eight of them)	0	1
Gravitational	[Graviton g]	[0]	[2]
Higgs boson	[H]	[>110]	[0]

Note: Brackets mean unconfirmed.
[a] In units of billions of electron-volts (GeV).
[b] In units of $\hbar = h/2\pi$.

muon (μ^-) is the positive muon (μ^+), and of the tau (τ^-) is the positive tau (τ^+). By the way, matter/antimatter annihilation takes place if and only if a matter particle comes into contact with its exact antimatter counterpart: a u quark will annihilate a $\bar{u}$ antiquark but not a $\bar{d}$ antiquark, nor a $\bar{\nu}_e$ antineutrino.

The bosons are their own antimatter partners. The antimatter partner of the photon is just the photon itself, of the W^+ is the W^-, and so forth. Antimatter fermions interact with each other through the exchange of the same field quanta bosons that matter fermions exchange—there are no special antifield quanta for the antimatter to use.

The mass-energies of all of the fundamental particles are listed in the tables 5.1 and 5.2 in units of billions of electron-volts, or GeV. Field quanta masses were discussed above; the heaviness of the W and Z bosons is an important aspect in the "phenomenology" (observed behavior) of the weak nuclear force.

The fermion masses increase as you go to higher generations. The choice we make to order the generations in terms of increasing mass is based on the observation that, when ordered in this way, the mixing between quark flavors (see the section "The November Revolution") is greatest for quarks from neighboring generations. We don't yet understand the source of this correlation between the masses and mixing properties of the fermions, but we suspect this correlation may well be an important clue.

While we now think that neutrinos have mass (because of the observation of the phenomenon of neutrino oscillation), none of the neutrinos has had its mass directly measured. Neutrino oscillation experiments measure the differences between the masses of the three neutrino types. Assuming that the electron-type neutrino is very light (just an assumption), the neutrino oscillation data seems to suggest a muon-type neutrino mass of several thousand electron-volts, and a tau-type neutrino mass of several million electron-volts (a little greater than the mass of the relatively light electron). However, it's premature to put much faith in these numbers.

Finally, as this book is being written, the Higgs has yet to be discovered, but were its mass-energy less than 110 GeV or so it would have been discovered; therefore, 110 GeV is a lower bound on its mass-energy.

As far as charges go, not all of the constituents partake in all of the four interactions, which is to say that not all the particles carry all four of the charges associated with each of the interactions. All matter fermions (and all bosons, for that matter) have mass-energy, and so all take part in the gravitational interaction (the first direct confirmation of Einstein's general theory of relativity was the deflection, by the sun, of photons reaching the earth from a distant star).

Likewise, all of the fermions carry weak isospin, the charge associated with the weak nuclear force. Strictly speaking, it's according to the value of the fermion's weak isospin that the pairs of fermions in a generation are arranged: the upper member of each quark or lepton pair always has a weak charge of $+\frac{1}{2}$ (in some units or other, which we'll ignore for now), while the lower member always has a weak charge of $-\frac{1}{2}$.

The leptons distinguish themselves from the quarks precisely because none of the leptons carry the strong nuclear (color) charge, while all of the quarks do. Finally, the electromagnetic charge is explicitly listed in tables 5.1 and 5.2; of the fermions, only the neutrinos have no electric charge. So, we see that quarks partake in all four forces, charged leptons in all but the strong nuclear force (which is lucky because if the electron had color charge, it wouldn't orbit the nucleus at such a relaxed distance, and we could forget about atoms), and neutrinos only have gravity and the weak force to concern themselves with.

To round out the picture for the bosons, we note first that the photon has no charge other than its gravitational mass-energy. Insofar as it is the field quantum of electromagnetism, the photon will form Feynman-diagram vertices with anything that's electrically charged, but it doesn't carry any electric charge. However, W and Z bosons, in addition to being the mediators of the weak nuclear force, carry the weak-isospin charge associated with that force. Moreover, the W bosons are electrically charged. Gluons carry color (strong-force) charge but are both electrically and weak-nuclear neutral. The Higgs carries weak isospin but is color and electromagnetically neutral. The graviton is expected to carry no other charges than mass-energy. As you can see, there's not much order to the list of which bosons carry which charges, but nevertheless, the Standard Model of the electroweak force does explain these photon, W, Z, and Higgs boson properties, and the corresponding theory of the strong force (quantum chromodynamics) explains why the gluons need to have color charge.

Whether a particle is stable or decays rapidly into other particles is not a fundamental property of the particle but, rather, follows directly from the other fundamental properties previously discussed above. Since everything that happens in the physical universe is thought to be brought about by the four known interactions, then the natures of the four interactions determine

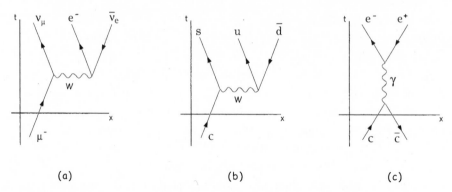

Fig. 5.7. Examples of particle decay: muon decay (a), charmed quark decay (b), and the decay of the J/ψ (c) through the annihilation of the two charmed quarks from which it is composed.

which particles will be stable and for those that aren't, just how long or short their lifetimes will be. Exactly how this plays out is shown, for example, in the three Feynman diagrams of figure 5.7, depicting muon decay (fig. 5.7a) and an example of c quark decay (fig. 5.7b), both through the weak interaction, and a typical J/ψ decay through the electromagnetic interaction (fig. 5.7c; recall that the J/ψ is composed of a c quark and $\bar{c}$ antiquark bound in orbit about each other).

Recall Feynman's rule—anything that can happen must happen—so these diagrams, which follow all our rules about the workings of the associated interaction, must take place. The muon, the c quark, and the J/ψ are all unstable. The individual particles in a sample of muons, for example, live for an average of 2.19703×10^{-6} seconds; the lifetime of the J/ψ is measured to be 1.26×10^{-20} seconds.

Not everything is unstable. To the best of our knowledge, electrons and protons live forever. Interstellar hydrogen atoms produced shortly after the big bang billions of years ago are still alive and well today. How does this square with the arguments of the previous paragraph: Why, for instance, can't one simply replace the symbol e for μ, and vice versa, everywhere in figure 5.7a? This would seem to produce a Feynman diagram for the decay of an electron into a muon, electron neutrino, and muon antineutrino, as shown in figure 5.8. Such a diagram is fine as far as the rules for weak interaction vertices go, but there's something else that we've forgotten to take into account—the principle of energy conservation.

Even if we ignore the neutrinos, the mass-energy of the muon alone is greater than that of the electron. So, there's no way that an electron, sitting at rest and minding its own business, can suddenly turn into a muon plus other stuff. Before the decay, the energy of the system under consideration is that of the electron's mass-energy. After the decay, you would have the mass-energy of the muon and the two neutrinos, plus whatever energy of motion is carried off by the muon and neutrinos. Energy is always positive (even for antiparticles) so all of these add together. The mass-energy of the muon alone is enough to guarantee that the energy after such a decay would be greater than the energy before the decay, so such a decay would violate the sacrosanct principle of energy conservation. Feynman's rule that anything that can happen must happen still applies. It's just that this particular (electron decay) Feynman diagram *can't* happen. When you add up the decay product energies, the sum will always be greater than that of the original electron. There is no electron-decay Feynman diagram that can be drawn for which energy is conserved, so electrons don't decay—they're stable.

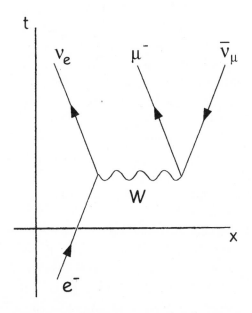

Fig. 5.8. A potential Feynman diagram representing the process of electron decay. This diagram obeys all the rules of the weak nuclear interaction. However, it does *not* obey the law of energy conservation, so it is forbidden. Electrons are thought to be completely stable, living forever unless annihilated by a positron, the electron's antimatter counterpart.

What would the world be like without stable electrons (and similarly, stable protons)? It's hard to imagine, but it wouldn't be a nurturing place. Energy conservation is a good thing.

So, it's the heavier forms of matter that are unstable, and it's the lighter forms that are stable, not because the forces treat the heavier generations in any fundamentally different way than the lighter generations, but simply because of energy conservation.

All of the neutrinos are light enough so that each by itself is stable. Other than this, there is not a single stable particle or combination of particles that contains anything other than members of the first generation (although the opposite is not true, there are many unstable particles that are composed only of material from the first generation). Also, there's nothing intrinsically unstable about antimatter. The positron (the electron's antimatter counterpart) is perfectly stable on its own. However, bring it into contact with an electron, and they'll both disappear in a flash of light (literally). An antimuon (μ^+) is exactly as unstable as a muon (μ^-), by which I mean that their short but precisely measured lifetimes are identical.

There's a little more to this stability-of-matter issue that we need to get out in the open. Say a top quark is produced, either in an accelerator experiment or in a natural cosmic ray event. If there's not an antitop quark that just happens to be lurking nearby to annihilate it (and there usually isn't), how are we going to get rid of it? It will have to decay into lighter quarks (as shown in fig. 5.9). Whatever interaction is responsible for this decay must have the following property: it must be "flavor-changing." It must allow the t quark, through the emission of a field quantum, to change spontaneously into a lighter quark (in fig. 5.9, this is a b quark, although an s or even a d quark is also possible, although less likely). Otherwise, you couldn't get rid of the t quark, and matter containing the top quark would be stable, in contradiction with observation.

There's only one interaction that allows for this change of flavor, and it's the weak interaction, specifically, the part of the weak interaction that is governed by W (rather than Z) boson exchange. Because the weak interaction behaves in this way, because of the mixing properties of the section "The November Revolution," the stuff of the universe is composed almost entirely of the constituents from the first generation of quarks and leptons, which happens to be the stuff from which we can create atoms and molecules. Again, we see that the weak interaction's renegade properties, rather than making life difficult, come to our rescue and render life possible.

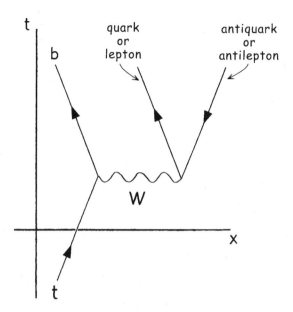

Fig. 5.9. The process of top-quark decay. To conserve energy in the decay, the top quark must change flavor (in this case, it changes into a bottom-flavored quark). The only interaction that allows quarks to change flavor is the charged weak interaction (the interaction mediated by the *W* boson), which is what is depicted here.

It's even more interesting than that because the weak nuclear interaction is, as its name implies, weak. This means that, because the decay of heavier quarks and leptons is controlled by the weak interaction, their lifetimes are not necessarily unfathomably short. The weaker the interaction, the less likely the decay is to take place at any instant of time and the longer the particle will live on average. On the other hand, particles that decay by way of the relatively strong electromagnetic interaction have lifetimes of 10^{-15} seconds or less, while particles that decay through the strong nuclear force tend to have only about 10^{-23} seconds to get their affairs in order.

Weakly decaying particles fair much better. For example, three of them—the c and b quarks and the τ lepton—all have a lifetime of roughly 10^{-12} seconds. Now, this is a very interesting number, for a particle with this lifetime moving close to the speed of light will travel a millimeter or so before decaying into a spray of more stable particles. Thus, if you can detect these short distances between the production and decay of these heavy forms of matter, you've got an iron-clad way to distinguish them from the more mundane lighter forms of matter. This lifetime of 10^{-12} seconds, short as it may seem, provides a potent tool for the detailed study of the properties of these exotic heavy forms of matter.

In the 1980s, when this picture began to come together, measuring these millimeter decay lengths lay just beyond the limit of the capability of particle physics detectors. Being trained to tackle just this sort of problem, experimental particle physicists dove in, and by the mid-1990s had developed detectors sensitive enough to detect these moderate lifetimes. The use of such a lifetime tag to identify particles containing these heavy forms of matter has become a standard capability of all the major experiments; the top quark, which almost always counts among its decay products at least one b quark, owes its discovery in 1995 to just such a tag.

In addition, it soon became clear that the resulting advances in particle detection that were developed in pursuit of this fortuitous lifetime scale of 10^{-12} seconds, in the realms of both detector technology and the associated sensitive electronic readout, have broad application throughout the field of imaging (medical imaging, microscopy, astronomical imaging, etc.). Today, many particle physicists can be seen working side by side with colleagues from these other scientific fields.

The Particle Zoo

While quarks are truly the fundamental, indivisible building blocks of nuclear and subnuclear matter, they suffer the empirical drawback that they are never observed freely in nature. It's impossible to isolate and study an individual quark as you can an electron or a proton. Quarks, which carry the color charge of the strong force, always come in groups that are color neutral. In other words, any observable particle that contains quarks must be composed of combinations of quarks for which the combined strong charge of the combination exactly cancels, just as the electric charges of the protons and electrons cancel in an atom, which is electrically neutral.

This property, known as *confinement*, is a consequence of the vast strength of the strong nuclear force, and the fact that the strong force-field quanta, the gluons, themselves carry the charge of the strong force. These conspire to concentrate the force field between two color-charged objects (e.g., two quarks) in such a way that the energy of their binding increases with their separation. This is very counterintuitive; usually, you would think that as two objects get farther and farther apart from each other, their influence on each other would wane rather than grow stronger. But such is not the case for the strong nuclear force.

As a result, the energy associated with this separation grows as the separation increases, until there's enough energy around to create some new quarks (out of the perpetually fluctuating vacuum) with color charges that cancel those of the two quarks you're trying to separate. Each of the new quarks pairs up with one of the old quarks, producing two new color-neutral objects.

As a result, if you try to separate the two quarks in a color-neutral object, the instant the two quarks get much more than a fermi (10^{-15} m) away from each other, you suddenly find yourself holding two color neutral objects composed of a total of four quarks. The energy required to create the two new quarks came from you because you had to exert energy to pull the original two quarks apart. It is impossible to pull a single quark out of nuclear matter and isolate it in the laboratory.

Strong-force (color) charge comes in three different types: red, blue, and green. While any electromagnetically charged object has some net positive or negative amount of electric charge, a color-charged (or "colored") object has some net positive or negative amount of red charge, some net positive or negative amount of blue charge, and/or some net positive or negative amount of green charge. To specify the color charge of an object, you need three numbers: one each for the red, green, and blue contribution to the overall color charge. If the object is strong-force charged, one or more of these numbers will be nonzero.

Electrons always have an electric charge of -1 (in units of the proton's charge), but quarks come in three varieties: red, blue, and green, each containing a positive unit of the appropriately colored charge. Antiquarks also come in three colors, each containing a negative unit of the appropriately colored charge.

Any particle that is composed of a color-neutral collection of quarks is known as a *hadron*. There are two observed (and theoretically understood) ways in which colored quarks can be gathered together so that the resulting combination is color neutral. The two different ways of making color-neutral assemblages of quarks lead to two different types of hadrons: mesons and baryons.

A meson is formed when a quark of a certain color is paired with an antiquark of the same color. Just as a system containing an electron and positron ("positronium") is electrically neutral because the electron's and positron's charges have opposite signs and cancel each other out, the color charge of a red quark is opposite to that of a red antiquark. So, a particle formed with a red quark and red antiquark is color neutral. Mesons are color-charge neutral combinations of a quark and an antiquark, bound together by the strong nuclear force.

The assembly of quarks into baryons exhibits a remarkable property of color charge. If you take one red, one blue, and one green quark (or one red, one blue, and one green antiquark), and put them together, the result is also color neutral. You combine three different types of positive color charge (or three different types of negative color charge if you're using antiparticles) and, amazingly, the result is neutral! This is counterintuitive but follows directly from the mathematical properties of the Lie group that underlies the behavior and properties of the strong force and its associated color charge. So, baryons are combinations of three quarks bound together by the strong nuclear force, which are color-charge neutral because the three quarks contain each of the three possible colors of color charge (note 5.8).

The *particle zoo* that so dismayed particle physicists before the advent of the eightfold way is just the myriad of such mesons and baryons that can be formed by combining the underlying quarks. In what follows, we'll not explicitly mention the colors of the quarks involved in these combinations; once we know the ways colors can be combined to form color-neutral hadrons (mesons and baryons), we can forget about color.

For example, let's make some baryons. The most stable possible combination we can imagine is one composed entirely of first-generation materials. Each of the three quarks is either a *u* or a *d*. Quarks have spin, internal angular momentum of $\frac{1}{2}\hbar$. As it happens, the lowest energy and most stable possible configuration of the quarks is one for which two of the spins are clockwise and the other is counterclockwise (or vice versa) and also where all the quarks are at rest inside the baryon, so there's no additional motion adding energy (and thus mass) to the overall baryon. Within these constraints, we have four choices: *uuu*, *uud*, *udd*, and *ddd*, with charges of $+2$, $+1$, 0, and -1, respectively.

For somewhat subtle quantum-mechanical reasons, the *uuu* and *ddd* choices are not allowed for this lowest-energy configuration of spins and internal motion. This absence of the *uuu* and *ddd* states confused particle physicists for some time. In fact, it was precisely the recognition that quarks come in three different colors—red, blue, and green—that made the quantum-mechanical arguments forbidding the *uuu* and *ddd* combinations work out. The absence of the *uuu* and *ddd* combinations in the lowest-energy (lowest mass) configuration of the baryons in the particle zoo was one of the first pieces of experimental evidence for the existence of color charge.

So, we expect two most-stable baryons: a *uud* combination, with electric charge $+1$, and a *udd* combination, which is electrically neutral (since *u* and *d* quarks have charges of $+\frac{2}{3}$ and $-\frac{1}{3}$, respectively). The positively charged *uud* baryon is the proton, while the neutral *udd* baryon is the neutron.

These particles are indeed quite stable, although not perfectly so; free neutrons, on their own out in the world, have an average lifetime of 14.8 minutes (note 5.9). For their part, free protons have never been observed to decay, although substantial effort has been expended searching for the phenomenon of proton decay.

Why do physicists suspect that protons might be unstable? If protons decay then hydrogen atoms (which are just protons with electrons orbiting about them) are not stable, and star formation would never have gotten started in the early universe. Protons are definitely quite stable; the question is whether protons have an immensely long, yet not infinite, lifetime. Attempts to unify the electromagnetic, weak, and strong forces into a single, all-encompassing theory of particle interactions as well as to understand the origin and evolution of the universe (our theories of "cosmology") tend to prefer that protons decay, albeit with an exceedingly long average lifetime. Current experiments (the Super-Kamiokande experiment in Japan being the most prominent) would have seen a proton decay if the mean lifetime were less than about 10^{31} years, and they haven't. So, these experiments have demonstrated that the lifetime of the proton is greater than 10^{31} years. How can they make the claim that the lifetime of the proton is at least 1 septillion (10^{21}) times the current age of the universe?

Simple. All you have to do is monitor vast quantities of protons all at once, with electronic sensors that are sensitive enough to pick up the signal from a single decaying proton. Such is the case for modern proton-decay experiments. Since the amount of water it takes to fill up the bottom of an abandoned mine shaft (far underground so that cosmic rays that fake proton decay don't penetrate) contains on the order of 10^{31} protons, and since none of these experiments has ever observed a proton to decay after watching for a few years, we now know that protons live longer than 10^{31} years. Such is the property of the random decay process. If any given proton will decay in 10^{31} years, then, on average, you would expect one proton out of a sample of 10^{31} to decay in any given year. Analogously, the chance of you getting struck by lightning on any given day is small, but in the population at large, it's a common occurrence.

You may recall that it was the Super-Kamiokande experiment, for want of something better to do after failing to observe proton decay (it's not their fault; protons just don't decay very fast, if at all), that mounted the most convincing evidence to date for neutrino oscillations. As is often the case in science, great discoveries (in this case, profound enough to garner a piece of the 2002 Nobel Prize in Physics) arise in unforeseen ways due to a combination of well-designed experimental apparatus and the creative scientific thinking of its proponents. The history of the advancement of science is peppered with such discoveries.

Beyond protons and neutrons, stability goes out the window. The various combinations you can make by replacing one of the u or d quarks with the next lightest quark, the s, are known as Λ and Σ baryons (these are the lowest energy uus, uds, and dds combinations). These decay through the weak interaction; although they are unstable, they are not horribly so by particle physicists' standards, with lifetimes of about 10^{-10} seconds. If, instead of swapping one of the u or d quarks for an s, you form a baryon of u and d quarks, but with all the quarks' spins in the same direction (all clockwise or counterclockwise), you get the Δ baryons. Or, if you leave the spins alone and instead give the quarks inside the baryon some orbital energy (get them turning in orbit around each other), you get the N baryons. In both situations, these new baryons can decay by way of the strong interaction by falling apart into a proton (or neutron) and a π meson (to be introduced below). Things that decay through the strong interaction are truly short lived: 10^{-23} seconds is typical.

Because there are six different quark flavors to use to make baryons and because for any given quark combination there are a number of different ways to configure the spin and orbital properties of the quarks in the baryon, you can see that the "zoo" of elementary particles, or at least the particles that were thought to be elementary before Gell-Mann's hypothesis of the eightfold way and the subsequent discovery of quarks, contains quite a number of exotic animals. The hypothesis and discovery of quarks was a tremendous simplification, as can be seen by the diminutive table of truly fundamental matter particles (table 5.1).

Nevertheless, it's not the bare quarks but, rather, their combinations in baryons (and mesons) that are directly observable. To this day, the number of species interred in the particle zoo continues to increase, overseen by the Particle Data Group, an energetic team of gamekeepers headquartered at the Lawrence Berkeley National Laboratory just up the hill from the University of California, Berkeley (an appropriate place, if ever there was one, to house a particle zoo). Their ever-updated publications, which comprise a sort of scripture for the congregation of particle physicists and their ilk, are available free to the general public and contain much more than dry lists of particles and their properties (note 5.10).

But what about mesons—nature's quark/antiquark approach to producing color-neutral quark combinations? As for the baryons, the most stable mesons are formed of quarks from the first, and lightest, generation: from a u or d quark combined with a $\bar{u}$ or $\bar{d}$ antiquark. The spins of the two quarks will be in opposite directions, one clockwise and the other counterclockwise, and the two quarks will be at rest within the meson. A quick bit of reflection reveals four possible combinations: $u\bar{d}$, $d\bar{u}$, $u\bar{u}$, and $d\bar{d}$, with electric charges of $+1$, -1, 0, and 0, respectively. The last two of these, with charge 0, get lumped together into two particles, each of which contain some $u\bar{u}$ and some $d\bar{d}$ (you can do this sort of thing in quantum mechanics); the resulting quartet of mesons consists of a trio of particles known as the π-mesons, or simply the pions: π^+, π^0, and π^-, and a neutral particle known as the η (eta). The difference between the π's and the η is that the pions possess nuclear isospin, while the η doesn't (we'll introduce nuclear isospin in chapter 7).

All of this is incorporated and subsumed, in an elegant and encompassing fashion, by the formalism of Gell-Mann's eightfold way. The pions and the η are understood to be part of a larger pattern of particles, a pattern that manifests itself in the mathematical space of the amount of electric charge (really, nuclear isospin, which is only incidentally related to electric charge), and strangeness (net number of strange quarks minus strange antiquarks) possessed by each particle in the pattern. This is as true for the mesons as it is for the baryons—there is no known particle formed from quarks and antiquarks that does not fall neatly into a prescribed and necessary place in one of the grand patterns of the eightfold way. Figure 5.10 below shows one such pattern—that of the so-called pseudoscalar meson octet; we'll revisit this pattern in a little greater depth in chapter 7.

Since the π^0 is composed of matter/antimatter quark pairs, its quarks eventually find each other and annihilate. When this happens, the π^0 al-

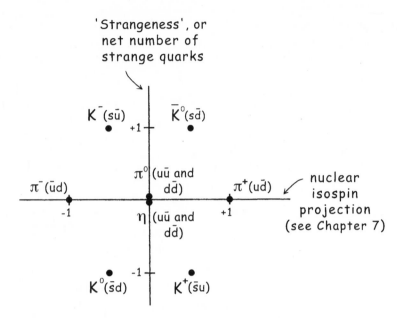

Fig. 5.10. The pattern of the pseudoscalar meson octet. The quark content of each of the pseudoscalar mesons is shown in parentheses.

most always decays to two photons through the electromagnetic interaction. Since the electromagnetic interaction is reasonably strong, the π^0 has a correspondingly short average lifetime of 8.4×10^{-17} seconds. The quarks in the π^+ and π^- are not matter/antimatter pairs and cannot annihilate each other. Instead, the charged pions must decay weakly (primarily to a muon and muon antineutrino), so they have a relatively long lifetime: 2.603×10^{-8} seconds. These charged pions are the most stable (longest-lived) mesons. The η decays both electromagnetically and strongly (it has a number of different possible decay modes), with a lifetime of about 10^{-18} seconds.

As with the baryons, you can make a staggering array of different mesons by swapping quark and antiquark flavors or by leaving the flavors alone and instead playing with the way the quark and antiquark spin and orbit each other. Doing the latter gives you sets of mesons with names such as ρ, η', ω, f_0, and a_1. If you take a pion and swap one of the u or d quarks or antiquarks for an s quark or antiquark, you get a set of mesons known as kaons, or K mesons. You can take the kaons and rearrange their quarks' spin and orbital properties, giving rise to the K^*'s, K_1's, K_2^*'s, and so on. Don't forget that there's also the charmed, bottom, and top quarks to play with. What a zoo!

The Particle Data Group's full-sized *Review of Particle Properties,* updated every other year, weighs in at a few pounds (in paperback).

Figure 5.10 shows the pseudoscalar meson octet—the eightfold-way pattern occupied by the pions, kaons, and the η—along with their quark content. Again, the η and π^0 are not the same: the π^0 possesses the soon-to-be introduced quality of nuclear isospin, while the η does not. The eightfold way knows all about isospin, and makes a sharp distinction between the two particles, even though they have the same electric charge and strangeness. That this eightfold-way pattern has precisely eight unique particle states is no coincidence. That's where the name "eightfold way" comes from.

A parting question: What role do the members of this vast array of nearly fundamental subnuclear particles play in the overall workings of the universe? Clearly, we need the protons and neutrons—*uud* and *udd* combinations arranged so that the resulting three-quark baryon has the lowest possible internal energy—to construct everyday matter. But what about all of this other junk: the unstable baryons and mesons that are jarred into a brief existence during one of the few earthly high-energy laboratory experiments or the occasional collision of high-energy cosmic radiation with the other, more stationary matter in the universe? Are they somehow a necessary component of the evolution of the universe from the chaos of the big bang to the eventual emergence of sentient beings?

My sense is that the answer is no. We know that we require three quark generations (six quark flavors) to allow for the mechanism of matter/antimatter asymmetry that provides for the necessary absence of antimatter in our evolved universe. We also suspect that the exact correspondence between the number of quark and lepton generations reflects a deep and necessary connection, not yet understood, between the strong and electroweak (unified electromagnetic and weak nuclear) interactions. However, the myriad states composed of various quark and antiquark combinations, all bound together by the strong nuclear force, seem to be more a by-product of the fundamental and necessary components of nature (quarks, leptons, and the forces that influence them) than to be necessary building blocks themselves. There may be some beneficial roles played by a small number of these particles; for instance, particularly violent cosmic ray collisions in our upper atmosphere produce a lot of pions, which quickly decay to muons, which keep us bathed in a fairly constant flow of ionizing radiation, and which may hasten the process of genetic mutation and the adaptation of species. Overall, however, I am hard pressed to think of truly fundamental and necessary biophysical roles played by these obscure and short-lived particles.

Yet, particle physicists spend long hours and substantial resources in the study of these particles. Their motivation is that, when it comes to quarks, we don't see the fundamental level—the quarks themselves. What we do see in particle physics experiments are large numbers of baryons and mesons. These composite particles contain the only clues we have about the fundamental world at work within them. Without a careful program of search and classification of these particles in the 1950s, for example, we never would have had an inkling of the presence of quarks. Things have not changed; to this day, one often relies heavily on our understanding of the properties of these particles, which come flying away in copious numbers from the violent collisions induced by modern particle accelerators, to unravel the fundamental physical behavior taking place at the core of those collisions. The taxonomy of the curious creatures in the particle zoo is no less central to the work of the particle physicist than that of earthly animals is to the evolutionary biologist.

With that said, it's time to leave the sullied warrens of the particle zoo behind us. Our next destination will be the ivory tower world of pure mathematics; please remember to clean your shoes well as we take our leave.

6

Mathematical Patterns

~~~~~~~~~~~~~~~~~~

Lie Groups

Much has been said about the ever-growing and almost miraculous interconnectedness between the abstractions of pure mathematics and the concrete study of natural phenomena. That the former provides such an essential tool for the pursuit of the latter is a continual source of inspiration and wonder for those who reflect on it. The ancient Greeks, whose developments in abstract mathematics retain a solid standing in the body of modern mathematical knowledge and who also put considerable effort into the interpretation of the natural world, made only limited headway in the application of simple mathematical principles toward the description of nature. The connection is just not that obvious.

It was with the rise of the western university system in the early renaissance that the connection between the abstract world of numbers and the infinitely complex and rich world of natural phenomena became firmly established. In the middle of the fourteenth century, academics at Merton College of Oxford University and at the Universities of Paris and Bologna developed the first rigorous descriptions of motion, including quantitative notions of speed and acceleration. This development reached a pinnacle in Isaac Newton's seventeenth-century formulation of calculus, and his application of this new mathematical tool to the description of celestial motion.

Mathematics has hardly stood still since the time of Newton, and as its development has led in increasingly more arcane and abstract directions, its application to the natural world has become all that more remarkable.

The specific example of group theory, and its application to the fundamental description of natural processes, provides an opportunity to convey a sense of this profound connection and to show how the musings of abstract mathematicians, sequestered in their ivy-clad garrets, have provided an essential component of the modern understanding of natural phenomena.

Group theory is one of the most gratifying topics in the domain of abstract mathematics. The requirements that an object must satisfy to be deemed a "group," in the strict mathematical sense, are deceptively easy to state. Yet the delineation and study of groups has entertained a lot of top mathematical talent over the years, particularly in the latter part of the nineteenth century, during which much of the theory of finite groups (groups composed of a finite number of objects) was laid out. Our interests, however, do not lie within the theory of finite groups but, rather, with that of infinite, or more specifically, continuous groups. In fact, our interests are focused even more narrowly than this—on a specific class of continuous groups known as Lie groups (note 6.1), first defined and studied in the 1870s by the Norwegian mathematician Sophus Lie.

Lie groups are at the heart of the description of a surprisingly large number of physical phenomena and enjoy broad application throughout the fields of natural science. In particle physics, Lie groups play such a central role that it is impossible to proceed further without their introduction. It wasn't until after Murray Gell-Mann's introduction of the eightfold way, an application of Lie groups to the categorization of the particle zoo (see chapter 5), that physicists realized that they had begun to speak the language of group theory. Nowadays, the connection is firmly established and in making this connection, physicists have benefited greatly from the abstract mathematicians' exhaustive treatment of Lie groups.

An Exercise in Abstraction: Mathematical Groups

To a mathematician, a group is any set of objects with an associated rule, or operation, that combines pairs of objects in the set. The obvious examples of operations and the ones from which the more general notion of operation was abstracted, are addition and multiplication. For example, if x and y are two numbers, say, the monthly bill for the two separate phone lines in your house, then $z = x + y$ is the combination of x and y that tells you how much you are obligated to enrich the phone company this month. The operation of addition takes the pair of numbers x and y and combines them, yielding a third number, z.

The nice thing about addition and multiplication is that, without any external reference, most of us know how to take any given numbers x and y and combine them into the result z. To a mathematician, however, the group's operation could be absolutely any rule that combines the objects in the set as long as the operation satisfies four specific requirements (described later in this section)—even if the only way to delineate that rule is to write out a complete table of pairs of numbers x and y and their combined result z. Generically, the process of operation is often represented by the symbol "$*$"; the expression $x * y$ denotes the combination of the objects x and y according to the rules of whatever operation you have chosen to associate with the group's set of objects, of which ordinary multiplication and addition are just two of many possibilities.

Not every set with an associated operation on its elements comprises a group, however. To form a group, the set and its operation must satisfy four criteria, the defining axioms of group theory.

First, the set of objects must be closed with regard to the operation. This is a shorthand way to say that if x and y are objects in the group, then the result z of their combination (under the operation associated with the group) must also be an object in the group. For example, the set of positive whole numbers 0, 1, 2, 3, 4, and so on is closed under the operation of addition: the sum of any two whole numbers is always a whole number. Similarly, we could consider the set of positive whole numbers with the associated operation of multiplication; again the set (whole numbers) exhibits closure under the chosen operation (multiplication). However, the set of whole numbers under the operation of division is not closed; for instance, one divided by four is ¼ or 0.25, which is decidedly not a member of the set of whole numbers.

The second axiom, known as the *associative law*, is the most obscure of the axioms. It won't play much of a role in our discussion, but we need to include it for completeness' sake. Let's say that you have not two but three elements, a, b, and c, that you wish to combine according to the rules of the associated operation, yielding some final element z within the set. The obvious thing to do is to combine two of them first, say b and c, and then combine the result with a: $z = a * (b * c)$, where the parentheses indicate which of the two indicated "$*$" operations should be performed first. On the other hand, one could also imagine combining a and b first, then combining the result of that operation with c: $z = (a*b)*c$. The requirement of the second axiom—the associative law—is that both of these double operations yield the same result z. One reason the associative law is relatively uninteresting is that it's hard to think of an operation that doesn't obey it. You can easily convince yourself that ordinary arithmetic operations such as addition and multiplication do obey it. Nonetheless, many of the rich and powerful results that form the mathematical theory of groups do require that the associative law hold true, and so it must be included as an axiom.

There are even more ways to combine the three elements a, b, and c, such as $z = (b * c) * a$ or $z = (c * a) * b$. These are all different combinations of the three elements a, b, and c. Shouldn't they give the same result z as above? The answer is, emphatically, not necessarily. For these two latter operations, the order of the elements themselves (and not just the order of the operations) has been changed relative to those of the previous paragraph. The operation associated with the group is not required to be the same when the order of the elements being operated on is switched. In other words, it is not necessarily true that $x * y = y * x$—that x and y "commute"—for all elements x and y in the set (matrices, for example, familiar to many from high school math, do not commute when combined together under the rules of matrix multiplication). Groups that possess this additional property are known as *commutative*, or *Abelian* groups. Most Lie groups, however, are noncommutative, or non-Abelian, a fact that will be seen to have rather profound physical consequences when we encounter their application to gauge theory in chapter 8.

The third axiom requires that the set possess the *identity element* of the chosen operation. This means that the set must possess an element, call it I, for which $I * x = x$ for *any* object x in the set. For example, for the whole numbers under the operation of addition, the identity element is "0": 0 added to any number just gives the same number back again. Similarly, the identity element of the whole numbers under multiplication is the number 1.

The fourth axiom associated with the definition of a mathematical group prevents the positive whole numbers, under either addition or multiplication, from forming a group. This axiom requires that for each and every element x in the set there must be one and only one element x', also in the set, for which $x * x' = I$, where I is again the identity element. For example, if the operation "$*$" represents multiplication (for which the identity element I is 1) the fourth axiom requires that for each element x, there be a corresponding element x' such that $x \times x' = 1$, or, equivalently, $x' = 1/x$.

The element x' is known as the *inverse* of x under the chosen operation. In these terms, the fourth axiom states that each element in the set must have an inverse under the chosen operation.

Consider the set of positive and negative whole numbers ..., $-3, -2,$ $-1, 0, 1, 2, 3, ...$ also known as the set of *integers*. The integers do not form a group under the operation of multiplication. For example, what number x' has the property that $3 \times x' = 1$? The answer is ⅓, which is not an integer. The integers do form a group when the associated operation is addition, with identity element 0. Once you recall that the operation of subtraction is really just the addition of a negative number, it's easy to see that the inverse of any integer n is $-n$.

To recap: A group consists of a set with an associated operation, or rule of combination. The set and operation must satisfy four axioms if the system of set and operation is properly to be called a group: closure, or the requirement that the combination of two elements must always yield a third element within the set; associativity, $(a * b) * c = a * (b * c)$; the possession of an identity element, or an element which, when combined with any other element, gives the other element back again; and finally, for each element in the group, the existence of an inverse which, when combined with the element, yields the identity element. If, beyond this, the group exhibits the property that $a * b = b * a$ for any elements a and b in the set, the group is Abelian. Most of the groups that will interest us, however, will be non-Abelian.

These four axioms are neither good nor bad. Concerned only with the ethereal world of abstraction, mathematicians are free to introduce this notion of a group and to require that anything that is a group satisfy these criteria. These axioms are neither right nor wrong; they are simply the rules that mathematicians have chosen to require of something that they have decided for some reason or other to call a "group."

Having somewhat arbitrarily specified what it means to be a group, however, one can then develop a theory of groups based on postulates that follow directly from the four defining axioms. This theory is no more or less arbitrary than the axioms underlying the definition of the group. All that one can say is that since the mathematical entities known as groups satisfy (by fiat) all four criteria, then they possess a number of other properties, and exhibit a number of other characteristics, that can be mathematically proven to follow inexorably from the four axioms. For example, two questions (of many possible questions) that you might hope your theory of groups might answer are, (1) how many distinct Abelian groups (if any) contain exactly 675 elements? or (2) for the group with 3,698 elements, how many distinct subsets of elements in the group's set form groups in their own right under the associated operation?

To the mathematically inclined, there is a deep beauty in the creation of the web of interlocking results that lead eventually to the solutions of basic questions one might ask about mathematical structures such as groups. It's like doing a crossword puzzle. When you're finished, you haven't produced anything that is likely to launch the next Fortune 500 company, but you have met and triumphed over an intellectual challenge.

There is a difference, however, between working crossword puzzles and the pursuit of higher mathematics. In the case of mathematics, you don't triumph over the capricious machinations of another human being (the designer of the puzzle) but, rather, over the absolute fabric of logical relations. The body of knowledge you have developed has the enviable characteristic of being demonstrably and absolutely true, given the set of assumptions (axioms) underlying your contemplations, irrespective of the foibles of your own human limitations, indeed, irrespective of the existence of humanity itself. And, as an added bonus, if it should so happen that the set of axioms on which your intellectual fortress is built is somehow relevant to the physical world, then you can even walk away with a deeper understanding of your natural surroundings. The wonder of group theory is that its relevance to the disciplines of both mathematics and natural science far exceeds the self-contained boundaries within which it was first developed.

The Mathematics of Clocks

There's nothing that illustrates an idea better than a concrete example, so here's a description of a particular group—in fact one of the smallest possible groups, which has only four elements in its set. This group is sometimes referred to by mathematicians as "Mod(4)."

The division of the terrestrial day into twenty-four hours is an arbitrary one. Let's consider a culture that instead chooses to divide the day into only four hours. Also, let's assume that this culture is sufficiently relaxed that its citizenry doesn't care about the minutes within the hour. Thus, clocks in this culture consist of dials with four markings—one through four—and a single hand that moves four times a day from one hour to the next. Figure 6.1 shows this clock.

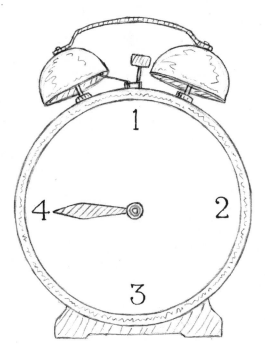

Fig. 6.1. A clock with but four possible readings: the hours one through four.

The four markings (the numbers one through four) will form the elements of our set. The associated operation will be that of adding elapsed time, in these long hours, to the current time of day to get the time of day after the elapsed interval. Thus, if it's now one o'clock, and we want to know what time of day it will be after two more hours have elapsed, we consult the rules of our operation to deduce that $1 * 2 = 3$. The operation represented by the "$*$" in this case behaves an awful lot like conventional addition, but it's not.

For example, what if the current time is two o'clock, and we want to know what time it will be after three more hours have elapsed. We consult our clock face to find that, after passing through four o'clock, the time cycles back to one o'clock. The day has changed, but that doesn't concern us because the elements of the set are just the four hours representing the time of day; the date itself is of no relevance. So, after 3 more hours have elapsed, the time of day is one o'clock: $2 * 3 = 1$.

It's easy to see that the set of numbers is closed under this operation, which is sometimes referred to as *clock arithmetic*. No matter how many hours you add to the current time, all you're going to do is spin the dial around to one of the four hours between one and four o'clock—the result of the operation of combining any two numbers between one and four is just a third number between one and four. In addition, no matter what time (*t*) of day you start with, if you add four elapsed hours to that time, you go through exactly one full day and get back to the original time of day: $t * 4 = t$, for any $t = 1, \ldots, 4$ in the set. Comparing this expression to that of the definition of the identity element above, we see that the number 4 is thus the set's identity element. Also, once you get the feel of clock arithmetic, it's easy to convince yourself that the associative law (the second axiom) holds.

That the fourth and final criterion (that every element in the set have a unique inverse within the set) is satisfied is demonstrated by considering the following three clock arithmetic operations: $1 * 3 = 4$, $2 * 2 = 4$, and $4 * 4 = 4$; you can verify that these operations are correct according to the rules of clock arithmetic with four elements. Recall that the inverse of an element is the element that, when combined with the original element according to the rules of the operation, yields the identity element. So, we see that the elements 1, 2, 3, 4 have the inverses 3, 2, 1, 4, respectively. And that's it; the numbers one through four, under the operation of clock arithmetic, form a group (note 6.2).

It can't be overemphasized that, to a mathematician, this particular group, Mod(4), or any other group for that matter, is purely an abstraction. The symbols 1, 2, 3, 4 are nothing more than labels for the elements of the group's set. The elements bearing these convenient labels could be, as far as the mathematician is concerned, absolutely anything: a pencil, a hammer, a bowl of spaghetti, and a glass eye, say, would work just fine. Then, to be the group Mod(4), the rule of combination would say that when you operate on a pencil and another pencil you get a hammer ($1 * 1 = 2$), on the pencil and the pasta you get an eyeball ($1 * 3 = 4$), and so forth.

As absurd as this seems, as long as you have the same number of elements as Mod(4), and the operation gives the exact same relations between all the elements as that of Mod(4), then the mathematician is happy to deem your odd assortment of objects and silly rule for combining them the group Mod(4). She might wonder why you went to all the trouble, but she would be compelled to grant you the designation you desire. Even more than that, she would exhibit tremendous insight in doing so, for the ability to abstract—to see the deep connections between things that on the surface seem to have nothing in common—is perhaps the single most defining characteristic of the human intellect (note 6.3).

Thus, rather than the nature of the objects themselves, it's the interrelations between the objects in the group—the pattern of relationships between the elements of the group that is given by the associated operation—that establishes the group's unique mathematical identity. It is through such patterns that the connection between the tangible world of particle physics and the ethereal world of abstract mathematics was first established in the 1960s.

Into the Continuum: The Lie Groups R(2) and U(1)

Groups need not have finite numbers of elements. Consider the set of integers, whole numbers from negative infinity to positive infinity, under the operation of addition. This set forms an *infinite group*—a group with an infinite number of elements.

Now, for each consecutive pair of whole numbers, add in all the fractions that fall between them. There's an infinite number of fractions between each consecutive pair of whole numbers and an infinite number of consecutive pairs of whole numbers. This infinitely infinite set of numbers, known as the *rational numbers*, also forms a group under the operation of addition; the sum of two fractions is always just another fraction, and so forth.

If we add in *transcendental* numbers—numbers whose decimal expansions (like that of $\pi = 3.14159\ldots$) do not terminate or exhibit repeating sequences in their decimal expansions (which all fractions do), we get the *real* numbers. Mathematicians have been able to show that, surprisingly, there are an infinite number of transcendental numbers between any two fractions that one chooses, no matter how close together those chosen fractions are. This is hard to imagine because it's possible to choose two fractions that are arbitrarily close together (since there are an infinite number of them between each two whole numbers, such as zero and one). But it's demonstrably true.

The infinitely infinite set of real numbers also forms a group under addition. Real numbers have a special berth within natural science, in that the result of the measurement of any physical quantity is expected to be a member of the set of real numbers. The set of real numbers forms a continuum of possible numbers. There are no gaps between successive real numbers, no holes between which you could squeeze another number that might be the result of a physical measurement. The set of real numbers under the operation of addition is a *continuous* group.

Similarly, all Lie groups are continuous. In fact, the set of real numbers under addition is one of the most straightforward examples of a Lie group, although not a particularly useful one for illuminating the special properties of Lie groups. For instance, most Lie groups are non-Abelian, but the group of real numbers under the operation of addition is Abelian because for any numbers x_1 and x_2, $x_1 + x_2 = x_2 + x_1$.

Without yet saying exactly what a Lie group is, let's discuss another relatively simple example of a Lie group. This Lie group will also be Abelian, but nonetheless will be of direct relevance to us in that it forms the basis of the *gauge theory* of electromagnetism, the most modern and up-to-date recasting of the quantum theory of electromagnetism. As such, it is also one of the two Lie groups underlying the unified Standard Model of electromagnetic and weak nuclear interactions.

Take a rectangular object—a piece of paper, a small box, or a book—and set it on a table in front of you. Place your finger on a corner of the object, and press down firmly, so that when you push on the object with your other hand, it will rotate around the corner that is fixed beneath your finger. If, in doing so, you hold your finger vertically above the object, your finger will represent the axis about which the object rotates, just as an axle represents the axis about which the wheels of a car rotate.

As you rotate the object about the axis defined by your finger, the object's position will change. Just how much the position changes depends on just how much you rotate the object. If you rotate it minutely, then its position changes some, but very little. If you give the object a real twist, the position of the object will change substantially (unless the twist is just about 360 degrees or a multiple of 360 degrees). In fact, the possible positions of the object after the rotation form an infinite continuum. For every possible angle between 0 and 360 degrees (i.e., for every real number between 0 and 360), there is a unique position at which the object comes to rest after the rotation. Note also that if you rotate the object through some angle and then again through some other angle, the result is just as if you rotated the paper once through some third angle, which is just the sum of the first and second angles.

This is starting to smell a lot like a continuous group. To see that this system does indeed obey the axioms required of a group, however, involves an interesting application of mathematical abstraction.

A group consists of both a set and a rule of combination, or operation, on that set. It's natural to think of an operation as some sort of action, so you might expect the act of rotation to be the operation in this case. However, this is *not* so. The act of rotation represents the *elements* of the group's set, not the operation. The elements of the set are all the possible rotations of the object about the axis defined by your finger; there's one such rotation for each angle through which you can rotate the object. So, just as the possible angles between 0 and 360 degrees form a continuous set of numbers, the possible rotations of the object form a continuous set of elements of this group.

The *operation*—the rule of combination—is the successive application of two rotations. A rotation of 32.4 degrees followed by a rotation of 19.1 degrees is the same thing as a single rotation of 51.5 degrees. If we represent with the symbol R_θ the element of the group's set corresponding to the counterclockwise rotation of the object through an angle θ, then we can write

$$R_{19.1} * R_{32.4} = R_{51.5}.$$

The result of the combination of the group elements corresponding to 32.4 and 19.1 degrees of rotation is a third element, also within the group, corresponding to a rotation of 51.5 degrees. If the two rotations are large, so that the total rotation is somewhat greater than 360 degrees, you will go all the way around and wind up with a total rotation that is identical to that of one of the rotations less than 360 degrees, like going all the way around the face of the clock in the clock arithmetic of the group Mod(4).

Once we make the abstraction that recognizes the different possible rotations as the elements of the group's set and the accumulation of the effect of two successive rotations as the operation associated with the group, we see that the axiom of closure (that the result of the operation must always lie within the set) is satisfied. There's an easily identified identity element: a rotation by 0 degrees. In other words, the group element R_0 has the property that $R_0 * R_\theta = R_\theta$; for any angle of rotation θ, the successive application of rotation through θ degrees followed by rotation through 0 degrees leaves you with your original rotation of θ degrees. That each element has an inverse is also easy to see: a rotation through an angle θ, followed by a rotation through an angle of $360° - \theta$, yields a combined rotation of $\theta + (360° - \theta) = 360°$. But going around by $360°$ (full circle) gets you right back to where you started and is the same as a rotation of $0°$, the identity element. Thus, the inverse of a rotation through the angle θ is just a rotation through an angle of $360° - \theta$. I'll leave it to you to show that the associative property (the second of the axioms we introduced) holds.

None of the rotations we've done in this section involve more than two of the three dimensions of space. As we turn the object around on the surface, the two-dimensional surface, of the desk, we need not concern ourselves with rotations that lift the object up into the air above the table. Our rotations involve only the two horizontal dimensions of the surface of the table, not the third, vertical dimension.

Accordingly, this set of rotations, which we now recognize to be a formal mathematical group, is known as *the group of rotations in two dimensions*, or R(2) for short. Since there is a continuum of possible angles of rotation in two dimensions, R(2) is a continuous group. Also, R(2) is Abelian. The order in which you perform the rotations (when combining them according to the rules of the operation) makes no difference whatsoever. A rotation through 32.4 degrees followed by a rotation through 19.1 degrees is the same as a rotation through 19.1 degrees followed by a rotation through 32.4 degrees. In both cases, you wind up with a rotation through 51.5 degrees.

It's convenient here to introduce another Abelian Lie group: the group U(1) of rotations in one *complex* dimension. Mathematically, U(1) is really just the same thing as R(2); to understand this, we need to pause for a moment to introduce the notion of complex numbers. This digression will be of real use to us further down the road because in quantum mechanics, the value of the wave function at any point in space is in fact a complex number.

Whenever you square a number (multiply it by itself), no matter what number you start with, the number you end up with is positive. This all boils down to the fact that two minuses, when multiplied together, make a positive, so if you square a negative number, the two minus signs combine to give a plus sign, and the result is positive. There's no way around this. Well, no way unless you're a mathematician and thus schooled in the art of making up whatever wacky thing you need to satisfy your latest whim.

In this spirit, let's invent a number i that, when multiplied by itself (squared) gives the result -1: $i \cdot i = i^2 = -1$. Thus, if b is any *real* number whatsoever, then the quantity $b \cdot i$ (the little dot is a more compact way of representing multiplication, which we also represent as "$\times$") has the property that its square is negative: $(b \cdot i) \times (b \cdot i) = (b \cdot b) \times (i \cdot i) = (i \cdot i) \times (b \cdot b) = -1 \times b^2 = -b^2$. Numbers such as $b \cdot i$, that have negative squares, are known as imaginary numbers. A complex number is a combination of a real number and an imaginary number. In other words, if z is a complex number, then we can always write $z = a + b \cdot i$, for some real numbers a and b. If a is zero, then $z = b \cdot i$ is an imaginary number; if b is zero, then $z = a$ is a real number. When you put together a real number and imaginary number, you get something a bit more complex, hence the name.

We know how to put real numbers in order. Given any two real numbers, anyone could say which one is bigger—which one they would prefer to receive, say, as the dollar amount of a gift from a benevolent uncle. But how do you discern the size of a complex number $z = a + b \cdot i$? Which is bigger: $100.2 + 22.6 \cdot i$, or $1.3 + 67 \cdot i$? By the whim of convention (albeit a very useful whim in this case), the size s of a complex number $z = a + b \cdot i$ is defined to be the square root of the sum of the squares of the sizes of its real and imaginary pieces: $s = \sqrt{a^2 + b^2}$.

The interesting thing about this rule is that two different complex numbers can have the same size. In particular, there is more than one complex number z that has a size of $s = 1$. You can easily verify on a calculator that the numbers $1+0 \cdot i$, $0.64+0.36 \cdot i$, and $^{12}\!/_{13} + ^{5}\!/_{13} \cdot i$ all have a size of one according to this rule for calculating sizes. In fact, there is a continuum of complex numbers of size one, since for any real number a between -1 and $+1$, you can find a real number b (also between -1 and $+1$) such that $\sqrt{a^2 + b^2} = 1$, so that the number $z = a + b \cdot i$ has size one.

Whenever you multiply together two complex numbers of size one, the resulting complex number also has size one. Thus, it turns out that the set of complex numbers of size one, together with the operation of complex number multiplication (note 6.4), forms a continuous group. This group is known as U(1), the group of complex numbers of size one under the operation of multiplication.

To get a better feel for the group U(1), let's go back to our exercise of rotating a rectangular object around on the surface of a table. Imagine drawing a 1-inch-long arrow—a size-1 arrow—along one edge of the rectangular object, with the base of the arrow lying at the corner where your finger

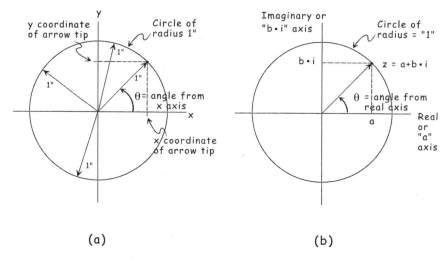

Fig. 6.2. The correspondence between the groups R(2) of rotations in two real dimensions and U(1) of rotations in one complex dimension (any point on *b* is specified by a single complex number $z = a + b \cdot i$, so the "complex plane" of *b* represents a single complex dimension). Any point on a circle of radius 1 in two real dimensions can be specified by the value of the angle θ shown in *a*. Likewise, any point of length 1 in one complex dimension can be specified by the value of the angle θ shown in *b*. In both cases, θ lies between 0 and 360 degrees. So, mathematically, R(2) and U(1) are equivalent.

presses. As the object rotates, while the base of the arrow remains fixed, the tip of the size-one arrow can come to rest at one of an infinitude, a continuum, of points, depending on which direction the arrow happens to point after the rotation. As shown in figure 6.2*a*, the arrow can point at any angle between 0 and 360 degrees on a two-dimensional plot. For all such angles, the points lying at the tip of the arrow form a circle. We label the point lying at the tip of any given arrow according to the point's coordinates along the x- and y-axes of a two-dimensional plot (see fig. 6.2*a*).

Now, from the point of view of figure 6.2*b* (note 6.5), since the elements of the group U(1) are complex numbers of size one, they are similarly represented by points at the tips of arrows with angles from 0 to 360 degrees. The graph in figure 6.2*b* is exactly the same as that of figure 6.2*a*, only instead of having an x-axis and a y-axis, we have a real, or *a*-axis, and an imaginary, or *b*-axis. With real numbers, any point on a circle with a radius of 1 inch is given by two numbers, *x* and *y*. However, the same point on the circle is given by a single complex number $z = a + b \cdot i$. In the ordinary world of real numbers, circles are two dimensional, but in the imaginary world of complex numbers, circles are only one dimensional.

Thus, the elements of the group U(1), the group of complex numbers of size one, correspond exactly to the possible angles θ of the two-dimensional rotations that form the group R(2). Furthermore, the multiplication of two complex numbers of size one is completely equivalent to the operation of combining rotations for R(2). If a size-one complex number z_1 is represented in figure 6.2*b* by an arrow at an angle of 32.4 degrees and z_2 by an arrow at an angle of 19.1 degrees, then the product $z_2 \cdot z_1$ is represented by an arrow at an angle of 32.4 + 19.1 = 51.5 degrees, just as following a 32.4 degree rotation with a 19.1 degree rotation is the same as a single 51.5 degree rotation.

At first blush, R(2) and U(1) look very different, but after some thought, we see that they are really the same. The fact that the two groups are based on different number systems (two real vs. one complex dimension) and employ entirely different operations (successive rotation vs. complex number multiplication) is immaterial. The mathematician recognizes that, while they are represented quite differently, the groups R(2) and U(1) are, in their essence, identical. This is an archetypical example of the workings of mathematical abstraction and a necessary step in the process of unveiling the connection between these mathematical entities and the structure of the physical universe.

The designation U(1) may be a bit obscure. For the set R(2) of rotations in two (real) dimensions, we read the "R" as "rotation in real space" and the "2" as "in two dimensions." In the case of the complex group U(1), however, we saw that we admit the possibility of something mathematically equivalent to rotation with just a single complex number—a single complex dimension. Furthermore, these "rotations" involve themselves solely with the complex numbers of size one—of unit length. Thus, for U(1), we read the "U" as "the set of unit-length numbers" and the "1" as "in one complex dimension."

Adding the Next Dimension: The Lie Group R(3)

The three-dimensional version of the group R(2) of rotations in a plane is, not surprisingly, the group R(3) of rotations in three real dimensions. The increase in complexity associated with the expansion from two to three dimensions leads to a number of interesting properties.

For one, the group R(3) is non-Abelian: The order in which you combine the elements of the group matters. This is a bit counterintuitive, but it's easily demonstrated. Another property of R(3) is that we can think of the infinite continuum of its elements as being *generated* by a finite number of prototypical elements of the group. This latter characteristic is the essence of what makes any continuous group specifically a Lie group. Both of these properties will have direct consequences on the nature of physical law when we eventually incorporate them into the formalism of gauge theory (see chapter 8).

To explore the properties of R(3), the set of possible rotations in three-dimensional space, find a cardboard box, a thick book, or some other such rectangular solid, and mark one of the eight corners with a pen. The object that you may have used above in exploring R(2) would be fine, as long as it's thick enough to prompt thinking in three dimensions. If you got by without the prop in the discussion of R(2), it might be helpful to actually go through the physical exercise in this case.

The marked corner will form the origin, the fixed point about which we will perform our rotations, similar to the corner that we held fixed in our discussion of R(2). The three edges that connect at the origin define three lines in three-dimensional space—the *x*-, *y*-, and *z*-axes of a three-dimensional Cartesian coordinate system. Label them *x*, *y*, and *z* (see fig. 6.3).

Figure 6.4 shows three basic types of rotations that we can do with this three-dimensional object. Practice rotating the box exclusively about the *x*-axis so that the edge you've labeled *x* remains in the same place as you turn the box back and forth (see fig. 6.4*a*). Rotate the box back to its original position, and then do the same exercise about the *y*- and *z*-axes (edges), again making sure you return the box to its original position before going on from the *y*-axis to the *z*-axis rotation. This is all just warm-up; the only thing you might want to notice is that, by the time you've gotten through all three exercises (the *x*-, *y*-, and *z*-axis rotations), there will be one, and only one, point on the box that hasn't moved the entire time—the corner you marked as the origin.

Now we're ready to begin our exploration of the group R(3). In what follows, we'll refer to these three types of rotations, by some unspecified angle but with one of the three axes (*x*, *y*, or *z*) fixed, as the three exercise rotations. Keep this terminology in mind because it will come in handy in later chapters.

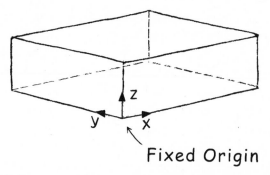

Fixed Origin

Fig. 6.3. The prop to use as an aid to learning about Lie groups.

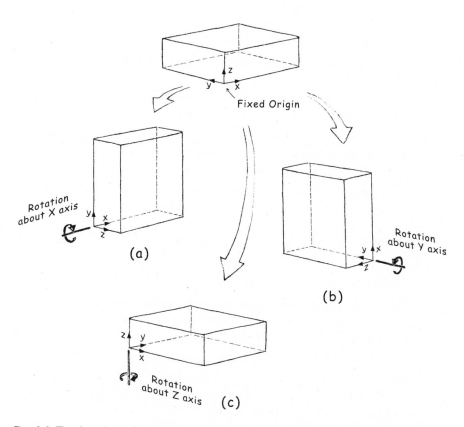

Fig. 6.4. The three basic "exercise" rotations about the x-axis (a), y-axis (b), and z-axis (c).

Place the box squarely in front of you, as before, with the origin of the box resting on the dot on the paper. Make a mental note of the position of the box. Now, pick up the box, and in the air above the table, spin the box around in some arbitrary way so that its orientation in space is random. Preserving this orientation, lower the box back to the table so that the origin again touches the dot on the piece of paper. (Because the surface of the table is rigid and won't let any of the box pass through it, not all possible orientations of the box will work, but there are enough orientations that do work that this doesn't really present a problem. Just pick one of the random orientations that does work.)

Now, compare this position of the box to the original position, of which you made a mental note. The two positions are related by an arbitrary rotation of the three-dimensional box about its marked corner—the fixed origin of the coordinate system. The set of all such possible rotations, including the ones you couldn't do because the table was in the way, form the elements of the group R(3). We can label these different elements of the rotation group R(3) by the angles at which the x-, y-, and z-axes end up after the rotation.

Just as for R(2), these actions, the set of possible rotations of the box, are the elements of R(3), not the associated operation. The operation, again, is the combination of two rotations by their successive application.

Note that the exercise rotations, the rotations exclusively around the x-, y-, or z-axis, by any angle between 0 and 360 degrees, lie within this set of elements. Also among these elements resides the possibility of no rotation whatsoever (just leaving the box sitting there unmoved); we still want to think of this as a rotation, albeit one that leaves the position of the box unchanged. This rotation of the box by nothing is the identity element of the group R(3), just as it was for the case of R(2).

The set of rotations R(3), in concert with the operation given by the successive application of rotations, forms a group. If you rotate the box, and then rotate again, what you always wind up with is just some other rotation. So, R(3) is closed under this operation. The identity element is the rotation of the box by nothing. The inverse of any rotation is the complementary rotation that undoes the given rotation by returning the box to its original position, so that the combination of the given rotation with its inverse leads to a net rotation of nothing—the identity element. The associative law holds, although again we won't care to show it.

We need to be able to specify the different rotations that form the elements of the group R(3). We'll do this in a way that gets right to the heart of its designation as a Lie group.

If the arbitrary rotation that you just performed was complicated enough, you will see that it is not possible to move the book from its original (resting) position to its final (held) position with just a single exercise rotation about either the x-, y-, or z-axis. However, any arbitrary rotation can be achieved by a succession of no more than three exercise rotations of the appropriate angle—one about the x-axis, another about the new y-axis (note that after the rotation about the x-axis, the orientation of the y-axis has changed, giving a new orientation to the y-axis), and a third about the even newer z-axis. Thus, instead of distinguishing between the different possible rotations by designating the angles at which the former x-, y-, and z-axes end up, we can instead simply designate the angles θ_x, θ_y, θ_z associated with the three exercise rotations necessary to produce the overall rotation.

This is the essence of what makes a group a Lie group. A Lie group contains an infinite continuum of elements; yet, its structure is delineated by a finite number of elements, known as *generators*, from which the continuous infinitude of elements are easily obtained. In the case of R(3), there are three such generators: the small exercise rotations about the x-, y-, and z-axes. Once you know what the generators of the Lie group are, all you need to do to specify any given element of the group is to figure out how much of each of the generators contributes to the formation of that element. In the case of the Lie group R(3), this amounts to picking the values (between 0 and 360 degrees) of the three angles θ_x, θ_y, θ_z that produce the rotation that you desire. Absolutely any element of R(3), any rotation in a three-dimensional space, can be produced in this way.

In our discussion of R(2), the group of possible rotations of an object in only two dimensions, we made no mention of generators. This is not because the group R(2) (which is a Lie group) has no generators. For the group R(2), the elements (rotations) were specified by the *single* angle θ through which we rotated the object. Thus, if you think about it, R(2) is a Lie group generated by a single basic exercise rotation: a small turn about the single available axis of rotation in the system. R(2) and U(1), which we argued is mathematically identical to R(2), are Lie groups formed from a single generator.

To recapitulate, Lie groups are continuous groups (composed of an infinitely dense succession of elements) whose elements are derivable from a finite number of generators. The properties of the generators alone, their number—three for R(3)—and their relations under the group's operation, definitively establish the properties and interrelations of the infinite continuum of elements in the full group.

Order Matters: The Lie Algebra

Since any Lie group is characterized by its generating elements, by exploring the properties of its generators alone we can answer the question of whether the order of combination of the group's elements matters.

To see that the generators of R(3) don't commute, that the order of application of the different exercise rotations does matter, place your labeled box once again squarely in front of you. Let the symbol X_{90} denote an exercise rotation of 90 degrees about the x-axis (which should still be labeled as such on your box). Similarly, let the symbols Y_{90} and Z_{90} denote exercise rotations of 90 degrees about the y- and z-axes, respectively.

Now, let's consider the behavior of two of these generators under the group's operation (note 6.6). The combination, or successive application, of the 90-degree exercise rotations about the x- and y-axis can be performed in two different orders:

$$R_{xy} = X_{90} * Y_{90},$$

$$R_{yx} = Y_{90} * X_{90}.$$

We'll adhere to the rather confusing mathematical convention that the order of combination proceeds from right to left, for example, the first of these expressions directs you to rotate by 90 degrees about the y-axis first, and then by 90 degrees about the x-axis (which will have changed position but will still be labeled with the x on your box).

Try these two combinations, and see whether you can confirm what is depicted in figures 6.5a and 6.5b—the two different orders of combination result in combined rotations R_{xy} and R_{yx} that are completely different. The order in which the elements are combined by the group's operation (successive rotation) does matter—the generators of R(3) and thus, more generally, the elements of R(3) that are constructed by taking different amounts of these fundamental generating rotations do not commute. The group R(3) is non-Abelian.

To make this point more clear, it's helpful to consider a counterexample:

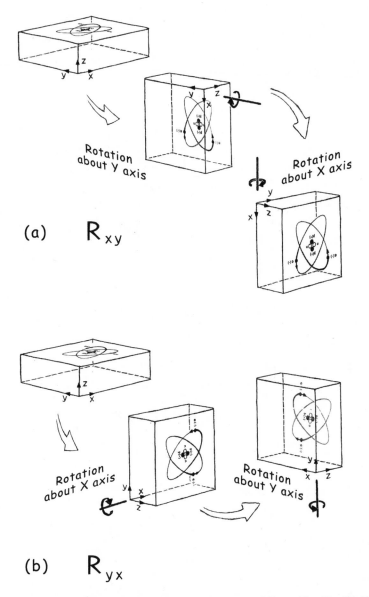

(a) R_{xy}

(b) R_{yx}

Fig. 6.5. The combined rotations $R_{xy} = X_{90}*Y_{90}$ (a) and $R_{yx} = Y_{90}*X_{90}$ (b). The order of the elements in the operation does matter!

a Lie group with three generators in which the generators do commute. Let's go back to rotation in two dimensions about a single axis—the system we used to introduce the group R(2)—and consider a case for which we have three separate objects stacked atop one another, lined up, and ready to rotate about the same axis of rotation (see fig. 6.6).

Rotations in this system are again determined by three angles: θ_L, θ_M, θ_U, the independent angles by which the lower, the middle, and the upper objects are rotated. The three generators of these rotations, from which all possible rotations of the system can be made, are just small rotations of each of the three objects (lower, middle, and upper) individually about the common axis of rotation. It's easy to see that in this case the generators *do* commute. It makes no difference which order you apply the separate rotations of the objects to compose any overall rotation in the group.

As we'll discuss in chapter 8, a physical theory based on this Abelian Lie

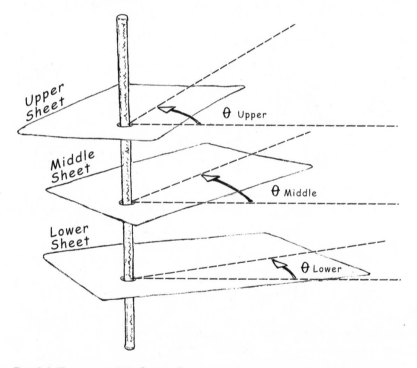

Fig. 6.6. The group $R(2) \otimes R(2) \otimes R(2)$, that is, three separate copies of R(2) (represented by the three two-dimensional planes) that rotate independently of one another. This group, like R(3), has three generators, but it's a different group than that of the rotations of an object in three-dimensional space. Among other things, this group is Abelian, while R(3) is not.

group, which is known as "$R(2) \otimes R(2) \otimes R(2)$" would have significantly different characteristics than one based on the non-Abelian group R(3), even though both of these groups have the same number (three) of generators.

So, it's not only the number of generators, but also their interrelation under the group's operation, that specify the identity of a Lie group. The exact nature of this interrelation—the precise difference between the combination of pairs of generators when combined in opposite order for all possible pairs of generators—is known as the group's *Lie algebra*. Even though R(3) and $R(2) \otimes R(2) \otimes R(2)$ have the same number of generators, they are entirely different Lie groups because all pairs of the three generators of $R(2) \otimes R(2) \otimes R(2)$ commute with each other, while none of the generators of R(3) do. To identify a given Lie group, you must specify both the number of generators and the Lie algebra of differences in the order of combination of all pairs of those generators.

When Lie groups are associated with the description of the natural world, as they will be in chapter 8, the specifics of the Lie algebra will be seen to have profound consequences. For example, were we able to somehow alter the Lie algebra of the group underlying the description of the strong nuclear force, we could do so in a way that would prohibit the formation of atomic nuclei. The dependence of physical law on the Lie algebra is not a subtle one.

The Lie Group SU(2)

We've already introduced the group U(1), the group of size-one complex numbers with the associated operation of complex-number multiplication. We argued that, mathematically, U(1) has precisely the same structure (number of elements and their interrelation under the group's operation) as the group R(2) of rotations in two real dimensions. Given this, we might ask two questions: Can one construct a group of rotations in two *complex* dimensions, and if so, is this group mathematically equivalent to a group of rotations in some number of real dimensions? The answer to the first of these questions is an unqualified yes; we can also answer the second in the affirmative, although in this case, with some interesting qualifications.

To see what is meant by rotation in two complex dimensions, let's first remind ourselves about rotation in two real dimensions. Take a piece of paper and mark a dot on it. Then, draw an arrow whose base is at the dot, extending in some direction or other on the surface of the paper (see fig. 6.7).

Press your pencil on the dot so that you can rotate the arrow about its base (the pencil represents the axis of this rotation). No matter how you rotate the arrow in the two real dimensions of the page, its size is always the same. Mathematically, we note that as the arrow is rotated, the x and y coordinates of its tip change (for example, to the coordinates x' and y', shown in fig. 6.7), but its size s, given by the relation $s = \sqrt{x^2 + y^2} = \sqrt{(x')^2 + (y')^2}$, is unchanged. Rotation in two real dimensions does change the x and y coordinates of any point (say, the point at the tip of the arrow) but in such a way so that the distance s of the point from the origin remains unchanged (note 6.7).

Rotation in two complex dimensions is not that different. Instead of a plane of two real coordinates x and y, we have a plane of two complex coordinates s_x and s_y (fig. 6.8). In real dimensions, x and y are the size of the real numbers that represent the values of the real coordinates. In complex dimensions, s_x and s_y are the size of complex numbers that represent the val-

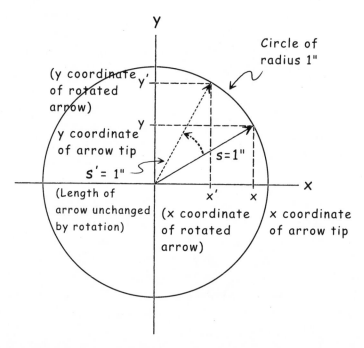

Rig. 6.7. Rotation in two (real) dimensions. When the arrow is rotated, the coordinates of its tip change from x and y to x' and y', but its length s stays the same.

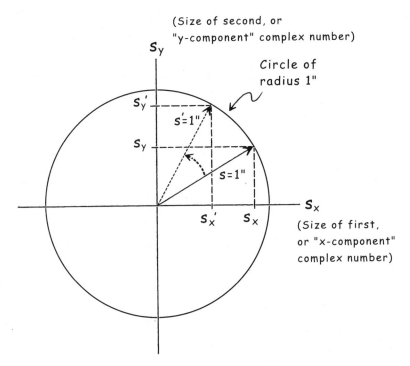

Fig. 6.8. Rotation in two complex dimensions. When the arrow is rotated, the complex coordinates of its tip change from s_x and s_y to s'_x and s'_y, but again its overall size s remains unchanged.

ues of the complex coordinates. The arrows of figures 6.7 and 6.8 don't look any different because they're not. One points to a location in a real two-dimensional plane and the other to a location in a complex two-dimensional plane. And, just as in two real dimensions, a rotation in two complex dimensions will preserve the overall size $s = \sqrt{s^2_x + s^2_y}$ of the two-dimensional complex number.

This seems to suggest that rotation in two complex dimensions is mathematically equivalent to rotation in two real dimensions: that the group of rotations in two complex dimensions is identical to R(2). But in the section "Adding the Next Dimensions" above, we argued that the group U(1) of rotations in one complex dimension is equivalent to R(2), so we must be missing something.

What we're missing is that each of the complex coordinates s_x and s_y is a complex number's *size*, not a complex number in and of itself. Take, for example, the coordinate s_x. This coordinate represents the size of a complex number $a_x + b_x \cdot i$, with the size $s_x = \sqrt{a_x^2 + b_x^2}$. Similarly, s_y represents a complex number $a_y + b_y \cdot i$ whose size s_y is given by $s_y = \sqrt{a_y^2 + b_y^2}$. But there are an infinite number of combinations of a_x and b_x that give the same size s_x and an infinite number of combinations of a_y and b_y that give the same size s_y.

The point is that, since *rotate* means *preserve the size of*, there are really three different ways to rotate in two complex dimensions. Analogous to rotation in two real dimensions, we can rotate the arrow of figure 6.8 so that we get a new arrow with new s_x and s_y coordinates but with the same overall size. We can, instead, simply pick a new a_x and b_x so that s_x (and s_y) remain unchanged. Or, we could instead pick a new a_y and b_y so that s_y (and s_x) remain unchanged. In all three cases, we've changed the complex numbers involved but have preserved the overall size $s = \sqrt{s_x^2 + s_y^2}$ of the arrow in the two-dimensional complex plane. The latter two of these three modes of rotation are unique to complex dimensions; rotation in two complex dimensions is much richer than rotation in two real dimensions.

Just as the set of rotations in two real dimensions forms a Lie group, R(2), the set of rotations in two complex dimensions also forms a Lie group, known as SU(2) (read "ess-you-two"). Again, the numeral 2 refers to the fact that two dimensions (in this case, complex dimensions) are at play. The letter U stands for the word *unitary*, which just means *size-preserving*, the bellwether property of a rotation. The letter S stands for the word *special*, and the explanation of its origin is not of interest to us.

The operation under which R(2) becomes a group is the successive application of two rotations, the result of which is a third, combined rotation that is always a member of R(2). Likewise, the result of the successive application of two complex two-dimensional rotations is always some other complex two-dimensional rotation. This operation, the successive application of complex two-dimensional rotations, also satisfies the other requirements listed at the beginning of this chapter, so SU(2), the set of rotations (size-preserving changes) in two complex dimensions is a group: a Lie group.

Because SU(2) is a Lie group, its infinitude of elements must derive from a small number of basic elements, or generators. For R(2), there was but a single generator: a small rotation about some fixed point (axis) on the two-dimensional plane. Similarly, the group R(3) of rotations in three dimensions has three generators: rotations about each of the x-, y-, and z-axes or, in terms of the prop we used, the three exercise rotations about each of the three different edges of a box that emanate from one of its corners. Any rotation in R(3), any possible final orientation of the three box edges, can be achieved by three rotations of the appropriate size about the three box edges. These three types of rotations, one for each axis or box edge, generate R(3).

Furthermore, the rotations of R(3) do not commute; the order in which you perform any two successive rotations *does* make a difference, leading in general to a different final orientation of the box, depending on which of two successive rotations is applied first. The precise way in which the rotations fail to commute—the difference between the successive application of two of the generating rotations in first one order and then the other—is a characteristic property of the Lie group. The list of these properties, the order differences in the successive application of each pair of generating rotations, is known as the *algebra* of the Lie group. It is this algebra that differentiates between different Lie groups with the same number of generators, such as the two three-generator Lie groups R(3) and R(2) ⊗ R(2) ⊗ R(2).

What about SU(2)? How many generating rotations do we need to be able to achieve any two-dimensional complex rotation? In two real dimensions, there is but one mode of rotation: turning in the plane about the given axis. So, R(2) has but one generator. However, as described above, there are three different ways to rotate in two complex dimensions, so SU(2) has just that many generators, three.

But we know not to stop here, for if we really want to compare SU(2) and R(3), we need to know more than just the number of required generators. We also need to compare the behavior of the combination (under the group's operation) of pairs of generators when we change the order of their combination. We need to compare the Lie algebras of the two groups.

Once you know how to represent the elements of SU(2) and R(3) mathematically (note 6.8), this is quickly done because there are only three unique ways to form pairs of two generators out of a set of only three generators. When you perform these three calculations, first for pairs of generators of SU(2), and then for pairs of generators for R(3), you find something rather surprising. The algebra of the two groups is precisely the same! In other words, the commutation, or ordering, difference between the first and second generators of SU(2) is precisely the same as that of the first and second generators of R(3), and so forth, for each of the three possible pairs of generators for each group.

At this point a warm feeling washes over us as we conclude, just as for U(1) and R(2), that the three-generator non-Abelian groups, SU(2) and R(3), are different manifestations of the same Lie group. It seems that the powers of abstract reasoning triumph over the pedestrian world of the merely apparent, revealing SU(2) and R(3) for what they are: the same individual, dressed in slightly different clothing. Mathematically, they are the same.

This epiphany is almost, but not quite, correct.

Let's go back to the intuitively accessible R(3) and the box we used as a prop to evince its properties. Consider any one of the three generating exercise rotations of the box, and rotate the box through 360 degrees of that rotation. You get back to where you started. A rotation of 360 degrees, or a multiple thereof, about any axis is the same as doing nothing; in other words, it is the group's identity element. Such is not the case for the group SU(2). For SU(2), it takes 720 degrees, twice 360 degrees, of (complex) rotation to get back to where you started! Requiring a rotation of 720 degrees rather than 360 degrees to restore a system to its original orientation may sound strange, but there's a relatively simple physical system that also has such a property (see fig. 6.9).

Place the palm of one hand near the cheek on the same side of your body, and face the palm straight upward (you'll need to be standing for this to work). Place a small box or book on your palm so that it rests comfortably with no other support than the palm itself, with the book resting near your ear like a waiter holding a tray in a fancy restaurant. Making sure the top of the box always faces upward, rotate your forearm about your elbow so that your palm (and the box) passes through 360 degrees of rotation about your elbow. If you've done it right, you'll now be in a mildly uncomfortable position, with the top of the box still facing up, but now *below* rather than *above* your elbow, with the book at about waist level and your palm pointing rearward. The system is in a different configuration than it was before the 360-degree rotation. A continuation of the rotation by another 360 degrees, for a total of 720 degrees, restores the system to its original configuration, with the book resting on your palm right near your ear, above the elbow. So, even though this system has little to do with SU(2), we see that it's actually not that out of the ordinary for a system to require 720 degrees rather than 360 degrees of rotation to get back to its original orientation.

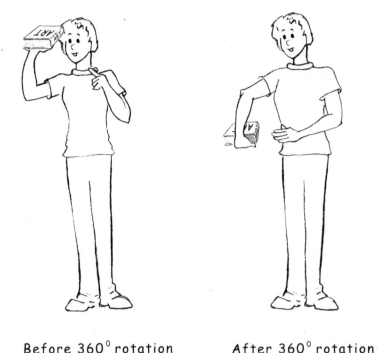

Before 360° rotation   After 360° rotation

Fig. 6.9. A system for which 720 degrees (not 360 degrees) of rotation is necessary to restore the system to its original position.

So, the mathematician hedges. She would say that, "locally," SU(2) and R(3) are the same group—the number of generators and the algebra are the same for both groups. But she would say that "globally," there are some differences; it takes 720 degrees, not 360 degrees of rotation to get back to where you started with SU(2), while for R(3), it only takes 360.

Physically, when we begin to apply Lie groups to our understanding of the natural world, it will be the *local* (Lie algebra) properties of the group that determine the basic behavior of the system governed by the structure of the Lie group. So, for example, SU(2) and R(3) both generally describe the behavior of systems with angular (rotational) motion. However, the *global* properties of the group establish what specific types of systems can be described. Thus R(3), with its mere 360 degrees of rotation, is more limited and can only describe systems with orbital angular motion. SU(2), with its 720 degrees of possible rotation, is more general and can describe systems with either (or both) orbital or spinning angular motion.

Lie Group Wrap-Up

Having introduced the group R(3) of rotations in three real dimensions, and the group SU(2) of rotations in two complex dimensions, we can wonder about rotation groups in higher dimensions, both real and complex. It turns out that the connection between complex and real rotation groups—the fact that R(2) and U(1) are one and the same mathematically and that R(3) and SU(2) are closely related—is fortuitous and is not a general property connecting the worlds of real and complex rotations. The next-higher-dimensional real rotation group R(4) (the group of rotations in four real dimensions) has six generators, while the next-higher-dimensional complex rotation group SU(3) (the group of rotations in three complex dimensions) has eight generators.

This lack of correspondence between real and complex rotations, however, is immaterial. The point is that we can determine, using the tools of abstract mathematics, exactly what the properties of these higher-dimensional rotation groups are. We can deduce the identity of their generators, work out the algebra of these generators, and even derive, if interested, the groups' global "topological" properties. Armed with this, we can fully explore the implications of physical theories based on each of these groups (what we mean by the statement "physical theory based on these groups" will be the subject of the next two chapters).

To this point, we've found that the groups that provide successful physical theories are U(1) and SU(2), together providing the basis for the theory of the electroweak force, and SU(3), which provides the basis for the theory of the strong nuclear force. So it's these Lie groups—the groups of rotations in one, two, and three complex dimensions—to which we'll need to return in the chapters ahead.

7

The World Within

———∿∿∿∿∿∿∿∿———

Internal Symmetries

It is often stated that the German Emmy Noether (1882–1935) was the greatest of all female mathematicians. Few would take issue with this claim, but it seems to me to sell her a bit short. There are few figures in the long history of mathematics, male or female, whose contributions stand clearly above those of Noether.

Like other great female scientific thinkers of her day (Marie Curie comes to mind), Noether's establishment in academic circles came only after a prolonged battle, championed, in her case, by her mathematical colleagues at Germany's University of Göttingen (David Hilbert and Felix Klein in particular) over the stodgy objections of the university administration. Against this inconducive backdrop, her achievements are all the more impressive, although there are also some male figures of similar renown (Karl Weierstrass being perhaps the most notable example) who overcame prejudice and other sorts of demons to advance and reshape mathematical thought (note 7.1). Mathematical genius has a way of bubbling forth inextricably from those few who possess it. Exposure to elevated discourse seems more critical an element than encouragement in the development of truly independent thinkers. Luckily, Noether happened to be the daughter of the respected mathematician Max Noether; one wonders how much innate mathematical genius has been squandered due to a simple lack of such exposure.

The abrupt and premature end of Noether's career in Germany was not due to her gender but, rather, to the accident of having been born Jewish in Hitler's Germany. Noether spent the last two years of her life as a visiting professor at Bryn Mawr College in the Philadelphia suburbs.

Noether's contribution to higher mathematics was primarily in the development and elevation of the study of mathematical systems known as *rings*, sets of elements on which, not one, but two interrelated operations are defined (note 7.2). While groups and rings are closely related, our interest does not reside in Noether's work on abstract algebra.

Instead, our interest in her work stems from her brief foray into the world of mathematical physics in the years surrounding 1915. In addition to work supporting the development of Einstein's general theory of relativity, Noether published a result showing the general relation between the sorts of invariance, or symmetry exhibited by physical laws and the existence of "conserved quantities"—measurable attributes of a physical system that remain unchanged over time, regardless of what other complex behavior the system may undergo. For instance, the principle of energy conservation, which has played several roles in prior chapters, is understood in this way to be due to the invariance of physical laws with respect to the passage of time. In other words, we believe that energy is conserved because the laws governing the behavior of the universe remain steadfastly fixed as time goes by. This notion of the fundamental origin of energy conservation is a specific application of Noether's more general result, published in 1915, and now referred to as Noether's theorem.

Noether's Theorem

To understand the statement of Noether's theorem (our only ambition; we won't attempt the proof here), we need to introduce the notion of symmetry, or invariance, a concept with widespread application throughout the quantitative sciences.

Suppose an introductory physics student does an experiment on static electricity in the teaching laboratory of some university, say, at the University of California at Santa Cruz, just to pick one at random. After a period of getting used to the apparatus, he is able to verify, to whatever degree of accuracy the apparatus permits, that the repulsive force between two charged objects is proportional to the static charge on either object and that the force diminishes as the square of the distance between the two objects.

He has provided experimental evidence in support of Coulomb's law of electrostatic repulsion.

In a phone conversation later that week with a friend enrolled at the National Autonomous University of Mexico, the student discovers that his friend has performed the identical experiment in her classroom in Mexico City. Because they are both good students working with quality apparatus, they both reach the same conclusion: Coulomb's law of electrostatic repulsion does an excellent job of describing the interaction of electrically charged objects. The laws of physics are no different among the redwoods in Santa Cruz than they are a thousand miles southeast and seven thousand feet higher in Mexico City.

This probably comes as no surprise, and to acknowledge it even sounds sarcastic: "You know, the laws of physics are no different in northern California than they are in Mexico City." To Noether, however, this was nothing short of remarkable, reflecting something about the underlying fabric of the space into which the workings of the physical world are woven. If one performs an experiment in northern California, then moves, or translates the experiment to Mexico City (or any other location) and performs the experiment again, one expects to find, and indeed does find, that the outcome of the experiment in either case is governed by the identical set of physical laws. The laws of physics are independent of location: one does not have to include in his or her description of the physical laws the exact location at which the laws apply; if they apply somewhere, then they apply everywhere identically. A physicist would say that physical laws are invariant with respect to translation, or symmetric with respect to translation.

Note also that the teaching lab experiments do not have to be redesigned every semester. Once an experiment is put together, as long as the apparatus remains in good repair, it will provide the same result year after year. You can't wriggle out of a poorly executed lab period by claiming that the laws of physics have evolved over time so that results achieved today might differ from those of a prior term. The laws of physics do not change over time; the prior measurements stand, and you had better prepare yourself for the humiliation of a poor grade. So, just as the laws of physics are invariant with respect to translations in space, they are also invariant with respect to translations in time. These are but two (actually, four; one for time, and one each for translation in the three independent spatial dimensions) of a number of symmetries obeyed by all known physical laws.

And now for Noether's theorem:

> For every symmetry exhibited by a physical law, there is a corresponding observable quantity that is conserved.

In other words, for any invariance exhibited by a physical law such as the electromagnetic interaction, there is some measurable quantity that remains unchanged over time, no matter how complex the behavior of the system being governed by the physical law. Thus, for example, since all physical laws are thought to be invariant with respect to translations in both space and time, there must be four quantities (one for each of the three possible independent dimensions of spatial translation, plus one for translation in time) that are conserved by all physical laws. What are these four conserved physical quantities?

Let's take one of these four symmetries, say, translation in the east-west dimension. What physical quantity is conserved, according to Noether, due to the fact that all physical laws, and really space itself, are invariant with respect to such translations?

The answer is precisely that portion of an object's momentum associated with east–west motion. The amount of momentum associated with the east–west motion of any physical system can never change on its own, without the influence of something external to the system. The reason for this, according to Noether's arguments, is fundamental: conservation of momentum in the east-west direction is a direct consequence of the fact that space itself appears perfectly uniform as you move east or west within it.

The claim that space is uniform as you move east or west within it may not sound correct. Certainly, the scenery changes dramatically as you move east from Santa Cruz into the interior of California. But that's not the issue. The point is that, independent of what you choose to clutter it with, space itself is unchanging as you move from point to point within it.

The earth moves continually through space as it orbits the sun and drifts along with the Milky Way, but the natural laws one derives from studying physics do not change as this motion plays itself out. Likewise, the momentum associated with motion in the north-south and up-down directions is conserved, again due to the uniformity of space in these directions.

What about invariance under translation in time—the fact that the result of any experiment is independent of when the experiment is undertaken? While spatial invariance leads to the principle of momentum conservation, time translation invariance, according to Noether, leads to the conservation of energy. Considering these four symmetries and their associated conserved quantities together, we can state that the principle of energy/momentum conservation is understood, thanks to Noether, to follow necessarily from an innate, intuitive property of space-time: the fact that the precise form of all physical laws that govern the workings of nature are independent of the location and time at which those laws find themselves at play. Such a profound implication from such a seemingly obvious and inconsiderable quality of the natural world!

The utility of the principle of energy/momentum conservation can be illustrated with a simple example. Consider a cigar box (taped shut) with some toads in it. To make it interesting, let there be not just one or two, but many toads, a whole squadron of them, kicking angrily about, wondering what malicious force of nature conspired to bring them together within the crowded and dingy confines of the box. Furthermore, let's assume that you and I know what this malicious force is, and that it's the funding arm of the National Institutes of Health, which has been twisted into action by a proposal to study a possible link between zero gravity and the spontaneous remission of warts.

Space shuttle mission leader Sally Ride, after packing the hapless amphibians into the box, gives them a small shove, sending them gliding slowly over to scientific mission specialist Steven Hawley. The system, box plus toads, has some specific amount of energy associated with its motion and a specific amount of momentum in a specific direction (from Sally to Steve). As the toads thrash, the box jerks wildly about. But, thanks to the miracle of energy and momentum conservation, at no time during this glide does the system, box plus toads, have an overall energy or momentum any different than it did just after Sally pushed it or just before Steve caught it. The amount of energy and the amount and direction of momentum of the system of box plus toads, taken as a whole, is precisely the same at each point in time during the glide, irrespective of how the incensed toads move about. As long as nothing external to the box of toads exerts an influencing force on it (which it doesn't, since the box glides through the space between Sally and Steve at zero gravity), its overall momentum and energy will not change. Sally can send the box off to Steve with no concern whatsoever about what the system of angry toads, isolated within the confines of the box, will do.

So, energy/momentum conservation is a powerful overarching principle, and, thanks to Noether's insight, one that we know to be attributable to the fact that space-time is invariant under spatial and temporal translations. If the toads were only aware of the beauty of the principles underlying their motion, perhaps they would be happier with their plight.

Conservation laws, of which energy/momentum conservation is just one example, play an absolutely central role in physics and its corresponding description of nature. If nothing else, the Schrödinger equation, the master equation of quantum mechanics, is nothing more than a statement of the meaning of energy conservation for objects that have wavelike properties (as all objects do). Noether's connection of energy/momentum conservation to the fundamental structure of space-time was a profound insight.

Noether's theorem establishes an essential link between the symmetries of a physical system and the rules that govern its behavior. It is precisely this connection that will take us beyond relativity, quantum mechanics, and quantum field theory to the next great leap in our understanding of the natural world: the development of gauge theory. The notion of symmetry and its associated mathematical underpinnings (Lie groups) will remain close at hand.

Lie Groups and Noether's Theorem

Lie groups are the mathematical entities that represent the full set of actions one can perform on a system exhibiting some given symmetry that, due to the symmetry, leave the physical properties of the system unchanged. For this reason, Lie groups are sometimes referred to as *symmetry groups*. For example, translation of a physical system through space leaves its overall properties unchanged; the set of all possible spatial translations forms a Lie group (although not one that we've discussed or named).

Recall that a Lie group is a group composed of a continuous infinitude of elements, each of which can be gotten by taking just the right amount of a small number of generating elements. So, for example, any of the infinitude of possible rotations in three dimensions, any member of the group R(3), can be achieved by the successive application of the appropriate amounts of the three generating exercise rotations: about the x-, y-, and z-axes of a Cartesian coordinate system (the axes defined by the three edges of a box that emanate from the corner of the box that is fixed by the rotation).

Similarly, any of the infinitude of possible translations of an object through space can be achieved by the successive application of appropriately sized versions of the three generating translations. If you want to translate your lab apparatus from Santa Cruz to Mexico City, you can do it by first translating the apparatus straight up by 7,000 feet, then straight south by a thousand miles or so, and then straight east by a few hundred miles. Translations in the east-west, north-south, and up-down dimensions are the three generators of the continuously infinite Lie group of possible translations.

Now, we've seen above that each of these generating translations corresponds to a unique, conserved physical quantity; in this case, the momentum in the east-west, north-south, and up-down dimensions, respectively. So, we make the following general conjecture: If a given physical law is symmetric, or invariant, with respect to a set of actions that forms a Lie group (such as the Lie group of translations in space), then Noether's theorem tells us there is a conserved physical quantity associated with each generator of the Lie group (such as the momentum in each of the three corresponding directions in space in the case of invariance with respect to spatial translation).

Not all symmetry principles observed by physical laws are represented by Lie groups in the way that translational symmetry is. One particular symmetry that will interest us is invariance under mirror-reflection, known also as *parity inversion* (note 7.3). In this case, there is no continuum of possibilities. You're either in the normal world or the mirror-reflected world. If you reflect once, you're in the reflected world. If you reflect again, you're back in the normal world. The group representing mirror reflection has but two elements: reflected and unreflected. If the laws of physics are the same in the reflected and unreflected world, then we have a symmetry, albeit a discrete yes or no–type symmetry rather than a continuous symmetry such as translation. Nevertheless, something analogous to Noether's theorem still applies, and there is an associated measurable physical quantity, known as parity, that is conserved.

The reason parity conservation is of interest to us is precisely because there is a particular law of physics that does *not* obey parity conservation—a law that is not invariant under mirror reflection. And that particular law of physics, you may have already guessed, is the law governing the behavior of the weak nuclear force, our favorite renegade. In fact, the failure of the weak force to conserve parity was one of the primary clues along the path leading to the development of the Standard Model. Today, measurements of the quantitative extent of weak-force parity violation comprise the most exacting tests of the ideas underlying the Standard Model. All this in good time.

Rotations and Angular Momentum

Let's head back to our introductory physics lab and return to our experiment on electrostatic repulsion. There's another aspect of the spatial disposition of our apparatus that we haven't yet considered; its orientation. Once again, we carefully measure the strength of the repulsive force between two electrically charged objects as a function of the distance that separates them. But now, we rotate the experimental apparatus about its center, so that the overall location of the experiment is unchanged, but it has a different orientation in space. We again perform the experiment and log the result. Intuitively, you would expect that our results will be the same. We don't expect the laws of physics to depend on the direction that a system governed by those laws happens to face.

Although intuition does sometimes have a way of leading you astray, this is not such a case. The laws of physics do seem to be independent of orientation, or in more formal language, invariant with respect to rotation. Thus, no matter which of the continuous infinitude of possible three-dimensional rotations we pick—no matter which element of the Lie group R(3) we choose—we find that Coulomb's law of electrostatic repulsion applies equally as well after the rotation as before. The group R(3) of rotations in the three dimensions of space is a symmetry of the law of electrostatic repulsion and of natural laws in general.

So, since physical laws are invariant under rotations, Noether's theorem tells us that there are one or more quantities associated with this invariance that are conserved. And since these rotations are related among one another according to the structure of a Lie group, R(3), we can figure out what these conserved quantities have to be. There's one for each generator of R(3)— one each associated with rotations about each of the three axes, x, y, and z, of a Cartesian coordinate system (each of the three edges emanating from one corner of a box).

The conserved quantity associated with the invariance of space with respect to translation in any of its three dimensions is the momentum associated with motion in that dimension. Accordingly, the conserved quantity associated with the invariance of space with respect to rotation about any of its three axes is the *angular momentum* (energy of orbiting or spinning) about that axis.

In our example above, no matter how the toads twist and spin as they ricochet off the walls of the box and each other, the total energy associated with *angular* motion (rotation) of the system of box plus toads, separately about each possible axis of rotation, is unchanged as the box glides from Sally to Steve.

Thus, another useful and overarching property of motion is understood in terms of its source deep within the structure of space-time. Toads or no toads, we begin to see that the universe possesses some remarkably beautiful organizing principles.

In fact, physicists like to believe that, in its essence, the universe is guided by a small set of such beautiful organizing principles—principles free of complex hypotheses that do little more than telegraph our ignorance, principles that are smooth and clean and devoid of the warts of contrivance and ad hoc conjecture. As we'll continue to see, our science has made great strides toward this goal.

A Surprising Paradox: The Story of Spin

The connection between Lie groups and the physical world is brought to the fore when we grapple with spin within the context of quantum mechanics. Planck's constant h has the units of angular momentum; its value of 6.626×10^{-34} kilogram-meters-squared per second represents a particular amount of energy/momentum associated with angular motion, a quality possessed by an object that is either in orbit about a fixed point (such as the earth about the center of the sun) or spinning steadily about its axis (such as the daily rotation of the earth about its poles).

Experience with the behavior of microscopic systems, such as individual atoms, for which the rules of quantum mechanics play an important role, tells us clearly that angular momentum is quantized. In other words, the amount of angular energy/momentum possessed by a system cannot be just anything but must instead be one of a discrete set of possibilities. For the case of orbiting motion, the angular momentum must be a multiple of $\hbar = h/2\pi$ — one of the values $0\hbar$, $1\hbar$, $2\hbar$, $3\hbar$, and so on. For spinning motion, however, there are twice as many possibilities: the angular momentum of a spinning object may be any half-integer multiple of $\hbar$ and may take any of the values $0\hbar$, $\frac{1}{2}\hbar$, $1\hbar$, $\frac{3}{2}\hbar$, $2\hbar$, $\frac{5}{2}\hbar$, and so on.

Spin in this context is something very specific: it is a fundamental prop- erty (quantum number) associated with each type of particle. Spin is the amount of intrinsic angular momentum possessed by a particle of any given type. The spin of a given particle is unchanging and unchangeable—as long as the particle exists, it will possess its characteristic amount of spin angular momentum—no more, no less. Depending on conditions it may also possess some amount of orbital-type angular momentum (which can be changed) about some other object. Whether or not it has orbital angular momentum, the spin will always be there, adding its fixed amount of an- gular momentum into any system containing the particle.

Most atomic nuclei, for example, possess angular momentum. For any given nucleus, some of the angular momentum is due the accumulated spin angular momentum of the nucleus's protons and neutrons as they turn about their own individual axes, while the remainder is due to the orbital angular momentum associated with the motion of the neutrons and protons as they orbit each other within the nucleus. For nuclei that possess no over- all angular momentum, it's not that the protons and neutrons within don't possess spin and orbital angular momentum, but, rather, that the accumu- lated effects of the angular momenta of the individual neutrons and protons conspire to exactly cancel each other out.

This picture is not quite correct. The angular momentum associated with spin, while indeed an intrinsic property of a given particle (in this case, of each individual proton or neutron within the nucleus), is not really caused by the spinning of that particle about its axis. Paradoxically, it's not really clear exactly what it is, physically, that spin *is* due to. This is an in- teresting point, and one that will play a role in the discussion of the enig- matic question of internal symmetry spaces that is soon to follow. For the moment it helps to regard the spin of a proton or neutron as being the re- sult of its intrinsic spinning motion about its own axis, even though we know this picture to be inaccurate.

Recall our distinction between *fermions* and *bosons* from chapter 4. Par- ticles whose intrinsic spin is a strict multiple of $\hbar$ are known as bosons, while particles whose intrinsic spin is an odd multiple of $\hbar/2$ ($\frac{1}{2}\hbar$, $\frac{3}{2}\hbar$, etc.) are known as *fermions*. Photons, the quanta of the electromagnetic force field, have a spin of exactly $\hbar$, and so they're bosons. Electrons have spin $\frac{1}{2}\hbar$, and so are fermions. For ease of expression, we conventionally drop the $\hbar$, since we know it's always there. So, we say that photons have spin-1 and electrons have spin-$\frac{1}{2}$.

To understand the paradoxical nature of spin, we need to introduce the notion of a *projection*. Suppose you find yourself happily bored on an equatorial beach at exactly noon on the equinox, in other words, with the sun directly overhead. For some reason, you are equipped with two metersticks, one of which you lay down so that it lies flat on the beach. You take the other and dangle it so that it hangs straight down; its bottom edge just touching the middle of the meterstick lying on the beach. Since the sun is directly overhead, the dangling stick casts no shadow on the prone meterstick.

If you were to lower the upright meterstick toward the one that's lying on the beach, keeping its lower end fixed to the same point on the middle of the prone stick, you would see a shadow develop on the prone stick (see fig. 7.1). The more you lower the upright stick, the longer the shadow on the prone stick becomes. We call the length of this shadow the projection of the upright stick on the axis defined by the prone stick—it's the length of the upright stick that's projected by the vertical light rays from the sun onto the horizontal line defined by the prone stick.

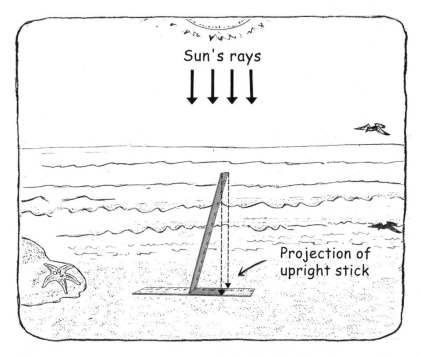

Fig. 7.1. The shadow represents the projection of the upright stick onto the horizontal stick.

If we define one direction on the prone meterstick (toward the end that reads 0 centimeters, say) to be negative and the other direction to be positive, we see that the projection of the upright stick on the axis defined by the prone stick can have a continuum of values, but it can be no more than 1 meter and no less than −1 meter (note 7.4). Such is the nature of projections. They can have any length whatsoever, as long as that length is no greater than the length of the projected object and no less than the negative of the length of the projected object.

Now think of a bicycle wheel—just a wheel in a bike shop, unattached to the rest of the bike. When the wheel spins, it spins about its axle, with the rod forming the axle defining an axis that can point in any direction in three dimensions. Similarly, the spin of a spin-½ particle, such as a proton or neutron, will also be directed along its axis (fig. 7.2). We can think of the spin angular momentum of a spin-½ particle as being like a stick of length ½ℏ, pointing along the direction of the axis about which the particle is spinning (note 7.5). Just as for our bicycle wheel, the direction of this axis, the direction associated with the particle's intrinsic angular momentum, can have any orientation in space.

Say that you want to measure the spin of this particle. Since it's a single submicroscopic particle, you can't very well hold it in your hand and mea-

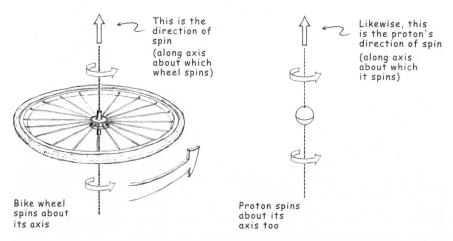

Fig. 7.2. The direction of an object's spin is defined as the direction of the axis about which it spins. In this case, the axis points up, but by manipulating the axle of the bicycle wheel (or the axis of the proton's spin), it can be made to point in whatever direction we choose.

sure its spin by watching it turn in a circle about its axis. The best you can do is to define a measurement axis (typically, given by the direction of a magnetic force field you conjure in the lab) and determine the projection of the particle's spin along that axis. For this measurement, the particle's spin angular momentum is like the upright stick on the beach, while the chosen measurement axis (given by the direction of the magnetic field) is like the prone stick onto which the upright stick is projected by the sun's rays.

In quantum mechanics, the result of such a measurement must come from a limited set of possible results, and all of these possibilities are separated from one another by some multiple of $\hbar$ (in fact, it is this property that is properly known as the "quantization of angular momentum"). For example, figure 7.3 shows the possible spin projection measurements along any given axis for a spin-½, spin-1, spin-³⁄₂, and spin-2 particle (a spin-0 particle will always be measured to have a spin projection of 0 along the chosen axis) (note 7.6). These possibilities are all separated by an amount exactly equal to $\hbar$.

For spin-½ particles there are only two choices: the projection, when measured, must have a value of either $+\frac{1}{2}\hbar$ or $-\frac{1}{2}\hbar$, corresponding to the particle's spin lying directly along or directly against the axis that you arbitrarily chose to project the spin onto when you decided to measure it. After you make the measurement, probing and disturbing the particle, the particle's wave function will have coded into its undulations one of these two possibilities: a spin projection of $+\frac{1}{2}\hbar$ along the chosen measurement axis, or a spin projection of $-\frac{1}{2}\hbar$ directly against the axis. It doesn't matter which measurement axis you pick; the result of your measurement will always be one of these two possible outcomes.

But what about the state of the particle before you disturb it by trying to measure its spin? Before you pick your measurement axis and force it to have one or other of the two allowable spin projections, it could be in a mixed state, with some fraction of it (or really, of its probability; remember that the wave function is an encoding of probabilities) with projection $+\frac{1}{2}\hbar$ along your eventual measurement axis, and the remainder with projection $-\frac{1}{2}\hbar$ against the axis. Until you measure it and force it into one or the other of these two, it can possess any arbitrary combination of these probabilities—as long as the total probability of finding the particle in one projection or the other is precisely equal to one, so that the particle will always be found in one or the other of these two possibilities when its spin is measured.

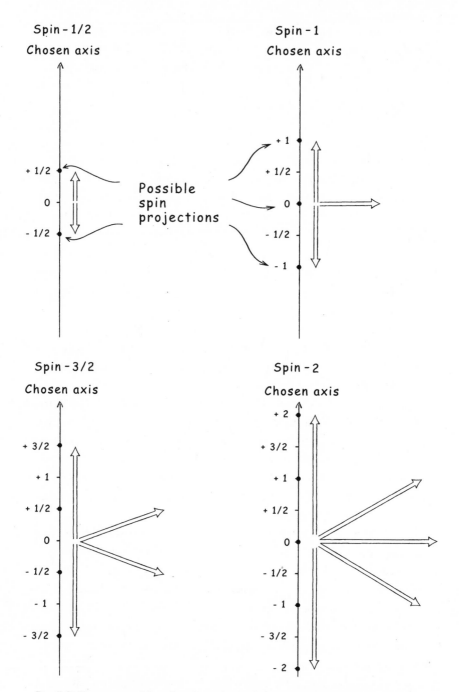

Fig. 7.3. The measurable spin orientations of objects with spin-½, spin-1, spin-³⁄₂, and spin-2. The possible projections of the spin orientation along the axis you are measuring it against are shown by the dots on the axis. The laws of quantum mechanics dictate that these dots are each separated by exactly 1 unit.

Let's formalize this notion by constructing an abstract two-dimensional space for which one dimension (the x-axis) is associated with the amount of probability that the particle has a spin projection $+\frac{1}{2}\hbar$ along your eventual measurement axis and the other dimension (the y-axis) is associated with the amount of probability that the particle has a spin projection $-\frac{1}{2}\hbar$.

There's nothing physically significant about this abstract space. You can take any two quantities and make a mathematical space out of them. One that comes to mind is a space where one dimension represents time and the other represents the magnitude of my checking account balance. Clearly, there's nothing physically relevant about this particular "space"; it's just a convenient way of conveying information. The inexorable downward march of the line, representing the time dependence of the balance, toward and beyond zero has an emotional force far greater than that of just looking at the raw digits on a succession of monthly statements. But there are no alien creatures inhabiting this balance versus time space, despite what I may tell my creditors in a moment of desperation. So that's all our two-dimensional spin projection state-space is, just a way to visualize, graphically, the amount of the wave function's probability associated with the two possible projections of the spin angular momentum.

Figure 7.4 shows two points — typical but not in any way special — representing two of the many possible choices for the spin projection probabilities encoded in the particle's wave function. The arrows pointing from the origin (axis crossing) of the space to the two points are just there to guide the eye. Point 1 is close to the $+\frac{1}{2}\hbar$ dimension axis, which means that when you eventually measure the particle's spin projection, you're much more likely to find it with spin projection $+\frac{1}{2}\hbar$ than $-\frac{1}{2}\hbar$. But it's not right on the $+\frac{1}{2}\hbar$ dimension axis, so there's still a small probability that you'll measure $-\frac{1}{2}\hbar$. Point 2, however, is about equally faraway from each of the axes, so you'll have about a fifty-fifty chance of finding it in either projection when you do the measurement. The arrows associated with each of the two points are exactly the same length, which shows that the total probability of measuring the particle with either one projection or the other is the same (one!) in both cases.

Now, if you were to write down on paper the mathematical expression representing the wave function of point 1 (it's not hard to do, but somewhat beyond the scope of our discussion), and you wanted to turn it into the expression representing the wave function of point 2, what would you do? The answer is intimated by the angle θ shown in figure 7.4. You perform a rota-

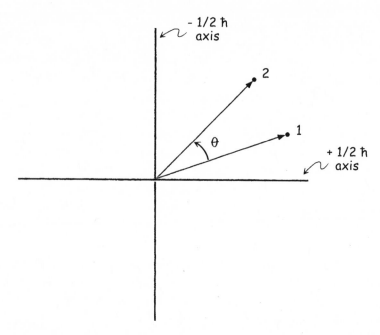

Fig. 7.4. The purely mathematical space in which we represent the orientation of an object with spin-½. If the point lies on the +½ axis in this mathematical space, then its spin lies along the measurement axis in physical (real) space, as shown by the upward-pointing arrow in the first diagram of figure 7.3. If the point lies along the −½ axis in this mathematical space, then its spin lies against the measurement axis in physical space, as shown by the downward-pointing arrow in the first diagram of figure 7.3. For points 1 and 2 in this figure, then, the orientation of the object's spin is partway between straight up and straight down. Note, however, that when the orientation is measured, it will be found as either straight up or straight down, and not somewhere in-between. The closer the point is to the +½$\hbar$ axis, the more likely it will be found with spin straight up when measured.

tion, a two-dimensional rotation in the abstract, but conceptually illuminating, space of spin-projection probability.

So, the set of mathematical manipulations that changes the spin-projection probabilities of the wave function of a particle with spin-½ seems to be the set of rotations in two dimensions. As we've mentioned in passing in previous chapters and will discuss in detail in the next chapter, wave functions are complex. To every point in space and time, the wave function associates not an everyday real number but a complex number. So this set of operations that change the spin-projection probabilities of a spin-½ particle is not

R(2), the set of rotations in two real dimensions, but, rather, is SU(2), the set of rotations in two complex dimensions. (By the way, this is how you make use of the concept of rotations in complex spaces, not in the context of physically meaningful spaces but, instead, in conceptual, abstract mathematical spaces, such as our spin wave function space.)

But wait again. There you are, rotating the spin of your particle around, in ordinary, three-dimensional space. Let's go back to the bicycle wheel example. The wheel spins about its axle, and the direction that the axle points is the direction that we assign to the spinning motion. If you want to change that direction, you rotate the axle through some angle in three-dimensional space until the axle points in the new direction that you want. So, if we listen to our intuition, we should forget about both R(2) and SU(2). When we rotate particle spins, you would think, you are rotating the orientation of the axis about which the particle spins through some angle in everyday three-dimensional space. To change the orientation of a particle's spin, it would seem, one must do it through some member of the group R(3) of rotations in three real dimensions.

But this time, our intuition does fail us. It is, in fact, the group SU(2), operating in the abstract complex space of spin-projection probabilities, that governs the physics of reorienting quantum-mechanical spins, not the group R(3) of rotations in our three real, everyday dimensions. And this is where the curiosity of quantum-mechanical spin lies.

Remember last chapter: as mathematical abstractions, the Lie groups R(3) and SU(2) are almost the same thing. Both have three generators, and the Lie algebra that relates the generators to one another is identical for R(3) and SU(2). The difference is that, while for R(3), as common sense would dictate, you get right back to where you started when you rotate by 360 degrees, it takes 720 degrees of rotation to go full circle in SU(2). If you don't take this into account when you work out the quantum mechanics of spin-½ particles, you get answers that disagree with well-established experimental results. Ask any physics graduate student; most of them have learned this the hard way. Thus, for our work, we are rewarded with something that almost makes sense, but not quite, for the system we develop to describe the angular momentum associated with spin has the hard-to-figure property that a rotation of 360 degrees doesn't get the system back to where it started. And yet it is physical spins—real, concrete, *observable* angular momentum axes being oriented and reoriented in our concrete, tactile three-dimensional space—that we are playing with. How could it be that when you take the axis about which an electron is spinning and rotate it through 360 degrees, you wind up with a wave function that differs, albeit subtly (by an overall factor of -1, to be exact) from what you began with? Is the space in which quantum-mechanical spin lives really the three-dimensional space of our common perception?

In fact, spin has other aspects that render it a bit difficult to understand physically. For all we know (and we've now tested this assumption down to sizes of less than 10^{-18} meters), electrons and quarks are pointlike particles that take up no space whatsoever. How can a particle with no spatial extent really be spinning about its axis and have energy associated with its spinning? If a particle has some extent, then the pieces of it on the surface have some speed as they rotate about its axis. If a particle has no spatial extent, then all points within the particle lie precisely on the axis about which the particle turns. But a point that lies exactly on the axis of a spinning ball does not move; so for a particle with no spatial extent, it would seem that there can be no motion, no angular momentum, associated with its spinning. Thus, it would seem impossible for there to be any energy/momentum associated with a fundamental particle's intrinsic spin. Yet, experiment shows clearly that spin-½ particles do have an angular momentum of $\frac{1}{2}\hbar$, so the very existence of the angular momentum of fundamental particles is mysterious.

On the up side, the first quantum-mechanical wave equation to incorporate Einstein's special relativity into the quantum theory (the Dirac equation; see chapter 4) specifically provides a description of the behavior of spin-½ particles (note 7.7). Relativistic wave equations describing particles with intrinsic spin other than $\frac{1}{2}\hbar$ followed soon after the groundbreaking development of the Dirac equation; each spin possibility has its own form of relativistic wave equation. As perplexing as the concept of intrinsic spin may appear, its existence does seem to be a necessary ingredient of a universe (such as ours) that adheres simultaneously to the laws of quantum mechanics and relativity.

So, we're stuck with the mysterious nature of quantum-mechanical spins, and the unnerving fact that the space in which particle spins exist is a space for which it takes not one, but two complete revolutions to get back to where you started. No matter how we may choose to represent this space in the abstract (such as the two-dimensional probability space of fig. 7.4), there's no question that this space is directly associated with the physical world, as the spins of electrons, quarks, protons, and so forth, are measurable quantities that play a fundamental and indispensable role in nature. But this space is subtly different from the space that we comfortably think ourselves as being part of.

So the question arises, what exactly is spin and this oddly construed spin-space in which it lives?

On the one hand, it's quite real, having associated with it the measurable physical quality of angular momentum. Furthermore, the angular momentum associated with ordinary orbital angular momentum is the same physical quantity as spin angular momentum: the total angular momentum of any physical system is just the sum of the various orbital and spin angular momenta of the components of the system. The fact that total angular momentum is observed to be conserved means, according to Noether's theorem, that physical laws are no less invariant with respect to orientation in spin-space than they are with respect to orientation in normal space.

On the other hand, a particle with no spatial extent shouldn't possess angular momentum, and the axis about which it spins shouldn't have to be rotated through 720 degrees to return the particle to its original state.

We don't really have a clue about the physical origin of spin. To describe spin as "intrinsic angular momentum" is like your best buddy describing how your car's differential works by explaining that it "employs a mechanical linkage"; the only useful information contained in this statement is that its author probably knows next to nothing about how a differential actually works.

The question of the origin of quantum-mechanical spin and the nature of spin-space is a conundrum that physicists have yet to solve. If you've understood, even vaguely, what you've read in this chapter, then your guess is truly as good as mine. And, interestingly, it gets even better (or worse, depending on your outlook) as we move on to other more arcane, but equally fundamental, properties of elementary particles.

Into the Within: The Story of Isospin

In 1911, working at Manchester University in Great Britain, New Zealander Ernest Rutherford demonstrated that the pattern of α radiation scattered off metal foils indicated that atoms contain a small and dense nucleus inside of which reside the atom's positively charged protons. Somewhat later, in 1932, working at Cambridge University's esteemed Cavendish Laboratory, Briton James Chadwick was able to demonstrate that when bombarded with α radiation, atomic nuclei sometimes release an electrically neutral particle with a mass about the same as that of the proton, which was promptly given the name *neutron* (note 7.8). Thus, the concept of the atomic nucleus as a dense collection of protons and neutrons was born, opening the way for great strides over the ensuing twenty to thirty years in what became the field of nuclear physics.

Since element type is determined by the number of protons in the nucleus, irrespective of the number of neutrons, a given element can have many different possible nuclei, corresponding to different numbers of neutrons bound into the nucleus with the given number of protons. Versions of the same element with different numbers of neutrons are known as isotopes (typically, only a few of these isotopes are stable with respect to radioactive decay).

For example, the nucleus of hydrogen, or at least that of 99.985% of the hydrogen you find in everyday things such as water, food, and gasoline, is formed of a single proton, with no neutrons. For the remaining 0.015%, the proton finds itself bound to a single neutron, forming a particle known as a *deuteron* (which is nothing more than just that: a proton and neutron bound together).

Deuterium, the 0.015% of earthly hydrogen that has a deuteron, rather than a proton, for a nucleus, has precisely the same chemistry as hydrogen and can be used instead of ordinary hydrogen to bind chemically with oxygen atoms, forming a substance known as *heavy water*.

The isotope of hydrogen with a nucleus formed of a proton and two neutrons, known as *tritium*, is not stable, decaying with a half-life (typical decay time) of twelve years to an isotope of helium plus an electron and a neutrino. This is why tritium is not a component of naturally abundant hydrogen.

Now, if you have a collection of positively charged particles, such as the protons in the nucleus of a heavy element, being held closely together, you had better have a pretty strong force holding them together (recall that like electric charges repel). This is the role that the strong nuclear force plays in the constitution of atoms.

As experimental results on many different atomic nuclei accumulated, nuclear physicists began a systematic comparison of "mirror nuclei"—nuclei for which the total number of "nucleons" (neutrons plus protons) is the same, but the numbers of protons and neutrons are switched. For example, the isotopes lithium-7, with three protons and four neutrons, and beryllium-7, with four protons and three neutrons, possess mirror nuclei. This study revealed that the strength with which mirror nuclei are bound—the amount of energy it would take to pull the nucleus apart into its individual nucleon components—is the same for both nuclei (after taking into account the extra electromagnetic repulsion associated with the extra proton in beryllium-7).

This, and similar evidence, suggested that the degree of attraction of the strong force that binds nucleons into nuclei is independent of whether the nucleon is a neutron or a proton. Two neutrons attract each other with the same strong-force pull as two protons, or as a proton and a neutron. The nuclear force is unprejudiced. It doesn't care, in fact it can't even tell, if the particles making use of its services are all protons, all neutrons, or some mixture thereof.

This circumstance was extremely suggestive to Werner Heisenberg and led him to introduce an otherwise unprecedented physical quality that he called *nuclear isospin*. This intuitive leap was to have tremendous unforeseen consequences, eventually playing a central role in the development of the scientifically compelling but philosophically unnerving notion of internal physical spaces.

Recall electron spin: given any axis, the measured projection of the electron's $\frac{1}{2}^-$ spin angular momentum along that axis has but two possibilities. Ignoring the factor of $\hbar$, which we'll remember is always there, the spin can lie along the axis ($+\frac{1}{2}$) or against the axis ($-\frac{1}{2}$). But, regardless of which of these two states the electron happens to find itself in, it will always behave like an electron when it's measured. Physics is rotationally invariant (in this case, as we've seen, in the enigmatic three-dimensional space in which spin lives), and a spin $-\frac{1}{2}$ electron is just a spin $+\frac{1}{2}$ electron that's been rotated by 180 degrees, and vice versa.

Heisenberg's insight was the following. The invariance of all physical laws with respect to rotations in space, as manifested by spin-$\frac{1}{2}$ electrons with their two possible spin projections, is very similar to the invariance of the physical law that binds nuclei together (the action of the strong force) with respect to the swapping of neutrons with protons. Purely by analogy, we might think of protons and neutrons as being in fact one and the same particle—the nucleon—possessing some abstract quality known as *isospin* with a magnitude, analogous to the electron's true spin, of $\frac{1}{2}$ (note 7.9). In this vein (see fig. 7.5), we hypothesize that protons are nucleons that are isospin up (note 7.10) in the abstract space of isospin (isospin projection $+\frac{1}{2}$ along some hypothetical vertical measurement axis), while neutrons are isospin-down nucleons (isospin projection of $-\frac{1}{2}$).

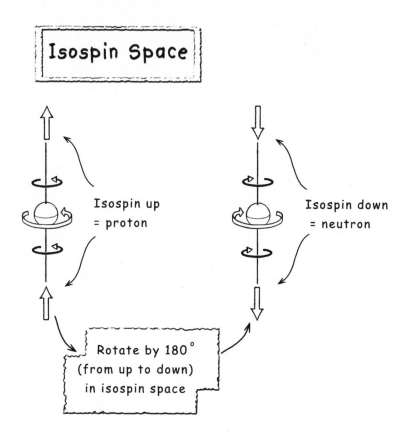

Fig. 7.5. Heisenberg asks us to think of a proton and a neutron as being one and the same particle—the nucleon. The nucleon possesses one-half of a unit of spin in a somewhat enigmatic space known as *isospin space*. If this isospin points up in isospin space, the nucleon manifests itself as a proton; if it points down, the nucleon is a neutron.

Electrons with spin up (true spin, not isospin) and electrons with spin down are all just electrons. Because an electron happens to find itself with spin up, while its neighbor finds itself with spin down, doesn't make either of them any more or less an electron than the other. Nor does one have any intrinsically different physical behavior than the other. The electron with downward-oriented spin follows precisely the same physical laws as the electron with upward-oriented spin. An electron is an electron, and that's that.

But, as we've just seen, as far as the strong nuclear force is concerned, protons and neutrons are similarly undifferentiated. As far as the strong force is concerned, a nucleon with isospin up (a proton) has the same binding properties as a nucleon with isospin down (a neutron). A nucleon is a nucleon, and that's that. The other three forces do care about the difference between a neutron and a proton (e.g., the proton has electric charge, while the neutron has none), but in regards to isospin, we're considering the strong force alone.

There is nothing at all physical or real about the space in which we rotate nucleon isospin, changing neutrons into protons, and vice versa. Isospin space is just another abstract space, like the spin wave function space of figure 7.4 or the bank balance–versus–time space discussed earlier. Rotations in isospin space have nothing to do with rotations in real, three-dimensional space or even the quasi-real three-dimensional space of real spin. The rotation in isospin space that changes neutrons into protons does nothing whatsoever to the physical orientation of the system that's being rotated. If I can convince the bank to wait a week before debiting my account for any check of mine it receives, then I have translated my graph of account balance versus time forward by one week, allowing my next paycheck to arrive before the graph crosses zero and the collection agency is called in. But, have I really moved time by one week? No, not in the physical world in which clocks run and trains need to be caught and diapers need to be changed. All I have done is to move time in my abstract space of account balance versus time. Likewise, if I rotate the wave function of a nucleon in the space of isospin, all I've really done is erase all the n's (neutrons) and p's (protons) on my page of calculations, and replace them with p's and n's. I haven't gone in and actually done anything physically to a proton and a neutron. So, the space in which isospin rotations take place is an abstract, unphysical one, associated with the task of keeping track of when I'm talking about protons versus when I'm talking about neutrons as I perform nuclear physics calculations. It's just a bookkeeping crutch. Or so it would seem.

It would seem that Heisenberg's analogy between true spin and isospin is nothing more than an academic reinterpretation of the observed behavior of protons and neutrons under the influence of the strong force; clever, but with no further use in deepening our understanding of nuclear physics. Let's not take this for granted; instead, let's pursue this question, making use of some things we learned earlier in the chapter.

Noether's theorem tells us that the invariance of any physical law with respect to a group of actions will have associated with it a physical quantity conserved by that law. If, in addition to invariance with respect to translations and rotations in ordinary space, the strong nuclear force is invariant with respect to the rotations in isospin space that swap neutrons with protons, what quantity associated with this new invariance will be additionally conserved by the strong force?

Recall that the invariance of physical laws with respect to rotations in ordinary space is associated by Noether's theorem with the conservation of angular momentum. It can be similarly shown, perhaps not surprisingly, that the conserved quantity associated with invariance with respect to rotations in the abstract space of isospin is isospin itself.

What this suggests, then, is that for any system under the influence of the strong force (the particular physical law that we propose to be invariant with respect to rotations in isospin space), the amount and orientation of the system's isospin will be unchanged over time, no matter what strong-force-induced ruckus goes on inside the system. So, if Heisenberg's analogy between true spin and isospin is valid, isospin conservation ought to be a principle of the behavior of the strong nuclear force.

The notion of isospin conservation is right up the alley of experimental particle physicists. Given the set of toys available to them in the 1950s and 1960s, the issue could be explored quite thoroughly. Experiments in which particles (say, two protons) were slammed together, producing several fragments (protons, neutrons, and mesons, such as the pions and kaons that we learned about in chapter 5) were able to demonstrate the conservation of isospin rather strikingly.

A particle that was studied thoroughly in those experiments was the Δ^+ baryon that, like the proton, is composed of two up quarks and one down quark. Rather than the proton's intrinsic spin of $\frac{1}{2}\hbar$, though, the Δ^+ has a spin of $\frac{3}{2}\hbar$, so it's a different particle than the proton. In addition, the Δ^+ is highly unstable, decaying via the action of the strong nuclear force into a nucleon and a pion with a lifetime of about 10^{-23} seconds.

Just as the Δ^+ baryon has a spin of $\frac{3}{2}$ to the proton's $\frac{1}{2}$, for quite independent reasons the isospin of the Δ^+ is also $\frac{3}{2}$, again to the proton's value of $\frac{1}{2}$; however, both the proton and Δ^+ possess an isospin projection of $+\frac{1}{2}$ on the "vertical" axis against which we measure things in isospin space. By the same measure, pions (π's) have an isospin of exactly 1, with a projection of $+1$ for the positively charged π^+ and of 0 for the neutral π^0.

We know all this because there are four Δ baryons: the $\Delta^{++}, \Delta^+, \Delta^0$, and Δ^-. These four baryons are the $+\frac{3}{2}, +\frac{1}{2}, -\frac{1}{2}$, and $-\frac{3}{2}$ projections of the generic isospin-$\frac{3}{2}$ Δ baryon, in the same sense that the proton and neutron are the $+\frac{1}{2}$ and $-\frac{1}{2}$ projections of the generic isospin-$\frac{1}{2}$ nucleon (see fig.

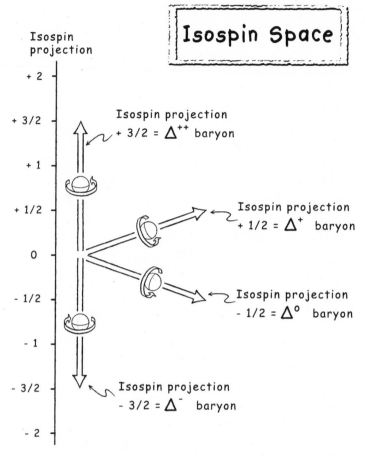

Isospin Space

+ 2

+ 3/2

Isospin projection
+ 3/2 = Δ^{++} baryon

+ 1

+ 1/2

Isospin projection
+ 1/2 = Δ^{+} baryon

0

- 1/2

Isospin projection
- 1/2 = Δ^{0} baryon

- 1

- 3/2

Isospin projection
- 3/2 = Δ^{-} baryon

- 2

Fig. 7.6. As for the proton and neutron manifestations of the nucleon, the quality that distinguishes the four Δ baryons (Δ^{++}, Δ^{+}, Δ^{0}, and Δ^{-}) among one another is the orientation of their isospin in isospin space.

7.6). Similarly, the π^{+}, π^{0}, and π^{-} are the $+1, 0$, and -1 projections of the generic isospin-1 pion.

Now, the Δ^{+} has the same electric charge as the proton, so when it decays into a nucleon and a pion, it has two choices. It can decay into a proton and an electrically neutral pion ($p\pi^{0}$) or an electrically neutral neutron and a positively charged pion ($n\pi^{+}$). Now the critical point: Since the Δ^{+} decays through the action of the strong nuclear force, then if the strong force conserves isospin, the magnitude and projection of the isospins of the nucleon and pion left over after the decay must combine to equal those of the decaying Δ^{+} baryon.

To avoid an impromptu lesson in quantum mechanics, I'll state without demonstration that the isospin axes of a proton and a π^0 can indeed be oriented such that, when added together, they give a total isospin of $\frac{1}{2}$, with a combined vertical projection of $+\frac{1}{2}$, the total isospin and isospin projection of the Δ^+. Likewise, the isospins of a neutron and a π^+ can be combined to yield the isospin ($\frac{1}{2}$) and isospin projection ($+\frac{1}{2}$) of the Δ^+. So the decays of the Δ^+ can indeed conserve isospin, as long as the isospin axes of the decay products have this proper orientation after the decay.

The more telling point, however, is that according to the rules of quantum mechanics, the first of these possible decays, $p\pi^0$, is exactly twice as likely to find itself with this isospin-conserving orientation as the second combination, $n\pi^+$. Thus, if isospin is conserved, the Δ^+ should decay exactly twice as often to $p\pi^0$ as $n\pi^+$: the decay of the Δ^+ should be $\frac{2}{3}p\pi^0$ and $\frac{1}{3}n\pi^+$. In fact, this is precisely what is observed.

Let's think about this for a moment. The Δ^+ baryon has two possible ways of decaying: to $p\pi^0$ and to $n\pi^+$. An obvious guess would be that the Δ^+ decays to both of these with equal likelihood, but nature dictates something quite different. The actual decay fractions—$\frac{2}{3}p\pi^0$ and $\frac{1}{3}n\pi^+$—are not at all what intuition would suggest, but they are a prediction of Heisenberg's notion of isospin invariance. Furthermore, no physicist has ever come up with an alternative explanation of why it is that these two decay fractions are unbalanced in this way. Isospin is not just a convenient analogy: it's an essential component of our basic understanding of nature.

For the case of regular spin, we had to take spin-space seriously because it was associated with a concrete, measurable, physical quantity—angular momentum. This was only mildly uncomfortable because, although spin-space has the somewhat hard-to-stomach property that you have to turn all the way around twice to get back to your original condition, it's otherwise pretty much like regular space. Isospin space, however, is completely abstract; it bears no relation whatsoever (other than through analogy) to anything we can grasp with our faculties of perception. How could rotations in such a space possibly have anything to do with the physical world? And yet the physical manifestation of the invariance of the strong force with respect to rotations in this space, the conservation of isospin, is a solidly established fact in the world of experimental science.

So, what then *is* isospin-space from a physical point of view? Physicists usually describe it as an *internal symmetry space*, but what's that, really? It's your old buddy again, telling you that your car's carburetion system "works on a vacuum principle." How's that going to help you to understand and fix the thing? It isn't.

Regarding the physical interpretation of the notion of isospin space, again your guess is as good as mine. Perhaps its experimental manifestations are hinting at some new and deeper truth about the universe that lies just beyond the current limits of our comprehension. Perhaps not. But one thing, however, is true: The introduction of the idea of internal symmetry spaces, of which isospin space was the first example, was an essential step forward in our understanding of the universe and the nature of the laws that govern it.

The Eightfold Way Revisited

Murray Gell-Mann's development of the eightfold way was one of the defining moments in the history of contemporary physics. It brought to our attention a particularly rich internal symmetry of the strong interactions, with a number of surprising predictions that could be (and were) addressed by the experiments of the day. While Heisenberg can be credited for introducing the notion of internal symmetry spaces, Gell-Mann can be credited with thrusting them into the forefront of our scientific consciousness in such a compelling way that they have been haunting our thoughts ever since.

If you know the properties of a Lie group's set of generators—how many there are, and what their algebra is (the interrelations between the generators associated with their dependence on the order in which they're combined under the group's operation)—you know pretty much everything you need to know to characterize the group. Any other group with the same number of generators and the same Lie algebra will be essentially the same Lie group describing the same underlying physics. The groups R(3) and SU(2) both have three generators, which exhibit the same ordering differences, and, correspondingly, they both describe rotations in three-dimensional space. The only difference is that while R(3) is adequate for systems that have angular momentum that is an integer multiple of $\hbar$, SU(2) must be called in if one is asked to consider the rotation of objects with half-integer multiples of $\hbar$—if one needs to consider particles' intrinsic spin. But nonetheless, R(3) and SU(2) both engender the set of rules for the same physical process: rotations in three dimensions.

Given this, one can turn the question around and probe the mathematics of Lie groups from what is essentially the opposite point of view. In this vein, the question one might ask is the following.

Given a Lie group's characterizing information—the number of generators and their associated Lie algebra—what are the concrete sets of elements, and rule for their combination, that manifest, or represent, the properties of the group? In the language of mathematics, what are the *representations* of the given Lie group?

Consider a system that possesses some amount of angular momentum. When this angular momentum is measured by determining its projection along a chosen measurement axis, the result must be one of a finite number of possibilities, each separated from the next by a unit of $\hbar$ of angular momentum. We can view this set of results as a pattern, a set of points along a line, representing the possible outcomes of the measurement. Figure 7.7 shows two such patterns of the many that we could depict: one for the case that the system possesses an angular momentum of $\frac{1}{2}\hbar$ (fig. 7.7a) and the other of $3\hbar$ (fig. 7.7b).

These patterns of points each form a *representation* of R(3) or SU(2) insofar as, when you take a state represented by any of the points in a given pattern, and perform any rotation from R(3) or SU(2), the result is a state represented by other points on the same pattern. Application of any element of R(3) or SU(2) moves you around within the pattern and doesn't take you outside of it to, say, a state with a larger angular momentum than you started with.

Rather than beginning with the hypothesis that the physics of angular momentum is governed by the Lie groups R(3) or SU(2), leading to observable patterns like those of figure 7.7, we might instead ask how the observation of such patterns might allow us to infer the underlying nature of angular momentum. Were we to look at systems possessing orbital angular momentum alone, we would see that the set of measurement possibilities falls within patterns such as that of figure 7.7b, containing an odd number of evenly separated points (with the exact number of points depending on the overall angular momentum of the system). A mathematician would then tell us that such patterns engender the properties of the Lie group R(3)—the set of all such odd-numbered patterns form the representations of the Lie group R(3). Similarly, were we also to consider systems with spin angular momentum, the observed patterns would be both even- and odd-numbered, and the mathematician would tell us that we were seeing the

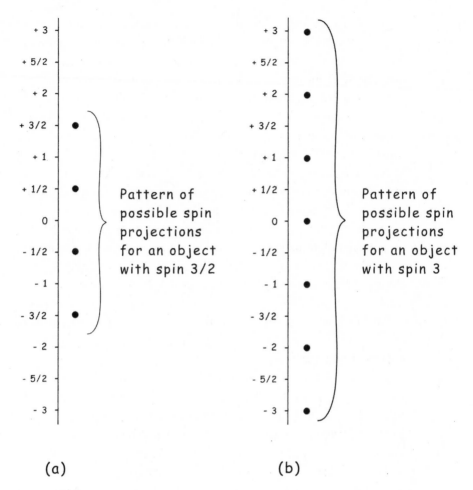

Fig. 7.7. The pattern of allowable spin projections onto any chosen axis of measurement for objects with spin-³⁄₂ and spin-3.

telltale signs, the representations, of SU(2). In either case, we have uncovered the underlying structure of the physical law. The pattern has revealed the paradigm.

Admittedly, in this case, the patterns are a bit uninspiring: a few dots on a line, the line being the projection axis, and the dots, all separated by an amount ℏ of angular momentum, being the possible projections of an object's angular momentum.

For the group SU(3) (the Lie group of rotations in three complex dimensions), however, the patterns are distinctive. This is largely because those systems that exhibit invariance under the transformations of SU(3) have not one but two separate axes on which to project the properties of physical states. So, instead of a progression of dots on a line, the representations of SU(3) are given by striking patterns of points in a two-dimensional plane. If you happen to observe states in nature that fall into some of these distinctive patterns, then it's abundantly clear: there's some (internal) space at play, and the physical law responsible for forming those states is invariant with respect to SU(3) rotations within that space.

Now, cut back to the early 1960s, by which time years of particle hunting at accelerator facilities had yielded, as mentioned in chapter 5, a daunting array of "elementary" particles that seemed to beg for some deeper understanding of the principles underlying their existence. In addition to the brute force task of finding each member of this lengthy list of particles, much clever thought and painstaking work had gone into identifying each particle's quantum numbers—its type (meson, hadron, antimeson, etc.), mass-energy, intrinsic spin, and so forth. It became clear that there were well-defined groupings of particles with similar quantum numbers, so physicists began to wonder what secrets might be hinted at by the properties of the particle states within these groupings.

It was Gell-Mann who recognized that these sets of particles with similar quantum numbers fell into suggestive patterns if one made use of two projection axes. On the x-axis, Gell-Mann plotted the magnitude of the projection of Heisenberg's isospin (the quantity depicted in fig. 7.6). On the y-axis, Gell-Mann plotted something of his own invention: a property he referred to as *hypercharge*, for which he concocted a quantitative formula whose value depended on whether the particle was a meson, a baryon, or an antibaryon and whether the particle possessed a somewhat nebulous quality known as strangeness. We now know that the value of a particle's hypercharge is related to the net number of strange quarks in the particle, but keep in mind that quarks had yet to be proposed; we'll get to them in a moment.

An example of a pattern produced in the mathematical space of isospin-projection versus hypercharge, that of the *pseudoscalar meson octet* introduced back in chapter 5, is shown in figure 7.8 (ignore the quark content labels for the moment). The particles that form this pattern are grouped together because they are all mesons (quark-antiquark combinations bound together by the strong nuclear force) and because they share two quantum numbers: each has an intrinsic spin of 0 and odd "intrinsic parity" (particles with odd intrinsic parity enjoy a change of sign of their wave function when they undergo mirror reflection, while particles with even intrinsic parity are unaffected by mirror reflection). Technically, a spin-0 particle is known as a *pseudoscalar* if it has odd intrinsic parity (and as a *scalar* if its intrinsic parity is even). Since there are eight of these spin-0 mesons with odd intrinsic parity, this set of particle states is known collectively as the pseudoscalar meson octet. Another technical point: for mesons, hypercharge and strangeness happen to be one and the same, so the hypercharge axis of this plot could also be called the strangeness axis, as it was in chapter 5.

In figure 7.8, each of the eight particles is represented by a dot; the *x* coordinate (position) of the dot is the particle's isospin projection I_p, and the *y* coordinate is its hypercharge Y. Now this is a distinctive pattern: a hexagonal shape, with a definitive separation in I_p and Y between each point, and with not one or three but precisely two particle states at the center point (note 7.11). Gell-Mann recognized this distinctive signature—this pattern that he referred to as the eightfold way—as one of the representations of SU(3) and, in doing so, established an entirely new way of thinking about

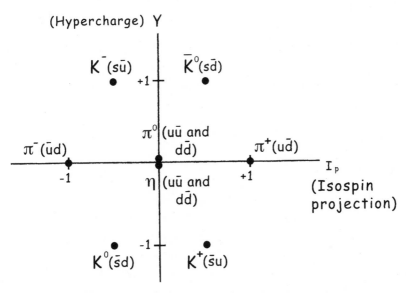

Fig. 7.8. The pseudoscalar meson octet. This diagram is almost identical to that of figure 5.10; the only difference is that the horizontal and vertical axes have been given the more formal labels of "isospin projection" and "hypercharge," respectively.

the natural world. Let's explore this notion, and its implications, a bit further.

One thing that Gell-Mann noted was that this eightfold hexagonal pattern was not the most basic representation (pattern) one could use to realize the abstract properties of the Lie group SU(3)—the pattern of the eightfold way can be constructed by combining two much simpler patterns. The first was a pattern containing just three points (see fig. 7.9a), which is the fundamental, or most basic, representation of SU(3). The second pattern (see fig. 7.9b), known as the antifundamental representation, is just the reflection of this first pattern, with all the I_p and Y values taken to their negatives. If you're willing to play with the graphs a bit, you can convince yourself that the eightfold way pattern of figure 7.8 can be formed by centering a separate copy of figure 7.9b around each point of figure 7.9a. When you construct the octet pattern in this way from the more basic patterns of figure 7.9, you're building your own meson states out of quarks and antiquarks.

Gell-Mann noticed that all the patterns formed by the groupings of elementary particles, not just the pseudoscalar mesons of the eightfold way but also the other sets of mesons or baryons for which the particles within the set all have similar quantum numbers, could be constructed in this way from various combinations of figure 7.9a and 7.9b. If you want to make mesons (which contain a quark and an antiquark), you combine a figure 7.9a pattern together with a figure 7.9b pattern in the way we just learned. If, however, you want to make the known baryons (which contain three quarks), you combine three figure 7.9a patterns. If you want antibaryons (which contain three antiquarks), you combine three figure 7.9b patterns. Thus, the three states contained in figure 7.9a struck Gell-Mann as the truly fundamental components of the known, dauntingly large, array of previously thought to be elementary particles. The states of figure 7.9b, being the same as those of figure 7.9a but with the signs of I_p and Y changed, represent the antiparticle states of figure 7.9a.

The literate Gell-Mann, recalling the line "three quarks for Muster Mark" from James Joyce's *Finnegans Wake*, dubbed these three states quarks. Given the (arbitrary) choice that the measurement axis in isospin space is vertical, the figure 7.9a state with $I_p = +\frac{1}{2}$ has its isospin pointing up (so that its projection on the vertical axis is positive). Thus, this state was dubbed the up (*u*) quark. Similarly, the $I_p = -\frac{1}{2}$ state was given the name down (*d*) quark. The third state, with $Y = -\frac{2}{3}$, has no isospin and doesn't appear in ordinary matter, which must have struck Gell-Mann as a bit

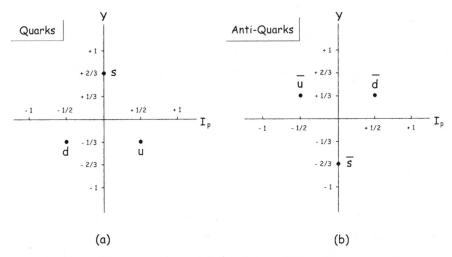

Fig. 7.9. The fundamental representation of SU(3), which houses the quarks in Gell-Mann's eightfold-way model. The antifundamental representation of *b* is that of the antiquarks.

strange, so it became known as the strange (*s*) quark. Correspondingly, the states of figure 7.9*b* are the anti-up ($\bar{u}$), anti-down ($\bar{d}$), and anti-strange ($\bar{s}$) antiquarks. None of the accelerators of the day (early 1960s) was powerful enough to produce heavier (charmed, bottom, top) quarks, so only these three quarks entered the picture at the time.

Having proposed quarks to explain the patterns observed in the properties of the elementary particles, and thereby organizing and vastly simplifying our perception of these basic states of matter, Gell-Mann nonetheless considered quarks to be no more than a convenient mathematical construct. If you can have internal spaces (such as isospin) dictating the way particles form themselves and behave, but having no real connection to the three-dimensional space of everyday life, then why not have a similar set of internal particle states? To Gell-Mann, the quarks could do their job well enough even if they weren't physical states that you can detect in the laboratory. But then came the groundbreaking scattering experiments at SLAC in the late 1960s (see chapter 5) that showed clear evidence for the presence of quarks within protons and neutrons. Quarks are no less real than the protons and neutrons (and pions and kaons and lambda baryons and so forth) that they comprise.

But where does the notion of symmetry work its way into all of this? Recall the original motivation for introducing isospin space and SU(2) rotational symmetry: two particle states, the neutron and proton, look entirely the same as far as the strong nuclear interaction is concerned. As far as the strong force is concerned, for a nucleus with some number of nucleons in it, it doesn't matter how many of the nucleons are neutrons and how many are protons: the strong force will hold them together with the same amount of attractive pull. No matter what SU(2) rotation we perform in isospin space, completely or partially swapping protons with neutrons (or, from the modern viewpoint, up quarks with down quarks) the strong force doesn't care; it stoically binds the whole mess together as if it couldn't tell the difference. This is what we mean by invariance under, or symmetry with respect to, SU(2) rotations.

Well, the same is true of SU(3) rotations in flavor space, except now, instead of just swapping up and down quarks, we're interchanging three quarks—up, down, and strange—among one another. If the strong force is invariant under the full set of SU(3) rotations in flavor space then, all other things being equal, the binding provided by the strong force should be the same regardless of what quarks are being bound together. The pseudoscalar meson of figure 7.8 that is formed from an up quark and down antiquark should be bound just as energetically, and thus have precisely the same mass (recall Einstein's relativity), as the pseudoscalar meson of figure 7.8 that is formed from, say, a strange quark and up antiquark. In fact, for the particles contained within any given representation, such as the octet, all other things are equal as far as the strong force is concerned; the only difference between the particle states is the flavor of the quarks they contain. So, all the particle states of any given representation, such as that of figure 7.8, should have the same mass. And this is what is observed, at least after you've taken into account the fact that the electromagnetic forces change a little when you interchange quark flavors (because the different quarks carry different amounts of electric charge) and that the different quarks themselves have somewhat different masses, so that combinations of different quarks can have somewhat different masses, even if they are bound with equal strength by the strong force (note 7.12).

Real space, spin space, isospin space, and so forth: In the physical world, there seem to be many different spaces (real or internal) at play. One of these spaces is the internal space in which rotations, governed by the rules of SU(3), swap quark types among one another; physicists needed to give this internal space a name. Somehow the different quark types came to be known as quark flavors, consistent, I guess, with the fact that quark is also a type of cheese, although few physicists seem to be aware of this. So, the internal space in which the three quark flavors (up, down, strange) are swapped around by rotations whose properties are governed by the Lie group SU(3) is known as SU(3) flavor space.

In any regard, the observation of "multiplets" of particle states with similar quantum numbers and roughly similar masses, arrayed in the characteristic patterns of the representations of SU(3), is abundant evidence that an internal SU(3) symmetry space is at play. And, with that knowledge firmly in hand, one can then use the properties of this space and of the group SU(3) and its representations to make testable predictions about the nature of the physical world.

In particular, if SU(3) governs the array of particle states through the patterns of its representations, then every particle must be part of such a pattern. But also, every pattern that has some particles in it must have a *complete* set of such particles, all with similar masses and quantum numbers. It's not enough to see that there are, say, five pseudoscalar mesons that lie on five of the eight points of figure 7.8. There have to be precisely eight such mesons, no more and no less, and they need to lie on each of the eight points of the pattern in figure 7.8, one particle state per point.

When Gell-Mann first proposed the eightfold way, one of the baryon (three-quark combination) representations was missing one of its members. In the frenzy of discovery during the 1950s that had prompted Gell-Mann's ruminations about SU(3), nine baryons with spin-$\frac{3}{2}$ and even intrinsic parity, and with masses all about $1\frac{1}{2}$ times that of the proton, had been discovered. For these nine particle states, the different values of isospin projection I_p and hypercharge Y—the physical quantities from whose values the SU(3) patterns are formed—fit nicely onto nine of the ten points of the decouplet, the SU(3) representation with ten points in it. The problem, of course, was that there shouldn't be just nine such particle states; if they are really states of the tenfold decouplet representation of SU(3), then there need to be ten of them.

Not only that, but the missing tenth particle needed to have exactly the properties associated with the remaining uncovered dot in the decouplet pattern, which, in this case, turned out to be an isospin projection of $I_p = 0$ and a hypercharge of precisely $Y = -2$. If you look back at the quark representation of figure 7.9a and ask how to get a total of $I_p = 0$ and $Y = -2$ by adding together three quarks, you'll realize that the only way to do this is

204 DEEP DOWN THINGS

if all three quarks are strange quarks (each strange quark has $I_p = 0$ and $Y = -\frac{2}{3}$). Confident that this particle must be out there, Gell-Mann boldly predicted the existence of this particle, the Ω^+, the strangest, if you will, of all possible particles.

Accordingly, this particle was discovered at Brookhaven National Laboratory (the Long Island, New York, atom smasher) in 1964. Two years after Gell-Mann's first papers on the subject, this discovery provided a dramatic confirmation of the notions of SU(3) symmetry and the eightfold way. It's one thing to come up with a model that fits all the existing data. It's quite another to use that model to unambiguously predict something new about the natural world—something as odd and unique as a particle formed from three strange quarks—and have the prediction confirmed by experiment. The discovery of the Ω^+, the confirmation that it had the correct mass, and that it was composed of three strange quarks (by studying the way it decayed into other, less exotic particles), plastered Gell-Mann's model of internal SU(3) symmetry solidly into the firmament of revolutionary twentieth-century physical theories. It didn't take long for the larger community to come to this recognition: Gell-Mann was awarded the Nobel Prize in Physics in 1969, just as experiments at the Stanford Linear Accelerator Center were providing the ultimate confirmation of Gell-Mann's model of the eightfold way by publishing the first evidence for the existence of the quarks themselves.

Two Further Thoughts

What, if anything, are internal symmetry spaces? Do they have any physical presence, in the sense of ordinary three-dimensional space, any "bulk" within which the motion and actions of physical objects play themselves out?

The sober answer is no. They are mathematical edifices, spaces with no more physical content than the space in which I plot my checking account balance against the unyielding advance of time. The employment of such a mathematical space can provide useful information and even help me live a more rewarding life, but the space of bank balance versus time is definitely not a space with physical extent, with a bulk that can support motion and interaction and all the things we associate with physical behavior.

It's easy for us to say that the various internal symmetry spaces—isospin and its extension to the space of SU(3) flavor transformations, and the internal spaces we'll meet in the next chapter—are equally unphysical constructs of the mathematical mind. Rather than being true physical spaces, they are (from this point of view) merely elegant ways of encapsulating the rules behind some aspects of the fundamental behavior of the natural world.

But then, what about spin-space? Rotations of the spin axis of spin-½ particles (a category into which many of our most favorite particles fall) don't follow the rules of R(3), the group of rotations that you and I can perform with an object in everyday three-dimensional space. Instead, these rotations follow the rules of SU(2), which operates within a somewhat different sort of space than that of R(3), a space for which a rotation of 360 degrees does not get you back precisely to where you started. And yet spin-space is very much a real space: When you rotate a spin-½ particle's spin by half a circle—180 degrees—about the x-axis in spin-space, it does go from pointing up to pointing down in our everyday, three-dimensional space, as physicists have demonstrated experimentally time and time again.

So perhaps this point of view is unnecessarily puritanical. An argument that might countervail it (if you will permit me a moment of untethered speculation) goes as follows.

We live in a world for which our perceptions, the mechanisms of our five senses, are governed by chemical behavior. Chemistry is really nothing more than the rich brand of physics for which information and actions are brought forth by exchanges that have an energy of the order of 1 eV. Chemical behavior was what was available to be made use of by life on earth as it developed, and it is what human senses and perceptions naturally and necessarily evolved to interpret.

Now, the energy scale of chemistry, of 1 eV and thereabouts, is far below that of high-energy physics. With exchanges of 1 eV of energy, you can't sense the behavior of the individual quarks inside the nucleons or even the nucleons within the nucleus or even the nucleus within the atom. You can't even rotate the spins of the spin-½ electrons that are orbiting about the nuclei in the atoms of tangible substances.

The point is this: The physical processes that we humans have evolved to make use of in probing the world around us, the various chemical behaviors that form the basis of our senses, are just not energetic enough to probe the behavior that we have been discussing in this chapter. Would humans have developed a sense of the bulk of ordinary three-dimensional space if we hadn't, through the chemistry of our senses, been able to invoke and perceive motion through that bulk?

Without the extension provided by sophisticated scientific apparatus, our senses, our biochemical senses, are just too anemic to invoke or perceive motion in the spaces we have introduced and discussed in this chapter. Might it be possible that the internal spaces of this chapter are in fact no less real than the three-dimensional space of our everyday perception? Could it be that, absent any way to influence or perceive motion in these realms, we simply lack the tools and motivation to evolve the ability to perceive them?

I'll leave you to mull this over while moving on to the second and final discussion of this section: that of an important qualification of the nature of the symmetry transformations we've been considering. So far, we've been restricting our thinking to what are known as global symmetry transformations.

When you hear the word "global," what is usually meant is something that applies equally and all at once everywhere in the world. That's more or less what's meant here, except the world of particle physics covers more than just the surface of our remarkable planet; it includes the whole universe; the full extent of whatever space the symmetry transformation happens to be operating within. Universal symmetry transformation may have been a better expression, but global conveys the notion well enough.

The point is that, when we apply one of the actions that is a member of the symmetry group of interest, we are applying it globally—equally and simultaneously everywhere in the space associated with the transformation (real space-time in the case of spatial or temporal translations and rotations, isospin space in the case of isospin transformations, and so forth).

Noether's theorem tells us, for example, that the conservation of overall angular momentum is due to the invariance of physical laws under rotations of the entire body of physical three-dimensional space. Conservation of angular momentum follows directly from the fact that, regardless of how the space surrounding you is oriented with respect to the lab equipment you just set up, the physical laws you derive will always be the same. When you rotate your laboratory equipment to test the hypothesis that physics is invariant under rotations, you are rotating the apparatus with respect to the entire body of space (or, equivalently, rotating the entire body of space with respect to your apparatus; there's no difference).

However, Noether's theorem doesn't follow from the consideration of the behavior of physical laws under rotations of small pockets of space. Anything reaching the boundary of such a pocket would exhibit a behavior as arbitrary and unpredictable as the rotation that was somehow imposed on the space inside the pocket. Angular momentum would certainly not be preserved for a system crossing such a boundary, the object would undergo a sudden and spontaneous reorientation at the point of crossing, quite independent of what forces were acting on it at the time. In doing such an operation only in a *local* region of space, you are doing more than rotating space, you are distorting it beyond recognition.

So, to get the same experimental result, to derive the same laws of physics, the entire space has to be reoriented, and all at once. The symmetry operation has to be global. In this light, there's obviously no merit whatsoever to the consideration of symmetry transformations that are not global, that are local, in the sense just described.

Unless you happen to be a particle physicist . . .

8

Physics by Pure Thought

————∿∿∿∿∿∿∿∿∿∿————

Gauge Theory

The role of the scientist is not to dictate natural law but, rather, to uncover and elucidate it. However, if it were up to physicists to choose the theoretical framework within which the physical world operates, gauge theory would be a promising candidate. It is through gauge theory that science makes its greatest inroads toward the reduction of the full spectrum of physical behavior into a single inevitable underlying principle of causation.

It's a bit of a shame, and a detraction from the elegance of the theory, that the Standard Model can't be presented at the end of this chapter as an example of a "pure" gauge theory. Instead, gauge theory is merely a component of the formalism of the Standard Model, and we'll need to move on to chapter 9 and its discussion of hidden symmetry to complete the picture. Nonetheless, much of the elegance of gauge theory, hidden or not, does carry over to the complete Standard Model, so the lessons of this chapter should not be discounted.

Gauge theory provides the intellectual cornerstone of the Standard Model; accordingly, this chapter really forms the crux of this book. I will therefore say, without apology, that it will be a long and involved chapter, covering a lot of ground while synthesizing a good deal of the material from previous chapters. We'll need to draw on and sew together threads from all of those chapters, including our discussions of quantum mechanics and the wave equation (chapter 3), field theory (chapter 4), the natures of the electromagnetic and weak nuclear interactions and their associated field quanta (chapters 2 and 5), Lie groups (chapter 6), and internal symmetries (chapter 7). All these topics were introduced in service of a single-minded purpose: the eventual introduction and description of gauge theory, and its connection to the workings of the natural world.

Phase Revisited

In chapter 3 we discussed the concept of the phase of a wave and spent some time emphasizing that the overall phase of the wave function of any system of objects can have no bearing on the physical properties of the system (recall our afternoon at sea, during which our naiveté about the irrelevance of phase put a significant dent into our social life). In this chapter, we'll take a deep look at this notion, and explore its surprisingly rich consequences.

At this point, we need to discuss exactly how it is that phase is treated in the quantum-mechanical representation of physical systems. Since any quantum-mechanical system is fully specified by its wave function, this just boils down to a discussion of how phase is incorporated into the wave function.

In several previous chapters it was mentioned, almost parenthetically, that wave functions are complex. A real (i.e., noncomplex) function is just a rule that assigns an ordinary real number to each point in space and time, so a complex function is nothing more than a rule that assigns a complex number to each point in space and time. For a quantum-mechanical wave function, the complex number assigned to any given point in space and time obeys the following rule: the square of the assigned complex number must be the relative probability of finding the object at that particular point in space at that particular time. Let's recall our discussion of complex numbers in the section "Into the Continuum" in chapter 6 in which we introduced the Lie group $U(1)$.

A complex number z is a number of the form $z = a + b \cdot i$, where a and b are any real (ordinary) numbers, and i is the make-believe number whose square is negative one: $i \cdot i = -1$. As mentioned in chapter 6, we know how to put real numbers in order: the number 10 is bigger than the number $\pi = 3.14159$, which is bigger than the number -12.5267, and so forth. However, which of the following two complex numbers is bigger: $z_1 = 12 + 3.1 \cdot i$ or $z_2 = 3.6 + 10 \cdot i$? The "real" part is larger for z_1 (12 for z_1 vs. 3.6 for z_2), but the imaginary part is bigger for z_2 (10 vs. 2.1), so it's not immediately clear which is bigger.

In chapter 6, we resolved this quandary by defining the size $|z|$ of any complex number $z = a + b \cdot i$ to be $|z| = \sqrt{a^2 + b^2}$ (by definition, we only use the positive square root). It's very important to note that since a^2 and b^2 are always positive real numbers (the square of any real number is always some other positive real number), then the size $|z|$ of the complex number z is always a positive real number (the positive square root of any positive real number is always some other positive real number). Since we all know how to order positive real numbers such as $|z|$, then this gives us a rule for ordering the underlying complex numbers $z = a + b \cdot i$ according to their size.

However, one quickly realizes that while no two real numbers have the same size, it's easy to find sets of complex numbers that, under this rule, have the exact same size. For example, you can easily verify that the complex numbers $5 + 0 \cdot i$, $3 + 4 \cdot i$, and $0 + 5 \cdot i$ all have the same size (which is 5).

To see how it is that this discussion of the nature of complex numbers relates to the phase of a quantum mechanical wave function, let's recall the discussion surrounding figure 6.2 in chapter 6. This recollection will be aided by figure 8.1, which is quite similar to figure 6.2b.

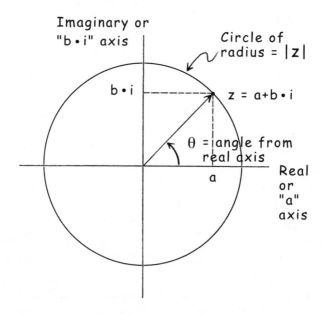

Fig. 8.1. A rerun of figure 6.2b. The phase of the wave function is represented by the angle θ in a.

Again, we choose to represent the complex number $z = a + b \cdot i$ as a point on a two-dimensional graph, for which instead of having x- and y- axes, we have a horizontal real axis and a vertical imaginary axis. We find the point corresponding to z by going out a distance a along the real axis, and then up a distance b along the imaginary axis, placing a dot corresponding to the complex number $z = a + b \cdot i$ at the point at which we end up.

If we draw an arrow from the origin (point at which the axes cross) to the point representing z, the length of the arrow is nothing more than the size $\sqrt{a^2 + b^2}$ of the complex number z (note 8.1). If this complex number z represents the value of the (complex-valued) wave function $\psi(x)$ of a quantum-mechanical system at some point in space and time, then the relative probability of finding one of the objects in the system at that point in space-time is just the square $(a^2 + b^2)$ of the length of this arrow.

Let's dwell on this for just a moment to make sure that one thing is clear. The position of the point z in the complex plane plot of figure 8.1 has nothing to do with location in real space. Instead, it's just a way of representing the value of the complex-valued wave function for whatever point in space and time you want to know it. When we discuss the wave function $\psi(x)$, x represents the point in space at which we want to know the value of the wave function. The value of the complex wave function, the complex number associated by the function $\psi(x)$ to the point x in real space, is just some complex number z, such as that represented by figure 8.1. Since complex numbers are, well, complex, you can't represent them by a position on a simple number line. Instead, they have to be represented by a point on a two-dimensional plot.

So, for any given space-time point, the representation of figure 8.1 provides complete information about the value of the wave function at that point. If we consider a different point in space-time, say, a point 2 centimeters to the right and 4 seconds earlier, we need to draw a new complex plane plot for the new complex number z_1 representing the value of the wave function at that space-time point. Thus, to represent fully the entire wave function in this way, we would need an infinitude of such plots, one for each of the infinite number of space-time points in the universe. It's important to keep this all in mind as we discuss what it means to change the phase of the wave function $\psi(x)$.

In any regard, we now need to note that in addition to the length of the arrow pointing to the complex number z we also need an angle (represented by θ in fig. 8.1) to specify exactly how to draw the arrow pointing to z. If we change the angle of the arrow to some new direction (new angle θ' in fig. 8.1), but without changing its length, the resulting complex number z' has the same length as the original complex number z and represents the identical probability of finding an object at the given point in space and time. This probability only depends on the length $\sqrt{a^2 + b^2}$ of the arrow pointing to z or z', and not on the angle θ or θ' that the arrow makes with the "real" axis. So the observable, the thing we can measure experimentally, is encoded into the length of the arrow representing the value of the complex-valued wave function at that point in space-time. Its angle (θ or θ') is unobservable; it's immaterial like the phase of the wave function is immaterial. And that's just the point, this angle is the phase of the wave function!

In chapter 3, we saw that it makes no real difference whether the sailor is at the bottom of a trough, the top of a crest, or anywhere in-between, at precisely noon. Similarly, there is no consequence to the physical state of an object if the complex value of its wave function is represented by an arrow (such as that of fig. 8.1) that points along the real axis, the imaginary axis, or any direction in between. What does matter is the length of the arrow.

We argued in chapter 3 that bobbing up and down on an endless series of waves is very similar to going around and around in a circle and that the phase of the bobbing motion at any given point in time is analogous to the angle through which you've turned on the circle. Here, we see that all complex numbers of a given length lie on a circle and again the notion of phase is provided by the angle through which you turn to get to the complex number z that represents the value of the wave function.

So, this is how the notion of phase is incorporated into the complex-valued quantum-mechanical wave function. The phase angle θ tells you where you are in relation to the crests and troughs of the quantum-mechanical wave: if θ is 0 or 360 degrees, you're at the top of the crest, if θ is 180 degrees, you're at the bottom of the trough, and so on. It's cast in a funny sort of way, in terms of the "real" and "complex" parts of the complex wave function, but it's the same thing.

This will be an essential component of the connection that we'll soon make between abstract Lie groups and the wave functions of physical objects. Remember that we introduced Lie groups as complete sets of rotations in a given space: R(3) for rotations in ordinary three-dimensional space, SU(2) for complex two-dimensional space, and so forth. The motivation for introducing complex Lie groups, such as SU(2), is that the complex nature of the Lie group allows it to change the phase of the wave function: the complex character of the Lie group's elements allows the phase of the wave function to be rotated around the circle of complex numbers (see fig. 8.1).

Global Irrelevance

The notion discussed in the previous section—that the overall phase of any quantum-mechanical system can have no physical consequence—is usually referred to by the rubric "global phase invariance." In this context, invariance means precisely the same as it did in the general discussion of invariance of chapter 7. Since the overall phase of the system has no bearing on its physical properties, the system is "invariant" under the (purely mathematical) operation of a change in its overall phase.

The term global in the phrase global phase invariance, contrary to its customary connotation, is a condition that actually *restricts* the nature of the phase transformations, the set of possible changes of phase of the wave function describing the system, under consideration. In this context, global doesn't mean "encompassing all conceivable possibilities." Instead, it implies that the transformations under consideration are only those that are identical throughout space and time—globally uniform. Thus, global phase transformations are purely mathematical operations by which the phase of the wave function describing a quantum-mechanical system is changed by the same amount everywhere in space and time.

In view of the above discussion of quantum-mechanical phase, this amounts to nothing more than picking some single angle between 0 and 360 degrees, and then rotating the arrow representing the (complex) value of the wave function at each and every point in space-time by exactly this amount. If the phase transformation is global, then this angle of rotation, whatever we chose it to be, must be *the same everywhere in space and time.*

It turns out that the notion of global phase invariance is, like everything else in quantum mechanics, built directly into the wave equation, which dictates the precise form of the wave function corresponding to any given physical system. We discussed a number of such wave equations; for example, the Schrödinger equation governs the behavior of nonrelativistic systems, while the Dirac equation provides us with the appropriate description of the behavior of relativistic spin-½ particles. Luckily, the discussion of phase invariance (both global and local) that is to follow is identical in essence for all forms of the wave equation, and thus we are free to carry out this discussion in the context of the more familiar Schrödinger equation.

In chapter 3, we presented the rigorous mathematical statement of the Schrödinger equation, for the special case that the object being described is moving in just one dimension (backward or forward along the x-axis). Here it is again:

$$-\frac{h^2}{8\pi^2 m}\frac{d^2}{dx^2}\psi(x)+V(x)\psi(x)=E\psi(x).$$

Let's reiterate some of the points that were made about this expression when it was introduced.

This equation explicitly represents the condition that the wave function must satisfy to represent appropriately the behavior of an object (of mass m) under the influence of one or more forces; the wave function itself is represented by the symbol $\psi(x)$. Wherever you are in space, whatever the value of x is, the (complex) number $\psi(x)$ is just that number whose square (size squared) is the relative probability of finding the object at that point.

The influence of the forces that are at play are introduced through the potential function $V(x)$ in the second term on the left-hand side of the equals sign. If an object is under the influence of some force or set of forces, it takes some amount of energy to move it to the point in space represented by the value of x (think of a car rolling against the force of gravity some distance x up a hill; it has to expend some amount of its energy to do this). If we let $V(x)$ be that amount of energy for all points x in space, we can plug this $V(x)$ into the Schrödinger equation above, and then solve the equation for the wave function. What we wind up with is the wave function $\psi(x)$ that describes the behavior of an object under the influence of the force described by the energy function $V(x)$. The E in the term to the right of the equals sign is the object's total energy—the potential energy $V(x)$ associated with moving it to x plus whatever kinetic energy the object has because it's still moving when it gets to the point x.

The first term—the one associated with the kinetic energy the object has because it's still moving when it gets to point x—is the one that we want to concentrate on because this term will make all the difference when we switch from the consideration of global phase invariance to the consideration of the more demanding and much more interesting requirement of *local* phase invariance. This first term involves a derivative (that's what the symbols $d^2\psi(x)/dx^2$ stand for), so let's talk about derivatives for a moment.

The function $\psi(x)$ gives the value of the wave function at any point x in space. If we change x a little (move to a slightly different point in space), then $\psi(x)$ can be expected to change some, too. The derivative $d\psi(x)/dx$ of $\psi(x)$ is nothing more than the rate of change of $\psi(x)$ when you change x. If $\psi(x)$ changes a lot when you move from one point in space to a nearby point, the derivative $d\psi(x)/dx$ is some relatively big number; if it changes only a little, then $d\psi(x)/dx$ is a smaller number. Actually, this first term involves a second derivative, which is the rate of change *in the rate of change* as you move from point to point in space—but that distinction is not necessary to grasp for the ensuing discussion.

Now, suppose that $\psi(x)$ represents a wave function that satisfies the Schrödinger equation for some potential energy function $V(x)$. This means that the rate at which the wave function $\psi(x)$ changes as you move from point to point in space (first term in the Schrödinger equation), when multiplied by the precisely known combination of numbers $-h^2/8\pi^2 m$, is just what you need so that when you add it to the potential energy (second term), you just get the object's total energy (term to the right of the equals sign). To satisfy the equals sign in the Schrödinger equation, there must be a delicate and precise balance between the rate of change of the wave function in the region surrounding the point x and the potential energy of the object at the point x, for each and every point x in space.

In other words, the total energy E on the right side of the equation is fixed for all points in space (due to energy conservation) so as $V(x)$ changes from point to point, the rate of change of the wave function $\psi(x)$ must itself change in just the right way such that the sum of the first and second terms in the Schrödinger equation is exactly the total energy E. If $V(x)$ is very small at some point x, then the rate of change of the wave function $\psi(x)$ in the region surrounding the point x must be correspondingly large, and vice versa.

Now, suppose that you arbitrarily decide to make a change of phase of this wave function—to change, at every point in space, the angle θ that the complex number $\psi(x)$ makes with the real axis in figure 8.1. Here's the critical point: If this phase change is *global*, if the amount by which you change the phase angle θ is the same everywhere in space, then this change of phase will not destroy this delicate and essential balance. Why is this?

First, by changing the phase of the wave function $\psi(x)$ you certainly don't change the potential energy $V(x)$ at all, so the energy of the second term in the Schrödinger equation is unchanged. The energy of the third term in the

Schrödinger equation, to the right of the equals sign, is also unchanged because the total energy E is conserved, and so is the same everywhere in space and time. But, in addition, if you change the wave function's phase by the same amount everywhere, you are also not changing the rate of change (derivative) of the wave function. So, the energy of the first time—the object's kinetic energy—is also unaltered by the global phase change.

To understand this, think of a slide on a playground. There is some function that represents the height of the slide at any point x along its length. We can label this function whatever we want to; so let's call it $\psi(x)$. Don't be confused here: We are using the height of the slide simply to give ourselves a way to visualize the value of a function $\psi(x)$ of position x; in this analogy, the height of the slide has nothing whatsoever to do with gravity and the potential energy function in the wave equation.

The slide has a certain slope, which is the rate of change of the height function $\psi(x)$ as you change x, that is, move from point to point along the length of the slide. So, this slope is the derivative of the height function $\psi(x)$.

If you move the slide to the top of a hill, being careful to change its height *by the same amount at every point along the slide*, the slope of the slide does not change. This *global* change of height, a change of height by the same amount at each point on the slide, leaves the derivative of the height function unchanged. Similarly, a global change in the phase of the wave function leaves the derivative of the wave function, and thus the kinetic energy (represented by the first term in the Schrödinger equation, which is the term with the derivatives), unchanged.

So, this *global* change in the phase of the wave function leaves all three of the terms in the Schrödinger equation unchanged, and the phase-changed wave function, call it $\psi'(x)$, still satisfies the Schrödinger equation. The new wave function $\psi'(x)$ still describes the behavior of an object under the influence of the force represented by $V(x)$. As far as observable, physical behavior goes (the only kind of behavior that counts), $\psi(x)$ and its globally phase-changed partner $\psi'(x)$ are one and the same wave function. In this sense, the notion of *global* phase invariance is built in to the Schrödinger equation, or any quantum-mechanical wave equation, for that matter.

The Real Trouble with Globalization

In the mid-1950s, two physicists, C. N. Yang of the Institute for Advanced Study at Princeton University, and R. L. Mills of Brookhaven National Laboratory, became deeply interested in the question of the phase invariance of quantum mechanics. What intrigued them most was that, on reflection, the idea of global phase invariance didn't quite wash with Einstein's notions of the nature of space-time. Yang and Mills were perfectly happy with the idea that the observable properties of a quantum-mechanical system should be independent of the phase of the wave function, as discussed at length above. What bothered them was that, to exhibit this independence of phase, one had to change the phase globally, by the same amount everywhere in space-time. We haven't yet demonstrated that changing the phase locally— by an amount that differs from point to point in space and time—disrupts the delicate balance of the Schrödinger equation, but we will in due course. In any regard, the need to require that quantum-mechanical systems be unaltered only by global changes of phase seemed to Yang and Mills to be very unnatural.

To understand Yang and Mills' queasiness with the notion of global phase invariance, let's recognize that, whatever the situation being described, the wave function needs to have some phase or other. The value of the phase we've said time and again is immaterial, but waves have phases, so we've got to pick something. In other words, to write down the wave function describing the quantum-mechanical state of an object, we have to choose its phase, but the choice that we make is arbitrary. It's like the slide on the playground: it doesn't matter whether the slide is installed down by the creek or on top of the hill. The choice of the height at which to install the slide is an arbitrary one.

However, once you put one of the legs of the slide down, then the heights at which the other legs rest is anything but arbitrary: they must be at just the right level, or the slope of the slide will be wrong; it will be either too steep to be safe or too shallow to be fun. Similarly, once you choose the phase of the wave function at one space-time point, the requirement of global phase invariance fixes it at all other space-time points.

In their 1954 paper in the *Physical Review*, Yang and Mills duly noted that the choice of the wave function's phase is an arbitrary one. They continued:

> As usually conceived, however, this arbitrariness is subject to the following limitation: once one chooses [the phase of the wave function] at one space-time point, one is then not free to make any choices at other space-time points.

It seems that this is not consistent with the localized field concept that underlies the usual physical theories. In the present paper we wish to explore the possibility of requiring all interactions to be invariant under *independent [choices of phase] at all space-time points.*

From the time of the first introduction of the idea of a force field, by Michael Faraday in the 1830s, up through the development of quantum field theory, great effort has been expended to avoid the disquieting notion of action-at-a-distance. Our innermost being tells us that if something is to be influenced by a force, the agent of influence had better be where the influenced object resides when it is acted on and not meters, or millimeters, or even microns away. If you're going to push on something, you've got to *push on it*, not just stand there flailing your arms some distance away!

Indeed, Einstein's relativity tells us that no object possessing mass-energy can travel faster than the speed of light, and since the exertion of a force by one object on another generally requires that energy be exchanged between the objects, no agent of force can act instantaneously at a point removed in space from that agent. Even more, quantum mechanics tells us that the very transfer of information from one point to another requires an exchange of energy (otherwise, how would the system of the observer be able to receive the transferred information?), so even information, in whatever form it may take, can be transferred from point to point at a speed no faster than that of light.

Given this, how could the value at point x of the universe's wave function, at exactly noon tomorrow, know what arbitrary choice was made for the phase of the universe's wave function at some distant point x' at precisely the same time? If the transfer of information regarding the choice made at x must obey the fundamental speed limit of 186,000 miles per second, what sense does it make to require that the arbitrary choice made at point x' be the same as that of point x? Information cannot be exchanged instantaneously between the two spatially separated points, so they can't know about each other; the two space-time points (x at noon tomorrow and x' also at noon tomorrow) are not *causally connected.* Nothing that happens at point x' at noon can influence point x at noon, since to do so, the information of what happened at x' would have to travel faster than the speed of light.

So the tenets of Einstein's relativity weigh against the possibility that the wave function at noon at point x can make itself aware of the properties of the wave function at the same time at point x'. In particular, the wave function at noon at point x won't be able to conform to whatever arbitrary choice was made for the phase of the wave function at noon at point x', since it can't know about the value of that choice of phase. The requirement that the arbitrary choice be global—unchanging throughout space and time— makes no sense because that's not the way nature seems to operate. This is what concerned Yang and Mills, as expressed in the quote from their seminal paper on gauge theory that we just read.

Local Irrelevance

Consider an electron somewhere out in the world, maybe under the influence of some force or set of forces, whose physical state is encapsulated in its wave function. Again, there's a certain freedom, or arbitrariness, in this wave function. We are free to change the phase of the wave function, in the manner of figure 8.1, as long as we change it globally, by precisely the same amount at every point in space and time.

But now, Yang and Mills admonish us that we shouldn't really talk about global phase invariance because not all points in space-time are causally connected. So, it makes no sense to require that the choice of change of the wave function's phase be the same everywhere in space-time. Instead, we must consider local changes of phase, that is, changes in the phase of the electron's wave function by an amount that varies from point to point in space-time.

But there's a major problem with this. As we've seen, the wave equation, to be satisfied by a given wave function, requires a delicate and precise balance between the rate of change of the wave function in any region of space (kinetic energy) and the value of the electron's potential energy in that region. These two must add up to the electron's given total energy, which is conserved; the total energy is the same no matter where the electron is and when it's there. However, if we take a wave function that originally satisfies the wave equation and change its phase by an amount that varies from point to point in space, we corrupt this delicate balance; what we wind up with no longer satisfies the wave equation. It no longer provides a valid description of any physical state of the electron, much less the state that we started with.

Let's make this point a little more accessible by assuming that the electron under study is not under the influence of any forces as it moves through space, that is, we are studying a free electron. In this case, the potential energy is zero everywhere in space (remember: we introduced the concept of potential energy precisely to allow the incorporation of external forces into the wave equation, so if there are no forces at play, there is no potential energy to worry about), and the wave equation becomes

$$-\frac{h^2}{8\pi^2 m}\frac{d^2}{dx^2}\psi(x) = E\psi(x);$$

because $V(x) = 0$, we can drop the $V(x)\psi(x)$ term that appeared in the wave equation.

Again, the term to the left of the equals sign, the mathematical notation for the rate of change of the wave function $\psi(x)$ times the well-known combination of numbers $-h^2/8\pi^2 m$—reveals the electron's kinetic energy, which is just the electron's total energy since, in this case, it doesn't have any potential energy. The term to the right of the equals sign is just the electron's total energy times the wave function. So, the free electron's wave function must still satisfy a precise and delicate balance, albeit a very simple one: the product of the rate of change of the wave function times the well-known number $-h^2/8\pi^2 m$, in any and all regions of space-time, must be precisely equal to the electron's energy.

Again, the essential point is that a change of phase of the free electron's wave function that varies from point to point in space will upset this delicate balance and render the electron's wave function unphysical. We can understand why this is with a fairly simple analogy.

Let's pack a sandwich and some juice and head back to the local playground, where we found the slide that helped us to visualize the notion of global phase invariance. As before, we'll think of the height of the slide as the value of the wave function $\psi(x)$ at the point x in space. The value of x just represents the point along the length of the slide at which you want to know its height; x corresponds to some point between the launching pad at the top of the slide and the bottom end of the slide from which wide-eyed children (and adults who should know better) come flying, only to be jarred to a halt in a cloud of dust.

In this analogy, the rate of change (derivative) of the wave function $\psi(x)$ is the rate at which the height of the slide changes as you move down it— the steepness of the slide. When you multiply this steepness by an appropriate factor, which includes your mass, the pull of gravity, the drag of friction, and so forth (analogous to the factor $h^2/8\pi^2 m$ in the wave equation), you get the kinetic energy E associated with your motion as you exit the slide. For a given energy E, there's one and only one slope, one solution for the function $\psi(x)$ for the height versus position of the slide, that will result in your motion as you exit the slide having precisely that energy. There's a delicate and precise balance between the rate of change of $\psi(x)$ and the energy E of your trip down the slide.

As for true quantum-mechanical wave functions, there's an arbitrariness associated with the specification of $\psi(x)$ analogous to the phase of the wave function. We talked about this a few pages back: it's the overall height of the slide above sea level. It doesn't matter whether you place the slide at the lowest or highest point on the park grounds. As long as you mount the slide in the same way, ensuring that the launching platform is perfectly level in either case, your trip down the slide will be characterized by the same kinetic energy E. All that matters is the slope of the slide; the choice of the slide's overall height (phase) is arbitrary.

So, the system of slide and slider is invariant under a change of phase of the wave function $\psi(x)$. If you arbitrarily decide to change this phase by relocating the slide at the top of the hill, you don't change the dynamics of the system. The amount of sand that gets lodged beneath your skin as you grind to a halt in the dirt below the slide is the same in either case. The energy E that you attain is invariant with respect to the overall height of the slide.

But as we've discussed, this is a global, not a local, invariance. In moving the slide from the bottom to the top of the hill, we changed the phase, the height of the slide, by an identical amount at each point x along the length of the slide. If the top of the hill was 5 meters higher than the bottom, then we changed the height of the entire slide—the launching platform, the bottom of the slide, and every point in-between—by precisely that 5 meters. The transformation of the slide's height was global.

What if, instead, we had performed a local phase transformation, by changing the slide's height by an amount that varied from point to point along the slide? Say, for instance, that we raised the launching platform by the full 5 meters but raised the bottom of the slide by only 4 meters? The top of the slide goes up by some amount, but the bottom of the slide goes up by 1 meter less than that, the middle of the slide by ½ meter less than the top, and well, you get the picture. The amount by which we raised the slide, that is, changed the phase of the wave function $\psi(x)$, varies from point to point in the space of distance along the slide.

Thus, we would have increased the slope of the slide, and we'd come off the bottom of the slide with even more energy. The transformed height function (wave function) $\psi'(x)$ does not satisfy the same wave equation that had the original, smaller energy requirement for its solution. The wave equation and its wave function solution are not, in this analogy, invariant under local changes of phase.

By changing the height function (wave function) by an amount (phase) that varies from point to point in space (i.e., locally), we alter the steepness (derivative) of the height function. This change in the derivative of the wave function is what destroys the delicate balance between the wave function's derivative and the object's energy and makes the wave function incompatible with the notion of local phase invariance.

It's at this point that Yang and Mills enter the scene, reminding us that, given what we know about the nature of space-time, it makes no sense to restrict ourselves strictly to global transformations, for which the slope of the slide remains fixed when it's moved.

We know that phase is irrelevant. The choice of the phase of an object's—or a system of objects'—or the universe's wave function is arbitrary. On the one hand, the hallowed wave equation tells us that the arbitrary choice of phase must be consistent from point to point in space and time or all hell breaks loose. On the other hand, as Yang and Mills point out, this requirement appears to be incompatible with Einstein's well-supported notions of the nature of space and time. The contradiction is direct and demands a resolution, so something has to give. And, as has so often been the case in modern science, in reconciling these two apparently incompatible points of view, Yang and Mills were led down a path that profoundly and fundamentally augmented the way physicists view the operating principles of the natural world (note 8.2).

The Relevance of Irrelevance: The Gauge Principle

As we've seen with our slide analogy, the aspect of local changes of phase that wreaks havoc with our wave function and its solutions is their effect on the slope of the function—the rate of change, or derivative, of the function. Here, once more, is the wave equation for a particle free of the influence of external forces:

$$-\frac{h^2}{8\pi^2 m}\frac{d^2}{dx^2}\psi(x) = E\psi(x).$$

The term to the left of the equals sign, involving the derivative operator, causes the trouble, so this was the term that Yang and Mills worked on. They hoped to resolve the conflict by *forcing* the wave equation to be invariant under local changes of phase. What they did was to add, completely by fiat, another term to the left side of the wave equation, as follows.

Whenever the phase of the wave function changes locally (by an amount that varies from point to point in space), the result of the derivative (rate of change) operation changes, introducing some "mistake" that causes the new wave function to no longer satisfy the wave equation. The thing—the only thing—we require of this new term we're going to add is that it commit precisely the same mistake, *but with a negative sign,* so that when the mistake from the term with the derivative is added to the mistake from the new term, they exactly cancel out, and everything is OK. In other words, Yang and Mills cheated; when nobody was looking, they added another term that got rid of the problem caused by the effect of the rate of change operation.

Think of a high-school student who solves a word problem with an expression that gives an answer that is too high by, say, three. Having found this out by looking at the answer in the back of the book, the student slyly adds another term into his expression: a term that subtracts three from the answer. Now the answer works out, although the student has done something arbitrary and logically indefensible in making it do so. But at least the answer is correct.

Yang and Mills' new term did exactly that: It precisely canceled out the "mistake" introduced by the derivative operator when the wave function's phase is changed locally, so that the phase-changed wave function still satisfies the (now somewhat modified) wave equation. The only difference between the high-school student's cheating and that of Yang and Mills is that the latter's cheating term varies from point to point because the amount of the mistake introduced by the derivative varies from point to point, so the cheating Yang and Mills do to make up for that mistake must also vary from point to point. Explicitly, this new term took the form

$$qA(x)\psi(x),$$

where $\psi(x)$ is the value of the wave function at point x in space, and q is just some fixed number whose importance we'll come to understand in due time. The critical factor in this expression is our fudge-factor function $A(x)$, which can be any function whatsoever as long as it satisfies one criterion: whatever damage the rate-of-change term in the wave equation causes when the phase of the wave function $\psi(x)$ is changed locally, $A(x)$ must have just what it takes to undo that damage.

On the face of it, this really doesn't sound so promising. Like the student who cheats on his math test, we've added an arbitrary term into the wave equation that (still forgetting about q for now) involves some function $A(x)$ about which we can say very little. And what we can say about $A(x)$ is not very inspiring: If the phase of the wave function changes, the function $A(x)$, whatever it may be, must change in such a way as to exactly cancel out the problems introduced by the derivative operator in the wave equation. So, even though the wave function still solves the (somewhat modified) wave equation when we change the wave function's phase locally, it only does so because we've changed $A(x)$ and thus changed the wave equation itself in such a way so that everything works out. We're after *invariance*—having the physical description provided by the wave function be unchanged by local phase changes. This cheating, by construction, does allow the phase-changed wave function to solve a wave equation, but it's a different wave equation than that satisfied by the original wave function because of the cheating term we added in. If the phase-changed wave function satisfies a *different wave equation* than the original wave function, then it must be describing a *different physical situation* than the original wave function. This doesn't sound much like invariance to me.

So, this seems thoroughly useless. Until, that is, you recognize one critical point—the point that is the central epiphany of gauge theory. This point involves the almost magical appearance of the most interesting source of natural phenomena that we know of—the electromagnetic interaction.

Take any electromagnetic force field that you can imagine—the field inside a television set, the field generated by a powerful radar gun, or that of a quasar in the depths of intergalactic space; it doesn't matter. For this particular force field (i.e., for this particular array of magnitudes and directions of the electric and magnetic force at every point in space), write down a function, and call it $A(x)$, which represents the potential energy of that force field at every point x in space. Now, let *this* function $A(x)$ be the one we put into the wave equation above so that things would hang together when we changed the phase of the wave function locally. Now, go ahead and make a local change, one that varies in amount from point to point in space, in the phase of the wave function. Then make the corresponding change in

$A(x)$ that you need to so that the phase-changed wave function still satisfies the wave equation.

The startling fact exploited by Yang and Mills was that, when you make this change to $A(x)$, despite the fact that it's a completely different function, *it still describes precisely the same electromagnetic force field*. Sure, you've changed the form of $A(x)$ in the wave equation in some funny way to compensate for the local phase change in the wave function but only mathematically. Physically, in terms of observable qualities (again, the only ones that matter), you haven't changed the wave equation at all. After the local phase change and the corresponding change in $A(x)$, the wave function satisfies a completely equivalent wave equation — the wave function describes the same physical situation after the local change of phase. The wave equation, and its wave function solution, are unaffected by — are invariant with respect to — local phase transformations. Thus, we can alter the wave equation in a way that satisfies our strong conviction that quantum mechanics be invariant under *local* changes of the wave function's phase.

Now, this begs some further explanation, for how is it that we can change the potential energy function $A(x)$ associated with a force field without changing the force field itself? Consider the following analogy.

It takes some amount of energy to climb a twelve-foot-high staircase — to increase your gravitational potential energy by twelve feet. However, it doesn't make any difference whether the staircase is at sea level or the last flight of stairs at the top of Chicago's Sears Tower, it takes the same amount of energy to climb those twelve feet of stairs. Only *differences* in potential energy matter: you're free to put the staircase itself at any elevation you please. No matter where you put it, although the potential energy (elevation) of the staircase is different, the energy exerted against the force of gravity in ascending the stairs is the same. You can change the potential energy function by any amount you want (by moving the whole flight up stairs up or downhill), but the force of gravity you experience on the steps is always the same.

So, there's a freedom of choice. No matter what value you choose for the potential energy function at the bottom of the stairs, the force associated with that function, the force you fight in going up the stairs, is always the same. By choosing a different value for the potential energy at the bottom of the stairs (by putting it at the top of a tall building rather than at sea level), you can change the potential energy function for the stairs. But the physical aspects of the system, the things you can feel and measure, are unchanged. The force of gravity is the same regardless.

Similarly, to compensate for a local change of phase, you are asked to change $A(x)$ in some particular way, depending on just how the phase is changed throughout space and time. But, analogous to the staircase example, the corresponding changes that you are required to make to $A(x)$ are unphysical. They have no bearing on the actual forces represented by the function $A(x)$. There's a freedom of choice in the way that you express $A(x)$, and this is exactly the freedom of choice you need so that $A(x)$ represents the same array of electromagnetic forces regardless of how you are asked to change it to compensate any given local change of phase.

So, the changes you need to make to $A(x)$ to compensate for the "mistakes" that are introduced by local changes in the wave function's phase are just those that are allowed by the freedom of choice one has in specifying the electromagnetic potential energy function for the electromagnetic fields that happen to be around. This quantum-mechanical requirement of local phase invariance, miraculously, just matches the freedom of choice one has in specifying the electromagnetic potential energy function for those fields. And so everything works out.

This freedom of choice that you have in writing down the potential function for a given field—be it the gravitational, electromagnetic, or even strong or weak nuclear force—is known as *gauge* freedom. It's from this term that the expression *gauge theory* arises. The precise nature of the gauge freedom for the case of electromagnetism is more complex and difficult to describe than it is for the case of earthly gravity that we discussed above, but the principle is the same; there are many different ways to write the potential energy function $A(x)$ for whatever electric and magnetic force field may be present.

To recapitulate: to make the wave equation and its wave function solution unaltered by local changes of the wave function's phase, you need to introduce a new term into the wave equation. This term contains an unspecified function $A(x)$ that itself changes when the phase of the wave function changes. But, if this function is the potential function of the electromagnetic field, then $A(x)$ changes in a way that doesn't matter in any concrete way, the wave equation and its wave function solution are essentially unaltered, and everything works out; we achieve the local phase invariance we so desperately need.

Beyond requiring $A(x)$ to change in a specific way when we change the phase of the wave function, we have said nothing about what the function $A(x)$ is; it could really be anything. But that's great! There are a lot of possibilities for the exact way the electromagnetic field can array itself in space, so no matter what particular electromagnetic field we find around us, we can pick an $A(x)$ that describes it, plug it into the new term in the wave equation, and we've described the effect of that field on the object.

In fact, one possibility is that the field is zero everywhere, which is just fine. There's a set of potential energy functions $A(x)$—that change into one another appropriately when you change the phase of the wave function—that all are associated with zero electromagnetic field. If we chose such an $A(x)$ for the "cheating" (phase-change compensating) term in our wave equation, what we have is a wave equation that, physically, is no different than the free-particle equation with no cheating term, except now, the wave equation is indeed invariant with regard to local changes of phase.

Thus, in addition to an object under the influence of any conceivable array of electric and magnetic fields, we can also describe a free object, one not under the influence of any force whatsoever, in this way. In other words, to make the wave equation invariant under local changes of the wave function's phase, we don't need the electric field to be present; we simply need there to be *such a thing* as the electric field! This point has rather interesting implications, to which we'll return toward the end of this chapter.

One more point needs to be made. The quality of an object that determines how forcefully a given electromagnetic field pushes it around is its electric charge q. If $q = 0$, then the object will feel no influence from the fields, while the bigger q gets, the more the object will be influenced. But look again at the term $qA(x)\psi(x)$ that we included. The factor q sets just how much of a contribution the potential energy function $A(x)$ makes to the wave equation. If $q = 0$, then the $qA(x)\psi(x)$ term is 0, and the object feels no effect from the fields. As q gets larger, the object feels a proportionally larger effect as $qA(x)\psi(x)$ gets bigger. So, in Yang and Mills' cheating term, q is just what it needs to be: the value of the object's electric charge. If $q = 0$, then the object is electrically neutral, and so the term containing the effect of the fields (the $A(x)$ term) will have no effect.

In summary, instead of being a burden, the cheating function $A(x)$ that we introduced to enforce invariance under local phase change is seen to be a great boon. It is *not* an arbitrary function that, when tacked on to satisfy our philosophical whim, renders the resulting wave equation meaningless. Exactly the opposite is true. What we have found ourselves forced to include is something that we are exceedingly thankful for. The requirement that the wave equation be invariant under local changes of phase has necessitated the introduction of electromagnetism, the physical force that lies at the heart of virtually everything we experience in our daily lives. Furthermore, the apparently discouraging requirement that the function $A(x)$ change when we arbitrarily change the phase of the wave function solution to the wave equation is seen to correspond to nothing more than an inconsequential change in the gauge of the electromagnetic potential function, that is, in the particular way we happen to choose to represent the physical electromagnetic force field in terms of its potential energy.

Our adherence to the notion that the phase of the wave function $\psi(x)$ is irrelevant—of no consequence to the physical, observable information encoded in the wave function—has led to something very relevant indeed: a precise and comprehensive incorporation of the phenomenon of electromagnetism.

Quantum Electrodynamics—Again

The above discussion was presented in the context of the Schrödinger equation: the wave equation of conventional nonrelativistic quantum mechanics. However, if it is to be of any use to us in our attempts to describe the natural world as we now understand it, the discussion must be recast within our true framework for the description of the behavior of the fundamental elements of nature: the relativistic quantum field theory of chapter 4.

With the hard work done, the development and interpretation of the gauge principle for the nonrelativistic Schrödinger equation, all we must do is apply the same principle to the appropriate relativistic wave equation. Everything we just learned about the relation between local phase invariance and the introduction of the electromagnetic interaction, through the cheating potential function $A(x)$, applies verbatim.

Recalling what we learned about quantum field theory in chapter 4, we should recognize that what the factor $A(x)$ really represents is the quantum of the electromagnetic field—the photon. The inclusion of $A(x)$ thus incorporates, within the field-theoretical description of the particle's behavior, the possibility that the particle emits or absorbs a photon, that is, the possibility that the particle emits or absorbs a quantum of the electromagnetic field. The probability that the particle does so, at any given point in space and time, is proportional to the *coupling strength q*, which is simply the magnitude of the particle's electric charge.

In the language of chapter 4, what the inclusion of the $qA(x)\psi(x)$ term

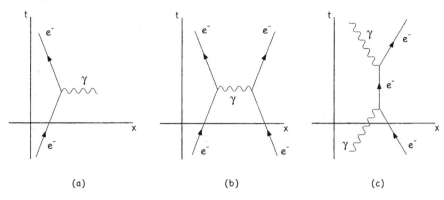

Fig. 8.2. A reminder of the nature of the electron-photon minimal interaction vertex (*a*) in quantum electrodynamics; *b* and *c* are the Feynman diagrams for electron-electron and photon-electron (Compton) scattering, two of many processes you can construct by cobbling together copies of the minimal interaction vertex of *a*.

has done is to introduce the *minimal interaction vertex* (fig. 8.2*a*) that connects charged particles (such as electrons) to the electromagnetic field quantum (the photon). As we saw in chapter 4, this then allows us to model all sorts of behavior associated with the electromagnetic interaction of particles and field quanta, such as the electrostatic repulsion of two electrons (fig. 8.2*b*), Compton scattering of light by electrons (fig. 8.2*c*), and so forth.

In fact, after including the function $A(x)$ into the relativistic wave equation, we find that the description of the minimal interaction vertex that arises is precisely the same, in every detail, as the description that was employed in the theory of quantum electrodynamics, the stunningly successful field-theoretical description of electromagnetic interactions that we discussed at length in chapter 4.

So, in view of the discussion in the beginning of this chapter, the inclusion of the theory of quantum electrodynamics into our relativistic description of the natural world is necessitated by a single, succinct requirement: that the quantum-mechanical description of an object's behavior, as represented by the wave equation and its wave function solution, be invariant under local changes of phase. If somehow the field of physics had blundered its way into the late 1950s without the development of quantum electrodynamics, this exploration of Yang and Mills', and their subsequent discovery of the gauge principle, could have allowed the entire theory of quantum electrodynamics to be developed essentially by a process of pure thought.

Electromagnetism, the prime mover behind so many phenomena that make the universe and life within it what it is, is understood, at the most fundamental level, to be a necessary consequence of the conviction that objects are described by wave functions and that this description should be invariant under local changes of that wave function's phase.

It is hard to imagine how a development in our understanding of the workings of nature could be more fundamental — and profound — than this. The ancients looked about them, beholding a world of such myriad complexity that it could only be attributed to the machinations of a large set of capricious and quarrelsome gods. Today, after millennia of lurching advancement, we now describe the full representation of the primary causative element in the world — electromagnetism — in terms of this single inexorable property of the wavelike nature of matter. All of electromagnetism derives necessarily, and with mathematical precision, from this single simple notion.

Finally, it should be noted that while Yang and Mills were uncomfortable with the requirement that the wave function be restricted to global (uniform) changes of phase, the locally invariant wave equations they developed were indeed invariant under global and local changes of phase. Any wave equation invariant under local changes of phase must also be invariant under global changes of phase. A local change of phase is one for which the amount by which you change the phase is a function of where you are in space-time; a global change of phase is, too. It's just that, for global changes of phase, that function happens to have the same value throughout space and time. So, global changes of phase are just special cases of local changes of phase, and if a wave equation is locally invariant, it will be globally invariant also.

Now, in chapter 7, we introduced Noether's theorem, which demonstrates that, for every global invariance of the wave equation, there's a corresponding physical quantity that is conserved. Our global phase invariance is such a case, and the conserved quantity is nothing other than the net electric charge q of the system. This, to the best of our current knowledge, is the physical origin of electric charge conservation — the requirement that, no matter what happens when bodies interact, the net electric charge must be the same before and after the interaction.

The notion of charge conservation is something that is probably, perhaps without your being conscious of it, fully ingrained in your own intuitive sense of natural law. In chapter 2, when we conducted the experiment on static electricity with the comb that became charged when we rubbed it through our hair, it was natural to assume that the electric charge that wound up on the comb was transferred from our hair to the comb. It likely would have struck you as strange if I had claimed that the process of rubbing a comb through one's hair magically creates an electric charge on the comb where before there had been none. This is because most of us intuitively expect that electric charge is indeed conserved, and the net charge (difference between the amounts of positive and negative charges) in any given system must be unchanged over time. So the charges on the comb had to come from somewhere (our hair) rather than just materialize out of nowhere.

Even in the more ethereal world of quantum field theory, with its virtual particles continuously popping in and out of existence in places one would think they would have no right to be, the virtual particles are always produced in pairs, with a net charge of zero, so even they don't violate the principle of electric charge conservation. To the best of our knowledge, the principle of charge conservation is completely inviolable.

Beyond Mere Phase 1: Lie Groups and Phase

Were the application of the gauge principle solely limited to this recasting of our understanding of the phenomenon of electromagnetism, the gauge principle would not be considered to be much more than an academic curiosity. As we saw in chapter 4, our preexisting theory of electromagnetism, quantum electrodynamics, does a more than satisfactory job of describing electromagnetic behavior. The theoretical developments that excite us most are not those that reinterpret past successes but, rather, the ones that propel our understanding forward into new realms.

With the application of the principles of mathematical abstraction of the sort discussed in chapter 6, we can understand how the gauge principle can be used to do just that and why it has become a tool in the development of essentially all new fundamental theories. It will be in this section that we revisit the mathematical constructs known as Lie groups and stitch them into our discussion of the gauge principle. The gauge principle, extended through the incorporation of Lie groups, will provide an essential step toward the incorporation of the weak nuclear force into our fundamental theory of causation. It will also provide the framework for our understanding of the strong nuclear force which, according to our current understanding, is provided by the pure gauge theory known as *quantum chromodynamics.*

Recall that ample experimental evidence on the behavior of atomic nuclei led to the realization that the strength of the force between any two nuclear constituents is independent of the identity of the participating particles. Thus, two protons interact in precisely the same way as two neutrons, or as a proton and a neutron. This fact led Werner Heisenberg to suggest that the neutron and proton are one and the same particle, the nucleon, that possesses an abstract property he called *isospin*. Just as the true spin of a particle can point either up or down (or perhaps, more sensibly stated, forward or backward) along any given direction in space, what we call a proton is really just a nucleon with its isospin pointing up along some chosen axis in the internal isospin space (isospin up), while the neutron has isospin down.

This is illustrated in figure 8.3; the double arrow represents the orientation of the given nucleon's isospin in isospin space. If the angle χ is 0 degrees, the nucleon is a proton; if χ is 180 degrees, it's a neutron. If it's somewhere in-between, as shown in the figure, it's a quantum-mechanical combination of a proton and neutron, with some corresponding probability that it will show up as either when a measurement is made on the system.

So, if in the course of some calculation, you want to change protons into neutrons or vice versa, all you need to do is rotate the given system by 180 degrees in isospin space: up goes to down and down to up, and you've done it. Since the strong nuclear force is unchanged (invariant) when protons and neutrons are interchanged with each other, we thus say that the strong force is "invariant with respect to rotations in isospin space." This notion leads to numerous experimental predictions, and in the vast body of experimental results, no data have ever been unearthed that disagree with any of these predictions.

If all this sounds familiar by now, it's because the preceding discussion was meant to serve as a reminder only; nothing new was said in the last five or so paragraphs that hasn't been introduced before. Now, with this in mind, let's consider one or two things that we might require of a quantum-mechanical description of nuclear behavior. Since, in the context of the strong nuclear force (the force that binds neutrons and protons together into nuclei), the proton and neutron are really just different manifestations of the nucleon, then our description will be one of how *nucleons*, and not protons and neutrons, behave under the influence of the strong force (note 8.3).

So, let $\psi(x)$ be the combined wave function of all the nucleons in some

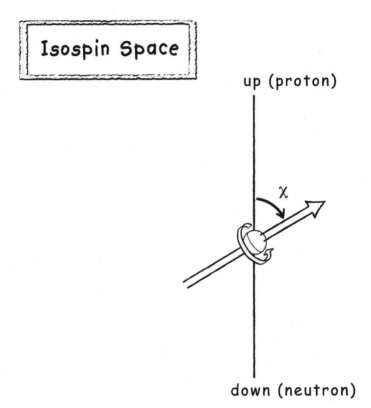

up (proton)

χ

down (neutron)

Fig. 8.3. The orientation of this nucleon's isospin in isospin space is neither purely up nor purely down, so it is a quantum-mechanical mixture of "protonness" and "neutronness." If detected, it will manifest itself as either a proton or a neutron, but until that measurement is done, it exists in a mixed state, with a certain probability of showing up as either. The smaller the angle χ, the more likely it is that it will be detected as a proton.

nucleus that's under study, moving around under their mutual strong-force attraction. Since this is quantum mechanics, this wave function and the wave equation it satisfies had better be invariant under changes in the phase of the wave function; this is always true for quantum-mechanical wave functions, regardless of the context. In addition, since these are nucleons moving under the influence of the strong force, the wave function must be, at the same time, invariant under rotations in isospin space that interchange the proton and neutron realizations of the nucleon among one another. What we ask the system to be invariant under is a generalized sort of phase invariance in which we can change either or both of the phase angle θ of the wave function (see fig. 8.1) and the angle χ that the nucleon makes with the proton/neutron axis in isospin space (see fig. 8.3).

Thus, to develop a quantum-mechanical description of nuclear behavior, we need to generalize, or *abstract*, the notion of quantum-mechanical phase. We must incorporate the additional possibility any given phase change includes, simultaneously with changes in the ordinary phase of the wave function, rotations in isospin space that transform protons into neutrons. For a system of nucleons within a nucleus, the quantum-mechanical description provided by the wave equation and its wave function solution $\psi(x)$ should be just as indifferent to this sort of a change as it is to ordinary changes of phase in the "wave" of the wave function.

To understand precisely what the nature of this abstraction is, we need to recall some things from the discussion of Lie groups from chapter 6. Recall that Lie groups are continuous groups, such as the real numbers on the number line, the elements of which form a continuum with no holes, and whose elements can be generated by the combination (according to the group's rule of combination) of an appropriate number of basic generating elements. The quintessential example of a Lie group that we held forth was the group R(3) of rotations in three-dimensional space. Any given orientation of an object in three-dimensional space can be achieved by rotations of appropriate magnitudes about the three axes $(x, y,$ and $z)$ of a three-dimensional coordinate system. Thus, the group R(3) has three generating elements, or generators, which are the exercise rotations about the x-, y-, and z-axes that we introduced in chapter 6.

In the section "Into the Continuum" of chapter 6, we introduced the Lie group U(1), which can be represented by the set of complex numbers $z = a+b \cdot i$ of size 1, that is, that satisfy the condition that $a^2 + b^2 = 1$. The operation associated with this group (remember that a mathematical group is defined in terms of both a set of objects and an operation governing their combination) is complex number multiplication.

Once again, take a look at figure 8.1. The arrow you see in this figure points to the location on the complex plain that represents the complex number z. If we operate on z with some member of the group U(1) (that is, multiply z by a complex number of length 1), we get a new number (call it z') of z (note 8.4). However, the orientation of the arrow pointing to the new z' will change, which means that we've changed the phase of the wave function. Thus, any member of the Lie group U(1), when used to operate on the wave function at a given point in space-time, just acts to change the phase of the wave function at that point. Different elements of the group U(1) move the angle through different amounts; in fact, we can characterize each of the elements of U(1) by how much angle it moves the arrow through or, alternatively, by how much it changes the phase angle θ of any complex number z. *The Lie group U(1) is just an abstract representation of the group of possible changes of phase of the wave function* at any given point in space-time!

For this reason, physicists refer to invariance under changes in quantum-mechanical phase as "invariance under U(1)," or, more succinctly, "U(1) invariance." And so quantum electrodynamics, formulated in terms of a gauge principle based on requiring the relativistic wave equation to be invariant under local changes of phase, is referred to as a "U(1) gauge theory." As we've seen, this succinct phrase encapsulates an appreciable portion of the natural laws that shape our day-to-day experience.

We should pause here for a moment and reflect on what we have just done. Earlier in this chapter, our discussion of the gauge principle allowed us to appreciate the profound relation between quantum-mechanical phase invariance and the origin and specification of the theory of electromagnetism. What we have done in the paragraphs just above is to step back and examine the nature of phase change in terms of its mathematical properties, not in the sense of the arithmetic of manipulating complex numbers but, rather, in the overall structural terms of the abstract mathematician. In discovering that this structure is that of a Lie group, we have uncovered an essential link between the world of abstract mathematical construct and the underlying structure of natural law.

Now, back to our attempt to develop a quantum-mechanical description of nucleons under the influence of the strong nuclear force. Again, the set of phase changes under which our nucleon wave function must be invariant is more general than the set of phase changes under which our electron wave function was required to be invariant. This is because, in addition to being invariant under changes in the conventional phase of the wave function, as described above, the wave function has to be simultaneously invariant under rotations in isospin space that swap protons and neutrons with each other. The critical step before us is to determine the mathematical structure of this more inclusive set of phase changes.

The beauty of the abstraction we made above for the electromagnetic force case of simple U(1) phase invariance is that we managed to connect the notion of phase change to the mathematical construct of Lie groups. So, we might suspect that the more inclusive set of phase changes for this more complicated case, involving both conventional phase changes as well as isospin rotations, will correspond to the members of some other, more complicated Lie group. What we need to do now is to figure out which Lie group it is.

Let's make a two-dimensional graph in another abstract mathematical space. Like the bank balance–versus–time space of the previous chapter, this space has no pretense whatsoever of being a true physical space. This space is *not* isospin space; it's just a bookkeeping space whose sole purpose is to help us keep track of whether a nucleon is a proton or a neutron.

This graph is shown in figure 8.4. The horizontal (x)-axis represents "protonness" and the vertical (y)-axis represents "neutronness." If a nucleon's wave function represents a proton, we draw an arrow lying along the x-axis; don't worry about whether it points to the right or the left; that's a subtlety that need not concern us. If the arrow lies along the x-axis, the wave function represents a proton.

Similarly, if the nucleon's wave function represents a neutron, we draw an arrow along the y-axis. Or, if the wave function is some quantum-mechanical combination of proton and neutron (i.e., a measurement of the properties of the nucleon would have some probability of revealing the nucleon as proton, but also some complementary probability of showing it to be a neutron), then the arrow falls at some angle between the x- and y-axes, which is the case shown in figure 8.4.

When you look at it from this point of view, what you see is that the rotation of the nucleon's isospin in isospin space, which swaps protons with neutrons, rotates the value of the nucleon's wave function around in the abstract space of this graph. If you start with a proton (which would be represented by orienting the arrow along the x-axis in fig. 8.4) and then go into isospin space and change it into a neutron (arrow along the y-axis in fig. 8.4), you would represent this operation in the abstract space of figure 8.4 by rotating the arrow by 90 degrees around its base.

We learn the following: the set of possible rotations in the isospin space of figure 8.3 is nothing more than the set of possible rotations of the nucleon's wave function in the two-dimensional space of protonness and neutronness of figure 8.4. However, the space of figure 8.4 is not a two-dimensional space of ordinary real numbers. Because wave functions are complex, this is a space of rotations in two complex dimensions. In isospin

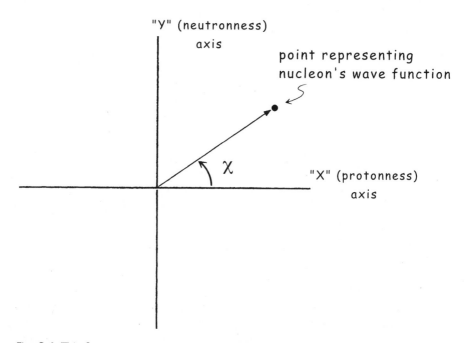

Fig. 8.4. This figure represents the same nucleon as that of figure 8.3, but in the abstract mathematical space of protonness and neutronness. This is a different, but equivalent, way to keep track of neutrons and protons that lends itself more readily to the mathematics that we take advantage of in particle physics.

space, we can rotate the wave function between protonness and neutronness, and we then keep track of our choice by pointing the arrow in figure 8.4 in the correct direction. But once we've done that, the values of the neutron and proton components of its wave function are two independent complex numbers, and we still have complete freedom to choose their phase. So, the wave function of a nucleon should be invariant under rotations in isospin space but also, independently and simultaneously, invariant with respect to changes in the complex-valued wave function's phase. But this is precisely what we mean by the expression "rotations in two complex dimensions." And, from chapter 6, we know that the set of rotations in two complex dimensions comprises the Lie group SU(2). The wave function of a nucleon under the influence of the strong force is invariant under isospin rotations and phase changes, the set of which has the mathematical structure of the Lie group SU(2).

To reiterate: We saw that the set of possible phase changes of an electron's wave function (no isospin) is equivalent to the Lie group U(1), which, as we see in figure 8.1, is really just the set of rotations in one complex dimension. When we instead consider a nucleon (with isospin), we need to consider two complex dimensions: one for the protonness and one for the neutronness of the wave function. Any such rotation, or transformation, will yield a new wave function that, as far as the strong nuclear force is concerned, is no different from the old one. The strong nuclear force is invariant under the group SU(2) of phase and isospin transformations.

Beyond Mere Phase 2: Lie Groups and the Gauge Principle

The wave function solutions of the wave equation that describe a nucleon must be invariant with respect to a more inclusive set of phase changes. This set is the group SU(2) of rotations in the complex two-dimensional space of isospin.

Fine. So, consider the value of a nucleon's wave function at some point in space and time. Pick an element of the Lie group SU(2), that is, an amount by which to rotate (transform) the nucleon from protonness to neutronness, as in figure 8.4, and amounts by which you want to change the phase of its neutronness and protonness, as in figure 8.1. Now, go to some other point in space and time. Again, pick an amount by which you want to rotate its isospin and amounts by which you wish to change its phase.

Since we subscribe to the philosophy of Yang and Mills, even if we choose different elements of SU(2) at the two separate space-time points, the transformed wave function should still describe the same physical state of the nucleon. This is what we mean by local invariance under SU(2), according to Yang and Mills, the only kind of invariance that makes sense to require.

As before, let's consider the wave equation for a free nucleon n, that is, a nucleon not under the influence of any external force, so that the potential energy function $V(x)$ is zero everywhere. This free-particle wave equation is no different than the one we wrote down for the electron, but here it is again:

$$-\frac{h^2}{8\pi^2 m}\frac{d^2}{dx^2}\psi_n(x) = E\psi_n(x).$$

The only difference is that, while the electron's wave function was asked to be invariant under changes in quantum-mechanical phase, a set of changes characterized by the Lie group U(1), the nucleon's wave function $\psi_n(x)$ (subscript n for "nucleon") is expected to be invariant under simultaneous changes of phase *and* rotations from protonness to neutronness in isospin space, a more complicated set of changes characterized by the Lie group SU(2).

Just as for the electron, we can try to transform the phase of $\psi_n(x)$ by picking some element or other from the Lie group SU(2) and performing the associated phase changes and isospin rotation. And, as before, if the amount by which we change phase and rotate, if the particular SU(2) element we choose changes from point to point in space, then the result of the rate-of-change calculation (derivative operation) in the wave equation changes. Just as before, this upsets the delicate balance between the left- and right-hand terms in the wave equation. This free-nucleon wave equation and its wave function solutions are not invariant under local phase/isospin transformations from the group SU(2).

We need to fix up the wave equation, and as before, we'll do this by adding in cheating terms whose sole apparent purpose is to correct the mistake made by the derivative operator when we make phase/isospin-orientation changes to the wave function that vary from point to point in space-time. To understand what happens when we do this, we need to remember a little about what we learned in chapter 6 about Lie groups and, in particular, about the Lie group SU(2).

In the section "The Lie Group SU(2)" of chapter 6, we discussed how the Lie group SU(2) is quite similar to the much more intuitively accessible group R(3) of rotations in three dimensions. Except for the fact that you have to rotate through 720 degrees to get back to where you started (recall the exercise of placing a book on your palm and rotating it above and below your elbow), SU(2) is identical in every way to R(3).

In particular, both groups have three generators, that is, three basic elements from which any other element in the group can be attained. In the case of R(3), these generators are just the exercise rotations about the three mutually perpendicular axes x, y, and z of three-dimensional space. It's fairly easy to convince yourself that, as required of Lie group generators, any rotation in three-dimensional space can be built up by the application of the right amount of each of these three basic rotations. The fact that SU(2), just like R(3), has exactly three generators—no more and no less—will be of immediate concern to us.

But, even more, the three generators of R(3) and SU(2) have the same

Lie algebra. Recall that we were able to demonstrate that the group of rotations in three dimensions is not Abelian; somewhat surprisingly, the order in which you apply the generating rotations does matter. If you rotate a box first about the x-axis and then about the z-axis, the box ends up with a different overall orientation than if you rotate first about the z- and then the x-axis. The precise difference between these two oppositely ordered combinations of generating rotations, for each possible combination of two of the three generators, specifies the Lie algebra of the group. Therefore, the statement that the Lie algebras of R(3) and SU(2) are identical means not only that they are both non-Abelian (that the ordering of the elements, or rotations, must be taken into account when combining them), but also that the difference between the application of the generators in opposite order is precisely the same for both groups.

Because the number of generators and the Lie algebra of those generators is the same, we are free to think, when it behooves us, of SU(2) and R(3) being one and the same Lie group. This should help us to maintain an intuitive feel for the group SU(2), since we already have a pretty good notion of what rotations in three dimensions are all about.

So now, it's time to go back to the free-nucleon wave equation and add in the cheating term with the intent of compensating for the problems caused by the derivative operator when we make SU(2) phase/isospin-orientation transformations that vary from point to point in space-time. When we considered the local phase, or U(1), invariance of the electron's wave function, we added in a cheating term of the form $qA(x)\psi(x)$, eventually discovering that q was just the value of the electron's charge and $A(x)$ the potential energy function of the electromagnetic field. Here, the cheating term we'll introduce will be $gW(x)\psi_n(x)$. This is exactly the same sort of term; we've just changed the names of the symbols because now we're considering a nucleon, with both phase and isospin orientation to worry about, rather than an electron, for which phase was the only invariance we had to pay attention to.

But things are indeed more complicated for our nucleon wave function, and this time, we don't get away with our cheating trick. With both phase and isospin orientation to worry about, the cheating function $W(x)$ cannot compensate for the full gamut of damage done by the derivative operation when we both change the phase and rotate the isospin of the nucleon wave function $\psi_n(x)$.

Say, for example, that our nucleon just happens to be everywhere in space-time, a proton with a quantum-mechanical phase of 0 degrees on the diagram of figure 8.1. We know that this nucleon's wave function $\psi_n(x)$ must be invariant under local changes of phase, so we change the phase of $\psi_n(x)$ by an amount that changes from point to point in space. As always, this changes the result of the derivative operation in the wave equation, but now our cheating term $gW(x)\psi_n(x)$ is there to compensate. Just as before, $W(x)$ changes (simply because we make it change) in just such a way as to cancel the error from the derivative operation. So far, so good.

But now, we also need to worry about *local rotations in isospin space*, by which we just mean changes that transform our proton into a nucleon with some neutronness in it and for which the amount of neutronness we end up with varies from point to point. And it's here that we get stuck because no matter how hard we might try, we simply cannot come up with a single function $W(x)$ that can compensate for both phase changes and isospin rotations. If $W(x)$ is engineered to compensate for local changes of our proton's wave function, then it can't compensate for the problems caused by local changes in the neutronness of its wave function. If, however, we require that $W(x)$ compensate for local isospin rotations, then it can't compensate for local phase changes.

So what do we do? Simple. We cheat more! If we can't do it with one function $W(x)$, then we try doing it with *two* functions $W_1(x)$ and $W_2(x)$. If that doesn't work, we try three functions, then four, and so on.

When we do this, we discover something very interesting. The number of $W(x)$ functions that we need to compensate for all possible phase/isospin changes to the wave function is three—*the number of generators of the Lie group SU(2)*. In fact, this is a general rule for wave equations for which we insist on these more complicated sorts of invariance. If the invariance we require has the mathematical structure of Lie group X, then the number of cheating terms that we need to introduce into the wave equation (to establish its invariance with respect to local changes of phase, isospin orientation, or whatever sorts of invariance we are requiring) is always equal to the number of generators of the Lie group X.

Note, for example, that we recognized (in retrospect) that the set of ordinary phase changes we considered for the electron's wave function has the properties of the group U(1). But U(1) has but one generator; you can generate any phase change in figure 8.1 by the repeated application of a single basic transformation: the change of the phase angle θ by some small amount. So, for U(1) invariance, we need only one cheating function, the $A(x)$ of the electromagnetic interaction.

But for the case of SU(2) invariance, instead of one cheating function $A(x)$, we use three cheating functions: $W_1(x)$, $W_2(x)$, and $W_3(x)$, one for each of the three generators of SU(2). Just as we included the cheating function in the electron's wave equation by adding in the extra term $qA(x)\psi(x)$, we now must add three extra terms into the nucleon's wave equation:

$$gW_1(x)\psi_n(x) + gW_2(x)\psi_n(x) + gW_3(x)\psi_n(x).$$

Just as we came to understand that $A(x)$ represented the potential energy function associated with the electron's electromagnetic interaction, the three cheating functions $W(x)$ together represent the potential associated with some interaction acting on the nucleon.

What interaction? Recall that we are exploring the consequences of requiring local invariance under rotations in isospin space, that is, in the swapping of protons with neutrons. But invariance under such transformations is a property associated with the strong nuclear interaction. So that must be the answer. We have taken a local symmetry associated with the strong nuclear force—invariance associated with changes of phase and rotations in isospin space—and required this of the free-nucleon wave equation. This necessitates the inclusion of new interaction terms into the wave equation—terms thus associated with the strong nuclear force. Without knowing anything more about the strong force other than its indifference to whether the nucleon it's influencing is a neutron or a proton, we have apparently derived a complete theory of its influence in nature. Such is the power of the gauge principle.

However, to avoid unnecessary confusion later on in the chapter, I need to state clearly and emphatically that this is in fact *not* a workable theory of the strong nuclear interaction. The reason for this is that, unlike electrons, nucleons are not fundamental particles. They are composed of quarks that, as we have discussed, are the truly fundamental constituents of matter that partake in the strong interaction. To derive a workable theory of the strong interaction, then, we need to think hard about the invariance of quark, not nucleon, wave functions. We'll come back to this in the section "Quantum Chromodynamics" where we'll introduce our true theory of the strong nuclear interaction.

In fact, what we're really learning about here is the *weak* nuclear interaction! As we'll discuss shortly, there is an SU(2) invariance exhibited by quarks and leptons, the truly fundamental constituents, with respect to the weak interaction. If you've seen one SU(2) theory, you've seen them all, so what we've just learned applies directly to the theory of the weak nuclear interaction that we're about to develop. Unlike the theory of the strong interaction that we just tried to put together, this theory of the weak interaction will not be a sham. It will provide an essential step toward the development of the full electroweak theory of the Standard Model.

True Gauge: The Theory of the Weak Interaction

Take a glance once again at the table of fundamental matter particles (see table 5.1). Recall that, in this table, the fundamental constituents of matter are arrayed in conspicuous patterns, known as *generational doublets*—pairs of associated particles, such as the up and down quark or the electron and its neutrino, that are related via the workings of the weak interaction. In particular, a quark or lepton partaking in the charged weak interaction (through the exchange of a W^+ or W^- boson) will be exchanged for its generational-doublet partner (e.g., look again at the Feynman diagram of fig. 5.2).

Now, just as the strong-force interaction between two nucleons is unchanged when neutrons and protons are swapped, the weak nuclear force is similarly unchanged when a lower member of one of the doublets (*d* quark, *b* quark, electron, muon, etc.) is swapped for its corresponding upper doublet partner (*u* quark, *t* quark, electron neutrino v_e, muon neutrino v_μ, etc.) or vice versa. The experimental study of the weak nuclear interaction being a bit of a challenge, the demonstration of this invariance of the weak nuclear force was somewhat indirect and retrospective, but it stands nonetheless, and the pairwise generational pattern exhibited by the fundamental particles provides one of its strongest pieces of evidence.

Thus, analogously to rotations in "conventional" isospin space that transform protons and neutrons into each other, leaving the strong-force interaction properties unchanged, we have a new symmetry space: that of *weak isospin*. Just as the strong nucleon binding force is invariant under proton/neutron (isospin-up/isospin-down) swapping rotations in conventional isospin space, the weak nuclear force is invariant under upper/lower doublet-member-swapping rotations in weak isospin space.

It must be emphasized that, other than their connection through this analogy, the respective internal symmetry spaces of strong-force isospin and weak isospin are wholly different notions. What we are introducing here is an entirely new internal abstract space, associated with the fundamental nature of the weak (as opposed to strong) nuclear force. However, while just an analogy, the strong-isospin/weak-isospin analogy is nonetheless a rigorous and complete one, so our discussion of local strong-isospin invariance from above is directly applicable to the discussion of local weak-isospin invariance that follows. For the case of weak-isospin invariance, however, we are considering an invariance associated with the behavior of truly fundamental particles—quarks and leptons—under the influence of the weak force, so what will emerge is, unlike our physically inaccurate model of the strong interaction between nucleons, a rigorous candidate for the theory of the weak nuclear interaction. It is a theory we'll need to reckon with.

That we can move so easily from the discussion of strong isospin to weak isospin and the theory of the weak interaction is a tribute to the power of gauge theory and, in particular, to its basis in the abstract mathematical theory of Lie groups. If you were to reread the previous section, you would confirm that the specifics of our "sham" theory of the strong nuclear interaction, the introduction of the three cheating functions $W_1(x)$, $W_2(x)$, and $W_3(x)$, were dictated solely by the identity of the underlying group SU(2) of symmetry transformations. As we further flesh out the properties of this theory (now in the guise of the theory of the weak interaction), we will continue to rely solely on the identity of this underlying symmetry group for guidance. Almost all of our work in constructing this theory will take place at this abstract level, for which we could as easily be talking about our "toy" theory of the strong interaction as our true theory of the weak interaction. Only at the very end, when we specify the nature of the charge that the interaction concerns itself with being that of the weak isospin (recall the discussion of the weak nuclear force charge in chapter 2), will the theory become specifically that of the weak nuclear interaction.

The first step in this development is to convince ourselves that SU(2) is the appropriate group of underlying symmetry transformations. Let $\psi_d(x)$ (subscript d for "doublet") represent the wave function of a particle from one of the doublets of table 5.1; just to be specific, let's choose the muon and muon-neutrino doublet. Just as, for the case of strong isospin, the nucleon wave function $\psi_n(x)$ could represent a proton, a neutron, or a mixture thereof, the doublet wave function $\psi_d(x)$ can represent a muon, a muon neutrino, or a state of mixed "muonness" and "neutrinoness." Our set of symmetry transformations must allow for rotations between muonness and neutrinoness in the two-dimensional weak isospin space analogous to that between protonness and neutronness in the conventional (strong-force) isospin space of figure 8.4. And, as always, the set must also include changes to the quantum-mechanical phase of figure 8.1, meaning that the muonness and neutrinoness dimensions in figure 8.4 are complex. As in the case of strong isospin invariance, the symmetry group has the structure of the set of rotations in two complex dimensions; it is indeed SU(2).

So now, when we ask for local invariance under changes in *weak* isospin orientation and quantum-mechanical phase, we have the identical set of problems to cure as for the case of strong-force nuclear isospin. Our rule — add one cheating term for each generator of the underlying Lie group of symmetry transformations — requires only that we identify the Lie group at play. Since that group is the same for weak isospin as it is for nuclear isospin, we need to add the same three cheating terms, one for each of the three generators of SU(2):

$$gW_1(x)\psi_d(x) + gW_2(x)\psi_d(x) + gW_3(x)\psi_d(x),$$

as we did for the case of nuclear isospin, thereby introducing the three potential energy functions $W_1(x)$, $W_2(x)$, and $W_3(x)$. These three terms represent the potential energy of the interaction of the doublet particle (muon, muon neutrino, or combination thereof) with the weak interaction force field. In other words, the wave equation (or at least its relativistic counterpart) modified in this way represents our theory of the weak nuclear interaction.

One important point is that, for the mathematics of the cheating functions to work out, all the g's must be the same. If we were to have different g's for each of the cheating function terms, we would find that the wave equation and its wave function solutions, even with the cheating terms, would not be invariant under local changes of SU(2) "phase" (real phase plus orientation in isospin space).

Identifying the physical meaning of this factor g is the next thing we need to do in developing our theory of the weak interaction. What is this g?

In the case of our discussion of quantum electrodynamics, we discussed symmetry under the Lie group U(1), changes in the phase alone of the wave function, and introduced the cheating function through the term $qA(x)\psi(x)$.

In this case, the number q turned out to be the electric charge, and it was the conserved quantity associated with invariance under changes of the phase of the wave function that is dictated by Noether's theorem. Now, for the weak nuclear interaction, invariance under isospin requires, again by Noether's theorem, that there must be an associated conserved quantity. Following Noether's prescription, we find that the conserved quantity is just weak isospin itself; the amount of spin that the particle possesses in weak-isospin space and which we are dialing around as we rotate the orientation of the particle's weak isospin in that space.

So, in this theory, the number g reflects the amount of weak isospin, the amount of the object's associated weak-interaction charge, possessed by the object. The greater the particle's weak isospin, the bigger g will be, leading to a correspondingly larger effect of the three weak-interaction cheating terms $gW(x)\psi_d(x)$. The greater the object's weak-interaction charge, the greater the effect of the weak interaction on the object; that makes sense.

In chapter 2, we claimed that to characterize a force, you need both to identify the nature of the charge associated with the force, as well as the overall strength of the force. We've done the first of these; another way to look at the second is to ask how much weak-isospin charge g is carried by a matter particle. As we just discussed, all the g's are the same, so it doesn't matter whether we ask this for quarks, charged leptons, or neutrinos. The answer we get will be the same.

There's only one way to answer this: empirically (deducing it, say, from the value $\tau_\mu = 2.19703 \pm 0.00004$ microseconds, the measured lifetime of the weak-force-decaying muon, which tells you precisely how readily the muon and its decay products partake in the weak interaction). With the value of g determined, our SU(2) gauge theory of the weak interaction is complete.

The measurement of the overall strength of the interaction, along with the identification of the underlying symmetry group, SU(2) in this case, are the only experimental inputs necessary to develop the complete gauge theory of a force of nature (note 8.5). Again, we underscore the power and economy of gauge theory.

To implement our development of the SU(2) gauge theory of the weak nuclear interaction, we need to figure out how this discussion plays out in the context of the quantum field theory of chapter 4, rather than that of the conventional quantum-mechanical wave equation we've considered so far.

In our discussion of the simple U(1) phase of the electron's wave function, the cheating function $A(x)$ that we introduced turned out, in the context of quantum field theory, to be a representation of the field of the photon, the quantum of the electromagnetic field. Thus, all we need to do is to recognize that the $W(x)$'s—$W_1(x)$, $W_2(x)$, and $W_3(x)$—are the potentials associated with the quanta, the exchange particles, of the weak nuclear force. It's these *three* field quanta that are tossed around by particles as they exert themselves through the weak nuclear interaction. In the language of chapter 4, each term $gW_i(x)\psi_d(x)$ (where the letter i can be either 1, 2, or 3) represents a *minimal interaction vertex* of the quantum field theory of the weak interaction. The factor "$\psi_d(x)$" is the wave function of the particle being influenced by the interaction, the factor "$W_i(x)$" represents the wave function of the field quantum that's responsible for the influence and the factor g sets the strength of that influence.

In other words, when we construct the Feynman diagrams that represent the exchange of a quantum of the weak force mediating the interaction of matter particles, we have not one but three minimal interaction vertices in our basket of building blocks. These three vertices are shown in figure 8.5; each consists of matter particles emitting one of the three field quanta. These diagrams are presented in terms of the second-generation lepton doublet of table 5.1, which contains the muon and its neutrino, but this choice was arbitrary. We could have chosen any of the six quark or lepton doublets.

There's a bit of formal quantum field theory necessary to get from the arguments presented above to the exact form of vertices in figure 8.5, but we can see that the emission of a W_1 quantum is associated with a transition

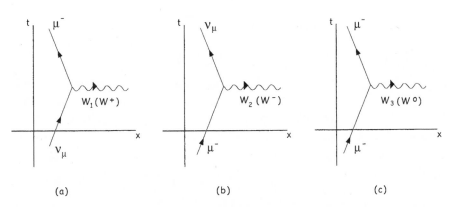

Fig. 8.5. The minimal interaction vertices of the three field quanta (W_1, W_2, and W_3, or if you prefer, W^+, W^-, and W^0) of the weak nuclear force's SU(2) interaction.

from an upper member of the doublet (weak-isospin-up v_μ) to a lower member (weak-isospin-down μ). Likewise, the emission of a W_2 induces a transition from a lower doublet member to an upper doublet member, while for the W_3 diagram the doublet member's identity is preserved.

You'll notice that we also have labeled the W_1 as the W^+, the W_2 as the W^-, and the W_3 as the W^0. The designation "+," "−," "0" has to do with the electric charge of the W field quanta in terms of the standard benchmark of the proton's charge. If you remember that the upper doublet particle always has an electric charge one greater than the lower doublet particle, the principle of electric charge conservation demands that the charges of the W's be such.

Now, the fact that the W_1 and W_2 minimal interaction vertices are associated with a transition between members of the participating generational doublet is a consequence of the fact that, in this SU(2) theory of the weak interaction, the W's themselves carry weak isospin, the very charge associated with the interaction they mediate. This is a very intriguing notion, having everything to do with the abstract mathematical structure of the underlying SU(2) symmetry group, and it leads to some very interesting behavior; we'll come to this in the section "Commute Issues."

For the sake of clarity, we should use these minimal interaction vertices to construct the Feynman diagrams for a couple of weak-interaction processes. To do this, it will help to recall one of the rules from the quantum field theory of chapter 4. This rule says that we can manipulate a minimal interaction vertex by swapping a particle for its corresponding antiparticle, as long as we change the trajectory of the particle so that the antiparticle we swap it for is represented by the corresponding particle traveling backward in time.

Figure 8.6 shows the minimal interaction vertices we get when we take the three vertices of figure 8.5 and swap the incoming particle for its antiparticle; recall that the antiparticle partner of the μ^- is written as μ^+, while the antiparticle of a neutrino is written as $\bar{v}$. Recall also that the arrows in the diagram point the wrong way for the antimuon and antineutrino since in the field theory antiparticles are represented as particles traveling backward in time. In other words, in the laboratory, the antiparticles of figure 8.6 are really flying away, forward in time, from the W field quantum they emit at the vertex (you observe the antiparticle traveling forward in time, not the particle traveling backward in time).

We can also represent the vertices of figure 8.6 in terms of absorption of

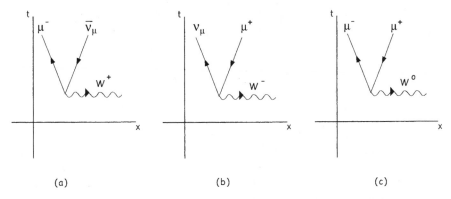

(a)   (b)   (c)

Fig. 8.6. The minimal interaction vertices of figure 8.5, with the incoming particle swapped for its antiparticle.

the W's (rather than emission), as shown in figure 8.7. In this case, we've just changed the direction of time for the W's, swapping each W for its antiparticle partner (remember from chapter 5 that the W^+ and W^- are a particle/antiparticle pair, while the neutral weak-interaction quantum is its own antiparticle), and so now the W's are absorbed at the minimal interaction vertex rather than emitted. You can confirm that the charge of the absorbed W boson is appropriately conserved by the leptons that fly away from the vertex.

Let's construct a diagram with, say, the vertices of figure 8.5b and figure 8.7a. In figure 8.5b, a muon emits a W^-, turning into a muon neutrino. Then, in figure 8.7a, the W^- decays into another muon and a muon antineutrino. This Feynman diagram of this two-vertex process is shown in figure 8.8a. The reason for the big "NO!" below figure 8.8a is that the mass-energy of what you end up with (a muon and a muon neutrino and antineutrino) is greater than what you started with (just a muon), so it's forbidden by the inviolable principle of energy conservation.

However, we could just as well have drawn the vertex of figure 8.7a with particles from the first lepton generation: that of the electron and electron neutrino. If we had, then the resultant Feynman diagram would be that of figure 8.8b, which has less mass-energy after the interaction than before — the difference being made up in the kinetic energy of the muon neutrino, electron, and electron antineutrino that you end up with. This process can and will happen. If you look at the diagram again, you will recognize this process as that of muon decay: $\mu^- \rightarrow e^- v_\mu \bar{v}_e$.

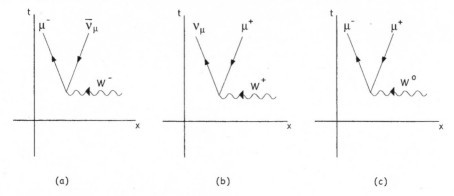

Fig. 8.7. The minimal interaction vertices of figures 8.5 and 8.6, with the additional swapping of the W field quanta for their antiparticles.

Two more interesting diagrams, out of the many that we could form from the variants of the three weak-interaction vertices, are those of figure 8.9, which we introduced in a different context in chapter 5. The first is a process by which a neutrino interacts with a down quark (in the nucleus of some material that lies in the neutrino's way) through the exchange of a W^+ field

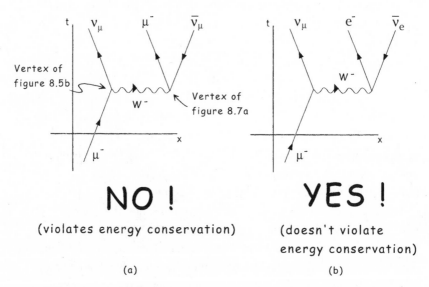

Fig. 8.8. Making use of the minimal interaction vertices of figures 8.5 and 8.7 to construct two Feynman diagrams that represent the process of muon (μ^-) decay. While consistent with all the rules of the weak interaction, a violates the principle of energy conservation, so it doesn't take place.

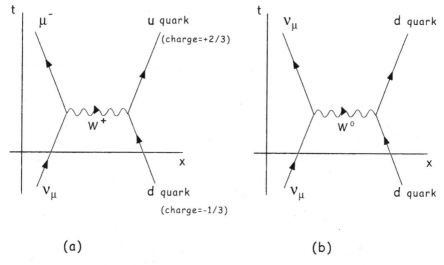

Fig. 8.9. Making use of the minimal interaction vertices of the weak SU(2) interaction to form Feynman diagrams for charged current neutrino-down-quark scattering (*a*; the charged *W* boson is exchanged in the interaction) and for neutral current neutrino-down-quark scattering (*b*; the neutral *W* boson is exchanged).

quantum. Since the W^+ is electrically charged, such a neutrino interaction is referred to as a *charged current* neutrino interaction (the motion of the W^+ being a current of electric and weak-force charge).

The second of these is a process by which a neutrino interacts through the exchange of a neutral weak field quantum: the W^0. If the diagram of figure 8.9*a* represents a charged current neutrino interaction, then that of figure 8.9*b* ought to represent a neutral current neutrino interaction; indeed, that is how it is known, which brings us to our next topic.

An Experimental Triumph: Weak Neutral Currents

The development of the theory of the weak nuclear interaction has its origins in the 1930s, when it was first recognized that the process of nuclear beta decay was not explicable in terms of the force responsible for binding protons and neutrons together into atomic nuclei. By the 1960s, as the gauge theory of the weak interactions began to emerge, the idea that weak-force processes such as beta decay and neutrino interactions should be mediated by the exchange of charged field quanta was already firmly established as a way to interpret the data that had been accumulated on such processes. So, the notion of charged field quanta—the W^+ and W^-—did not originate with gauge theory. What is unique and was not predicted by any data accumulated throughout the 1960s, was the possibility of processes, such as that of figure 8.9b, mediated by the neutral W^0 weak interaction field quantum. Such processes were a clear and bold prediction of gauge theory.

The diagram of fig. 8.9b was drawn quite often in the early 1970s, particularly by a pan-European group of neutrino physicists known as the Gargamelle collaboration. Their interest in this particular weak neutral current diagram stemmed from the fact that this process would produce a striking experimental signature, which would clearly confirm the contribution of the W^0 to the workings of the weak nuclear interaction.

Let the incoming neutrino be a member of a beam of high-energy neutrinos from a particle accelerator, and the d quark be one of many quarks in a nucleon deep within the sensitive region of a particle detector. Since neutrinos are very shy (the only charges that they possess are those of the feeble weak and gravitational interactions), they interact only infrequently. However, when they finally do interact, they transfer a good deal of their energy to the target, just like any other high-energy particle in an accelerator beam. Thus, were the interaction represented by this diagram to occur, what the detector would record is a burst of energy (due to the mightily jarred d quark) appearing suddenly in its midst, and nothing else. No trail leading into the interaction, and no trail leading out because if the neutral W^0 is responsible for the interaction, the scattered neutrino remains a neutrino and is thus very unlikely to interact again in the detector. All that would be observed would be an isolated splash of energy in the depths of the target, at the correct time to be associated with the passage of the neutrino beam through the detector. The observation of such events would be a sure sign of W^0 exchange.

And so it was, at the CERN laboratory in 1973, with the observation of about 100 such events in the Gargamelle bubble chamber. Weak neutral currents, processes mediated by W^0 exchange, were observed as predicted, providing a major triumph for the notion that the weak interaction is represented by an SU(2) gauge theory and for gauge theories in general. Let's reflect on the magnitude of this accomplishment.

The study of electromagnetism dates back hundreds of years and, during that time, occupied some of the greatest minds ever to address problems in physics. The emergence of the current, exceedingly successful, theory of electromagnetic interactions (quantum electrodynamics) in the middle part of the twentieth century, great achievement as it was, reflects the result of this tremendous effort, made possible in part by the vast base of experience that had been amassed with respect to the everyday workings of electromagnetism.

The weak nuclear force was only recognized as a distinct natural phenomenon in the middle of the twentieth century, and our realm of experience with its consequences is all but nonexistent, amounting to little more than the observation and measurement of the properties of nuclear beta decay (electron and positron emission by unstable atomic nuclei) in reactors and physics laboratories. There is absolutely nothing from the world of everyday human experience that is related to or predicated on the workings of the weak nuclear interaction.

And yet, exploiting the paradigm of gauge theory, informed by nothing more than the simple doublet pattern observed for the elementary matter constituents (fermions) of table 5.1, we arrive at a theory of the weak interaction that is every bit as complete a description of the weak interaction as quantum electrodynamics is of the electromagnetic interaction. In addition, the resulting $SU(2)$ gauge theory of the weak interaction strongly suggests that the weak interaction is more complex and more richly endowed with different types of essential behaviors than the electromagnetic interaction. An additional class of weak-interaction behaviors that are entirely absent in the case of the less intricate electromagnetic interaction—those due to the non-Abelian nature of $SU(2)$—will be the topic of the next section.

This is the tremendous power and inspiring beauty of gauge theory. We start with very little more than a vague notion of the form and structure of some underlying internal symmetry space, a two-dimensional complex space rotated around on itself by the elements of the Lie group $SU(2)$, and wind up with a complete, quantitatively precise, fundamental description of a vast set of weak-interaction phenomena even richer than the staggeringly diverse set of electromagnetic phenomena of everyday experience.

Now, as we'll see in chapter 9, this gauge theory of the weak interaction is not quite correct, although it's a big step in the right direction. As we already know, the electrically neutral quantum of the weak interaction is really known as the Z^0, not the W^0, and it has subtly different properties than those we've discussed here. It was in fact evidence of the exchange of the Z^0 that the Gargamelle group sought, and found.

It's not that there's anything wrong with gauge theory; rather, this basic SU(2) theory is just not quite the complete gauge theory for the weak interaction. However, as we'll see, the SU(3) gauge theory of the strong nuclear interaction is, in its pure and unadulterated form, the correct and complete theory of that force of nature. The paradigm of gauge theory is, without a doubt, one of the notable achievements of twentieth-century science.

Commute Issues

In the previous section, we succeeded in presenting what is perhaps the central theme of this book: the deep connection between the ethereal world of abstract mathematics and the concrete world of the interactions that underlie the behavior of the natural world. The discussion, however, is so far incomplete in that we've neglected to consider one aspect of the nature of the Lie group SU(2) that we dwelt on at some length when we introduced the group in chapter 6: the fact that SU(2) is non-Abelian. At that time, I promised that this would have profound consequences when we began to understand the connection between Lie groups and the behavior of natural processes.

Recall that a group is labeled as non-Abelian when its elements fail to commute with each other, when we can identify at least two elements a and b of the group for which $a * b \neq b * a$ when combined under the operation "$*$" of the group. In other words, a group is non-Abelian when the order in which you combine its elements matters.

In chapter 6, we discussed how it is that the Lie groups R(3), of rotations in three real dimensions, and SU(2), of rotations in two complex dimensions, are almost one and the same. In both cases, you can generate the group from combinations of varying amounts of three basic elements. In the case of the Lie group R(3) of rotations in three real dimensions, these were the three exercise rotations: simple rotations about the x-, y-, and z-axes of a Cartesian coordinate system (or, if you prefer, about the axes defined by the three edges emanating from one corner of a box).

For the Lie group SU(2) of rotations in two complex dimensions, the three generating elements can't be so easily described (it's hard to envision a space with complex dimensions), but the relationship between the group's elements and the three generators is identical to that of R(3). The only difference is that, for SU(2), you need to go around twice—rotate by 720 degrees rather than 360 degrees—to get back to where you started.

The reason for making this connection here is that, for R(3), it's easy to see that the group is non-Abelian—that it matters in which order you combine the group's elements (rotations) using the group's operation (successive application of those rotations). As discussed in chapter 6, if you rotate a box first about the x-axis and then about the y-axis, the box ends up with a different orientation than if you instead start with a y rotation and follow it with an x rotation. The total rotation you end up with (the element of R(3) that you end up with) is different for the two cases. The order of combination matters, so R(3), and by extension SU(2), is non-Abelian. In both cases, the elements of the group fail to commute.

Recall from earlier in this chapter that the gauge principle—the need to include interaction terms in the free-particle wave equation to ensure local invariance under some group of possible changes to the wave function—is all about cancellation. The new "cheating" terms, the interaction terms, that are introduced are put there solely to cancel the deleterious effects caused by the local changes to the wave function, that is, changes for which the chosen member of the Lie group of possible changes varies from point to point in space and time. The fact that these cheating terms represent interactions—that they have exactly the form they need to represent the interaction of the object with an external force field—was an exceedingly welcome by-product of the exercise of establishing the local invariance of the wave equation and its wave function solution.

Now, the types of cheating terms that one needs to add and the nature of the corresponding force-field interaction intimately depends on the nature of the Lie group of possible changes to the wave function. If this group is the Lie group U(1) of simple phase changes to the wave function, one only needs to add a single minimal interaction vertex, which turns out to have precisely the properties of the minimal interaction vertex of a photon (electromagnetic field quantum) with an electrically charged particle, and the theory we end up with is nothing other than quantum electrodynamics, our quantum theory of the electromagnetic force. If the Lie group is more complicated, then one needs to add more minimal interaction vertices, as many as there are generators of the Lie group. If we require that the wave equation and its wave function solution be invariant under a set of changes (e.g., phase/isospin-orientation) that form the Lie group SU(2), we wind up with cheating terms corresponding to three new interaction terms since SU(2) has three generators. Each of the three terms corresponds to a new minimal interaction vertex of the object with a new field quantum. The set of three new field quanta are just the quanta of the new force that we needed to introduce.

Here's the interesting point: If the Lie group of possible changes to the wave function is non-Abelian, if the order of combination of the group's elements matters, then it is not sufficient to just add one cheating term (minimal interaction vertex) for each generator of the Lie group. You do need those terms, but then, to take care of the extra complication caused by the fact that the elements of the group don't commute (I'm afraid that you need to work through the math to see this), you need to add in even more cheating terms. Indeed, after playing around a bit, you can find a set of additional cheating terms that do work to cancel the remaining problems caused by the local changes to the wave function—the additional problems caused by the fact that the elements of the Lie group don't commute with each other.

When you do this, and then look at the result from the point of view of quantum field theory, what you discover is that you added more minimal interaction vertices into the wave equation. However, these are not minimal interaction vertices between the object and the field quanta but, rather, between the field quanta themselves! For example, three of the minimal interaction vertices between the W field quanta of our weak-isospin SU(2) interaction are shown in figure 8.10. These vertices don't represent a complete set but are rather a few examples. Perhaps you can think of a few more yourself.

Remember, the field quanta will only interact (form minimal interaction vertices) with objects that carry the charge of the interaction they mediate. Since, in these vertices, the field quanta are forming vertices with themselves, we are led to the surprising conclusion that, for forces arising from local invariance under non-Abelian Lie groups, *the field quanta carry the charge associated with the force*: the field quanta themselves are charged. And since they carry the charge associated with the force, they are subject to the influence—the interactions—of that force.

For example, figure 8.11 shows three possible interactions that you can form by connecting together some of the minimal interaction vertices of figure 8.10. In both cases, you start with two SU(2) field quanta and end up with the same SU(2) field quanta but moving off in new directions; the two field quanta have bounced off each other.

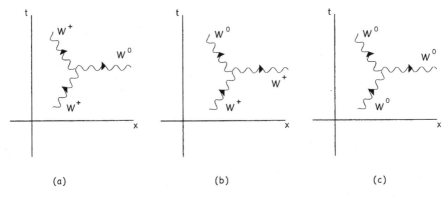

Fig. 8.10. Some examples of minimal interaction vertices of the weak SU(2) interaction that involve the W field quanta alone, and lead to interactions between the field quanta themselves, as shown in figure 8.11.

If this doesn't strike you as a bit odd, consider (or perform) the following experiment. Take a flashlight and switch it on inside of a dark room, preferably one that's a bit dusty. The beam of light that you see is nothing more than a beam of photons, composed of quanta of the electromagnetic field. Now, take a second flashlight, turn it on, and point it so that it passes through the beam from the first flashlight. What you observe is just that: the two beams pass through each other. That is, after all, what waves do. What did you expect?

Well, if you had learned about non-Abelian gauge interactions, such as our weak-isospin-charge SU(2) interaction, before having this experience with photon beams, you might well have expected something quite different. If instead of being the quantum associated with an Abelian, U(1), gauge interaction, the photon was a quantum associated with a non-Abelian gauge interaction, then photons would carry the charge appropriate for that interaction, and they would interact with each other. The quantum field theory representation of that interaction would just be that of figure 8.11c but with photons rather than W^0 field quanta. So, were this the case, the two flashlight beams would not pass benignly through each other but, rather, would bounce off each other and scatter throughout the room.

It's pretty obvious that the world would be a vastly different place if photons bounced off each other. For one thing, there would be no mechanism for the biophysical phenomenon of vision. If we go outside on a sunny day to play softball, we can only see the ball because photons in sunlight

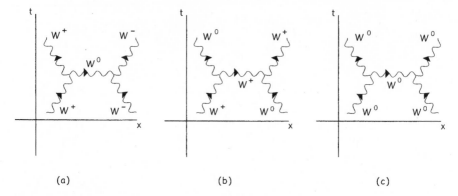

Fig. 8.11. Some Feynman diagrams that can be constructed from the minimal interaction vertices of figure 8.10. These diagrams, which are consistent with all the rules of the weak nuclear interaction, as well as the principle of energy conservation, represent the interaction of the W field quanta among themselves.

bounced off the charged particles on the surface of the ball and made their way directly to our eyes. If after bouncing off the ball, the photons then bounced again off all the rest of the photons continually streaming from the sun, very few of the photons carrying information about the ball would reach our eyes, and those that did would be too disturbed by the journey from ball to eye to do any good. In fact, you wouldn't even have eyes because there would have been no evolutionary advantage to having developed them. This is a truly striking consequence of abstract mathematicians' recognition that the generators of a given Lie group don't commute with each other.

But the photon is the gauge quantum of electrodynamics, a gauge interaction that arises from local U(1) invariance, and U(1) is Abelian, so light beams pass through each other, and fancy things like eyes are of some use to us (note 8.6). Luckily, the most ubiquitous (at least as far as everyday experience goes) interaction is electromagnetism, a simple Abelian gauge interaction.

While we can now draw the numerous new minimal interaction vertices between the three SU(2) force-field quanta (of which the vertices of fig. 8.10 are several of the possibilities), there's also the question of what it is that determines the precise properties of these vertices.

Just drawing all the different possible vertices is enough to get an idea, qualitatively, of what types of field-quanta interactions are demanded by the requirement that the wave equation be invariant with respect to SU(2) trans-formations. But, as we've seen, quantum field theory provides an exceed-ingly precise prescription for the modeling of the behavior associated with any given interaction. To take advantage of this, we need to begin with an equally precise knowledge of the properties of each of these minimal inter-action vertices.

What we need to know is the exact magnitude of the strength of each of these vertices, which is just the quantitative statement of how likely the oc-currence of each of these minimal interactions is. The answer to this ques-tion lies in the structure of the Lie algebra of the underlying Lie group SU(2).

The need to introduce the unusual vertices connecting the field quanta to themselves was caused by the fact that the SU(2) is non-Abelian — the fact that the elements and, in particular, the generators of SU(2) don't commute with each other. Now, as we've discussed, each of the three field quanta W^+, W^-, and W^0 is associated with one of the three generators of the Lie group SU(2). We also just argued that the new minimal interaction vertices con-necting the field quanta to each other arise because the SU(2) generators don't commute with each other; they arise because, when you combine any two of the generators of SU(2) according to the group's operation, the order in which you do so matters.

So, it stands to reason that the precise way in which the generators fail to commute with each other will establish the properties of these vertices. If we take SU(2) generator A and combine it with SU(2) generator B, we get some element of SU(2). However, if we start with generator B and then combine it with generator A, we get some other element of SU(2). There's a difference — a very well defined difference — between these two elements. The set of such differences, for each pair of generators A and B, is what pre-cisely establishes the strengths of all the vertices that connect the field quanta to each other.

But, flipping back through the pages of chapter 6, we are reminded that this set of differences, for all pairs of generators, is nothing other than the group's Lie algebra. We now appreciate that the precise form of the group's Lie algebra has a direct and unavoidable consequence on the nature of any physical interaction associated, through the gauge principle, with that Lie group.

In chapter 6, we discussed the claim of a nineteenth-century abstract mathematician that to specify a Lie group, to determine, abstractly, which

Lie group you happen to be considering at any moment, you need to know two things about the group: the number and the identity of its generators and the Lie algebra of those generators. We have now seen how both of these characteristics play a profound role in shaping the fundamental characteristics of the natural world. The number and identity of the generators establishes, with definitive accuracy, the types of interactions between charged objects (objects carrying the charge associated with the interaction) and the interaction's field quanta. The Lie algebra then establishes, in an equally precise fashion, the existence and nature of the interactions between the field quanta themselves. If the forces of nature do indeed arise from the requirement of local invariance under various Lie groups of transformations within enigmatic internal symmetry spaces (such as that of weak isospin), and we now have strong evidence that they do, then these abstract Lie group properties, the nature of the Lie group generators and their associated Lie algebra, have everything to do with the way we live our lives. In fact, they have everything to do with the very fact that we have lives to live in the first place.

We'll close this section with an observation, tying up a loose end that might otherwise be a point of confusion.

You might ask: Well, which will it be then? Should an object interact in the manner of electromagnetism, as established by requiring local invariance under $U(1)$ changes to its wave function's phase, or rather should it interact in the manner of the strong nuclear force, as established by requiring local invariance under $SU(2)$ changes to its wave function's phase/isospin-orientation? How does the object know which of these principles it needs to pay attention to?

The answer is that it doesn't need to decide. It needs to adhere to both of these principles continuously and simultaneously. These wave function transformations, the $U(1)$ set of changes in the wave function's phase and the $SU(2)$ set of changes in the wave function's phase/isospin-orientation, act on the wave function in entirely different domains. For example, the transformations from the Lie group $SU(2)$ that change the orientation of the object's weak isospin do so only in the internal symmetry space of weak isospin, an entity that is completely divorced from the "real" four-dimensional space-time in which we apparently live our everyday lives. If the object carries the charge associated with the interaction (in this case, weak isospin), then the wave function must express itself within this weak-isospin space to completely incorporate the physical properties of the object.

The question (unanswered to date) of whether internal symmetry spaces such as isospin space are in any sense real is not of particular relevance. In addition to familiar space-time, the wave function of an object possessing some amount of both the electromagnetic and weak-interaction charges must extend, at least mathematically, into both the one complex–dimensional U(1) space of changes to the phase of the wave function, and into the two complex–dimensional SU(2) space of changes to the orientation of the object's weak isospin in weak-isospin space. The object, represented by its wave function, must obey the rules of local invariance in both of these spaces, simultaneously and independently. Thus, finding itself spreading into these internal symmetry spaces, the wave equation governing the behavior of the object must exhibit the interactions associated with local invariance in these spaces, those of electromagnetism and the weak nuclear force, as imposed according to this chapter's gauge principle.

So, the number and nature of the independent interactions that are available to influence physical objects, the set of so-called forces of nature, is definitively established by the number and nature of such internal symmetry spaces. The nature of each of these internal spaces is definitively established, in turn, by the identity of the associated Lie group of transformations that reorient the wave function within the internal symmetry space and under which transformations we demand (precisely because the space is an internal symmetry space) that the wave equation and its wave function solution be invariant.

Thus, it seems that we now understand that interaction—really, the essential phenomenon of causation—arises quite specifically because, to represent the wave function of objects possessing a given type of charge, that wave function must express itself in an appropriate independent, abstract symmetry space. Enigmatic as they may be, our current notion of the origin of fundamental interactions relies intimately on the existence, at least mathematically, of these internal symmetry spaces.

Further Plaudits: Gauge Theory and Renormalization

In chapter 4, we discussed the framework for modeling fundamental interactions in a way consistent with both quantum mechanics and Einstein's special theory of relativity. This framework, known as relativistic quantum field theory, was based on the notion of the minimal interaction vertex between matter particles and the force-field quanta associated with the interaction in question. From these minimal interaction vertices, one can construct Feynman diagrams representing any physically permissible process mediated by the interaction. Each such Feynman diagram represents an explicit calculation, which when carried out gives the corresponding interaction or decay probability of the process as well as the precise way in which, on average, the particles that are left over after the interaction or decay array themselves in energy and in direction of motion.

Compelling as this may be, after motivating and presenting this framework in the context of the specific theory of quantum electrodynamics, we were nevertheless confronted with a serious caveat: the theory gives results that are utter nonsense. Any single diagram representing a given process, uniquely identified by the number and character of initial and final particles, yields a sensible calculated result. However, when one considers the full set of Feynman diagrams associated with the given process, reflecting the infinitude of different ways in which the process can be influenced by the lively and ever-fluctuating vacuum of quantum mechanics, the results of the calculation for the process become infinite and therefore nonsensical.

In the section "The Living Vacuum" in chapter 4, we remedied this shortcoming by introducing the technique of renormalization. This scheme was based on the following insight: that any process we would use in the lab to measure a physical quantity, such value of the electron's charge, must necessarily be representable by such a set of Feynman diagrams.

We know from the basic tenets of quantum mechanics that to measure is to probe—to influence the observed object in a way that it yields information about itself—and the only way to influence objects is through one of the forces of nature. So, if you want to measure some electromagnetic property of an object (e.g., the electric charge of an electron), you do so by probing it with the electromagnetic force, the probing of which is represented by some set of Feynman diagrams, which in turn are cobbled together from the minimal interaction vertices of quantum electrodynamics. These diagrams range from the very basic, with only one minimal interaction vertex (fig. 8.12a) to the very complex, for which a large number of such vertices connect together to represent an interaction beset with a considerable amount of activity in the vacuum (e.g., fig. 8.12b).

The result of the measurement is inarguably finite and sensible (measuring the charge of the electron is a popular laboratory experiment in the undergraduate physics curriculum); yet, by doing the calculation, we would

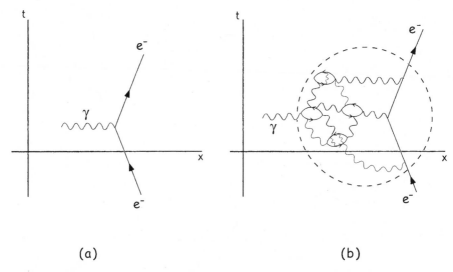

Fig. 8.12. Measuring the charge of the electron by shining light on it; *a* is the basic process of the absorption of light by the electron; *b* would be observed as the same process but incorporates substantial vacuum activity.

find that the combination of all the different diagrams representing our measurement process gives a result that is nonsensical.

We got out of this mess by speculating that the electron whose charge we are measuring is not the "bare" electron represented by one of the incoming legs on the Feynman diagram, but, rather, incorporates the myriad of possible vacuum fluctuations that cause us the trouble in our naive calculations. The real electron, the electron that we are probing with our experiments in the lab, is not the entity represented by the incoming and outgoing straight lines in figure 8.12*a*. Instead, the electron that we probe in the lab is the combination of all the activity contained within the circles of diagrams such as those of figure 8.12*b*. The electron we observe contains, as a part of its intrinsic substance, all of the possible vacuum fluctuations of the sort shown in figure 8.12*b*. It is *this* measurement—of the electron and its vacuum entourage—that leads to the sensible value for the electron's charge (1.602×10^{-19} coulombs, for what it's worth) observed by the students in the lab.

Thus, what we need to do is to renormalize, or simply adjust, the value of the bare electron charge that enters and exits the diagrams, such as those of figure 8.12, that represent the process of measuring the electron's charge.

We just set this value, artificially, to whatever it needs to be so that when we calculate and combine all the possible diagrams like those of figure 8.12, we get the result that we measure in the lab.

More cheating! What value is a theory such as this? It seems as if we've lost all of our predictive power by playing with the properties of real, tangible matter (electrons) just so that our theory will work out. It seems like any theory at all can be made to work in this way. Any set of rules for calculating the behavior of some physical process can be made to give the right answer if you're allowed to adjust its inputs artificially to make it agree with experiment.

The critical point, we argued, is this: If we can, through the renormalization of a finite number of parameters (charge, mass, etc.) render the resulting full calculation finite for any and every physical process permissible by the interaction and at every possible interaction energy, then the theory is indeed quite useful. By artificially adjusting a few parameters, we arrive with a theory that works in an infinitude of possible situations. For example, the magnetic properties of the electron, the measurement of which gave us our stunningly precise confirmation of quantum electrodynamics at the end of chapter 4, are not members of this set of arbitrarily renormalized parameters; they are a true prediction of the theory.

Such a theory, one definable in terms of a finite number of artificially renormalized parameters, is known as a *renormalizable* theory. If such is not the case, then the theory is *nonrenormalizable*, and can at best only be a crude approximation to a true theory of the interaction.

The mathematical process by which a theory is demonstrated to be renormalizable (or not) is very technical and quite challenging. In fact, what is found is that, beyond quantum electrodynamics (the most straightforward possible implementation of quantum field theory), it is exceedingly difficult to engineer theories that are renormalizable. Early attempts at descriptions of both the strong and weak nuclear forces, while partially successful in describing observable phenomena, notoriously lacked this quality.

The remarkable, perhaps even miraculous, thing about gauge theories, that makes them such a potent weapon in the theoretical particle physicist's arsenal, is that there is a deep connection between local gauge invariance and renormalizability. Gauge theories tend to be renormalizable and are about the only implementations of quantum field theory that are so. Quantum electrodynamics is renormalizable, but after reading the section "Quantum Electrodynamics—Again" we now appreciate that this is only because quantum electrodynamics *is* a gauge theory, albeit one based on the simplest possible Lie group: the group U(1) of changes in the phase of the wave function.

For a more complex theory, such as the SU(2)-based theory of the weak interaction, it's only a remarkably delicate balance between the characteristics of the various minimal interaction vertices that establishes its renormalizability. If any of these minimal interaction vertices are ignored or even have their characteristics altered slightly, the balance is ruined and the resulting theory is not renormalizable.

Thus, for gauge theories based on non-Abelian Lie groups, for which the resulting field quanta are predicted to have the unusual property of interacting with themselves, these interactions are essential to the health—the renormalizability—of the theory. If we leave out any of the minimal interaction vertex ingredients associated with these self-interactions, renormalizability is forfeited. Or, even more tellingly, if the characteristics of the minimal interaction vertices involving the interaction of the field quanta with themselves are altered slightly from the characteristics that are rigorously and uniquely established by the Lie algebra of the underlying symmetry group, renormalizability is again lost. Trying to establish a complex fundamental theory, such as the quantum theory of the weak interaction without the guidance provided by the principle of local gauge invariance would be like looking for a needle in a haystack.

Quantum Chromodynamics

The thrust of this book is the presentation of the conceptual basis of the Standard Model of electroweak physics. We've talked so far about separate theories of the electromagnetic and weak force based on underlying internal symmetry groups and the application of the gauge principle; the fully unified gauge theory of the electroweak interaction, however, awaits its presentation in the next chapter and that chapter's discussion of spontaneously broken, or hidden, gauge symmetry.

However, there is another force for which a successful quantum theory exists—the strong nuclear force. We won't say too much about this force, but what we do want to discuss is appropriately interjected here because the quantum theory of the strong nuclear interaction—quantum chromodynamics—is a pure gauge theory. Thus, the theory of the strong nuclear force is completely prescribed by the principles we've discussed in this chapter.

In this light, the development of the underlying theory of quantum chromodynamics involved two steps. The first was the identification of the nature of the charge associated with the strong nuclear force, the so-called color charge (hence, the "chromo" in "chromodynamics") possessed by the strong-force-abiding quarks. The leptons—electrons, muons, tau leptons, and their associated neutrinos—do not possess color charge and, as a result, do not take part in the strong interaction.

The second step was the identification of the associated Lie group—the group SU(3) of rotations in an internal symmetry space (color space) of three complex dimensions. As for the theory of the weak nuclear force and for gauge theories in general, the identity of this underlying mathematical structure of the strong force was revealed by the patterns evidenced by the properties of the fundamental participants in the theory—by the color-charge properties of the quarks.

With this in hand, the identification of the set of strong-force minimal interaction vertices and their precise properties followed in an orderly and prescribed way from the application of the gauge principle. The final step in specifying the theory was the determination of the single empirical (experimental) input to the theory: the overall strength, g_s of the color charge.

The assignment of credit for the development of the theory of quantum chromodynamics is not completely cut and dried. The introduction of the concept of quark color, as far back as 1965, is generally attributed to the estimable Japanese theorist Yoichiro Nambu of the University of Chicago. The first conjecture that the strong nuclear interaction should be described by a gauge theory based on SU(3) transformations in the internal symmetry space of color charge seems to have come in 1973 from Harald Fritzsch and Murray Gell-Mann of the California Institute of Technology, the latter already quite familiar to us because of his invention of the eightfold way.

However, for quantum chromodynamics, the task of employing the rules of the minimal interaction vertices in the calculation of laboratory processes is a particularly thorny one. Much credit is given to David Gross and Frank Wilczek (Princeton) and H. David Politzer (Harvard) for demonstrating later in 1973 that the theory of quantum chromodynamics provides a fundamental explanation of why it is that quarks only manifest themselves when nuclear matter is explored with an energetic probe. Unlike protons and neutrons, quarks cannot move freely about; they can only be freed (and then only for exceedingly short periods of time) if they are struck very hard by one of the force-field quanta. This property, known as *asymptotic free-*

dom, is the reason the discovery of quarks eluded us until relatively recently and is perhaps the single most characteristic property of the strong nuclear force. The appearance of asymptotic freedom in the theory of quantum chromodynamics is intimately connected with the mathematical properties of the Lie group SU(3), which we'll return to briefly below.

To kick off our discussion of quantum chromodynamics, let's recall that the fundamental arena of the strong force is not the interplay of nucleons (protons and neutrons) and their binding into nuclei but, rather, the binding together of quarks into individual nucleons, as well as into other (unstable) particles that are made up of quarks.

Quark-containing hadronic matter particles come in two types: *mesons*, which are bound quark/antiquark pairs (such as $\pi^- = \bar{u}d$) and *baryons*, which are triplets of three mutually bound quarks (such as the proton, $p = uud$). In the 1960s, when people were trying to make sense of the properties of hadrons in terms of Gell-Mann's eightfold-way quark model (see the section "Three Quarks for Muster Mark" in chapter 5), it was found that it was simply impossible to construct, given the presumed properties of the three constituent quarks, a wave function that represented the known properties of the proton and neutron. Only by hypothesizing, with no other motivation than to make these wave functions work out, that quarks carry a novel property known as color (recall the discussion of color in chapter 5) was the quark model rescued.

According to the theory of color, each quark can come in one of three colors: red (R), blue (B), or green (G). No one, of course, is claiming that quarks actually come in three decorative shades; it's just that, mathematically, when an extra factor is added into the wave function to account for this possibility, the rest of the wave function suddenly makes physical sense. Modified in this way, the quark model led to a number of significant predictions regarding the production of quarks and the properties of exotic, unstable baryons that were later confirmed in experiments at particle accelerators. For example, in producing quark/antiquark pairs through the annihilation of high-energy electrons and positrons, the collision rate was exactly a factor of three higher than one might otherwise expect: for any given flavor of produced quark/antiquark (say, down/antidown), the reaction could produce any of the three colors: a red/antired, a blue/antiblue, or a green/antigreen version of the down/antidown pair, leading to the factor of three enhancement in the annihilation rate. However, while this very strongly supported the notion of color, the physical meaning of this new property of matter was not understood at the time these measurements were first performed (note 8.7).

Time passed, and the origin of the property of quark color remained a curious mystery. However, in 1973, Fritzsch and Gell-Mann proposed that color, or more specifically, rotations in an abstract color space that swap the three colors among one another, may be responsible, through the gauge principle, for the strong nuclear force experienced by quarks.

Recall that rotations in the two complex dimensions of the Lie group SU(2) that swap upper and lower weak-isospin doublet members (v_e and e^-; u and d quarks, etc.) among one another, were responsible, through the gauge principle, for the generation of the weak nuclear interaction. In other words, when we ask that the free-particle wave equation and its wave function solution be invariant under local rotations between upper and lower doublet members (rotations by amounts that vary from point to point in space and time), we found that we needed to introduce the $W^\pm$ and the W^0 field quanta that are the mediators of the weak nuclear force.

For the SU(2)-based theory of the weak interaction, the amount of upper-doublet particle in the complex wave function represents the first complex dimension in weak-isospin space (represented by distance along the x-axis in complex two-dimensional weak-isospin space). Similarly, the amount of lower-doublet particle content represents the second complex dimension (represented by distance along the corresponding y-axis).

According to Fritzsch and Gell-Mann, the group of transformations that rotate the three quark colors into one another must be the group of complex three-dimensional rotations: one complex dimension for each quark color, just as for the weak force we had one complex dimension for each of the two members of the weak-isospin doublet. The strong force doesn't care which of the three quark colors is at play, just as the weak force doesn't care which doublet member, upper or lower, is involved in the interaction. Thus, any wave equation and corresponding wave function solution describing a quark had better be invariant under transformations swapping the colors among one another, that is, under any rotation belonging to the group SU(3) of rotations in the three complex dimensions of the internal symmetry space known as *color space*.

That there are three quark colors is, in and of itself, not quite enough to specify that the underlying internal symmetry group is necessarily SU(3). There are other possible Lie groups operating in three complex dimensions that could conceivably serve instead of SU(3) as the underlying symmetry group in our internal color symmetry space. For example, we could restrict ourselves to rotations in color space that leave the amounts of "redness," "blueness," and "greenness" in the wave function unchanged but change the complex (ordinary quantum-mechanical) phases of the red, blue, and green components of the wave function by some amount.

However, the nature of the Lie group SU(3) has something interesting to say about the pattern of interrelations between the color charges, should the three color charges indeed be related by SU(3) rotations in the complex color space. To be specific, this pattern of interrelations provides that the combination of all three color charges, in equal amounts, yields a net color charge of zero (note 8.8)! This property is quite unusual. It's somewhat akin to combining three particles with positive electric charge and winding up with a combined object that is electrically neutral — only the three particles have equal positive amounts of three different kinds of strong-force charge: red, blue, and green. Recalling the discussion of the section "The Particle Zoo" of chapter 5, this property is essential: The existence of baryons, among which we count protons and neutrons, is predicated on the fact that the three quarks that form them, when combined, have no net color charge.

This property of color charge — its pattern of interrelations — is uniquely characteristic of charges related through rotations from the group SU(3) and hence our conviction that SU(3) is the true underlying symmetry group of the strong nuclear interaction. Again, the patterns exhibited by nature, in this case the three colors of strong-force charge and the interrelations between the three colors of charge, provide the essential clue to the underlying mathematical structure that organizes its fundamental behavior.

Thus inspired by the thought that the strong nuclear interaction arises from the invariance of the wave equation with respect to SU(3) rotations in this internal color symmetry space, the next step is to apply what we've learned in this chapter to construct the corresponding gauge theory of the strong interaction. Again, the arguments of Yang and Mills compel us to think of this invariance as a local symmetry, and in forcing it to be so, we are thus required to entertain the notion that the quarks are subject to the influence of gauge fields $A_c(x)$. The subscript c reminds us that these gauge fields arise from the requirements of local invariance under transformations in this SU(3) color space, as opposed to the SU(2) space of weak isospin or the U(1) space of quantum electrodynamics. Here's a summary of how this works; nothing in the paragraph below is new; it's just a recap to remind you of what's involved.

Again, the gauge fields, or interactions, $A_c(x)$ need to be introduced into the free-particle wave equation to patch up the damage caused by color-space rotations that swap the quark colors among one another by amounts that differ as you move from point to point in space-time. Over there, you rotate, or transform, a blue quark into a half-green, half-red quark, while over here, you transform a blue quark into a pure red quark (and maybe change the quantum-mechanical phase of the resulting wave function by some amount also). These differing transformations alter the way the wave function changes from point to point in space. This, in turn, destroys the delicate balance, required by the wave equation, between the rate of change in the wave function around any point in space and the energy of the object described by that wave function. We thus find it necessary to introduce gauge field (interaction) terms of the form $g_s A_c(x)\psi(x)$ as "cheating terms" that act to restore this delicate balance and thus make the wave equation and its wave function solution invariant with respect to local SU(3) color-space rotations.

As before, the $A_c(x)$ turn out to be the fields associated with a new force — in this case, the color force between quarks — that we hypothesize must just be the strong nuclear force. The resulting theory, detailing the workings of quark interactions — quark dynamics, as opposed to the statics of Gell-Mann's eightfold way — under the influence of the color force is known as quantum chromodynamics, or QCD.

As we emphasized above, this theory, being a gauge theory, provides a complete and definitive specification of the rules by which the strong-force interactions of quarks play themselves out. If you're fortunate enough to be well schooled in the techniques of relativistic quantum field theory, you can use the theory of QCD to calculate the properties of any process that involves the interaction of quarks via the strong, or color, force. While these calculations lie beyond the scope of this book, we do now know enough to say something about the sorts of minimal interaction vertices that QCD sets forth to use in building up the Feynman diagrams that represent these processes; we leave the maddeningly complex task of calculating such diagrams to those who actually get paid to do so.

We know that there will be not one interaction term but, rather, exactly as many different interaction terms $g_s A_{c,1}(x)\psi(x)$, $g_s A_{c,2}(x)\psi(x)$, $g_s A_{c,3}(x)\psi(x), \ldots, g_s A_{c,8}(x)\psi(x)$ as there are generators of the Lie group

SU(3). As mentioned in chapter 7, the group SU(3) has eight generators, so this means that there are thus eight gauge fields, which is why the above list ran from one to eight. The factor g_s, common to all eight of the interaction terms, just represents the overall strength of the color-force (strong nuclear) interaction, the one and only input to the theory that must be determined by precise experimentation.

Now, this is quantum field theory, so these eight interaction terms really represent the introduction of eight separate field quanta: the strong force has eight photonlike particles to run around and express its will, which is simply to have quarks influence one another. Again, this influence, or interaction, between the quarks is exactly what causes them to be able to bind together into useful things like protons and neutrons; these field quanta are the glue, if you will, that binds the quarks together into everyday matter. Hence, physicists have chosen, somewhat tongue-in-cheek, to refer to these eight field quanta as gluons. There are indeed eight of them, although they're so similar, differing only in the type and amount of color charge that they each carry that we just call them collectively "the gluon." This gluon (really, each of the eight gluons) forms a minimal interaction vertex with any given quark in the usual fashion (see fig. 8.13a). Note that by conven-

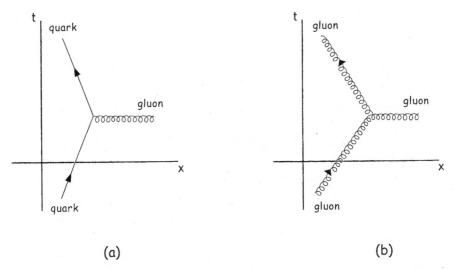

(a)   (b)

Fig. 8.13. Two minimal interaction vertices of the strong SU(3) interaction: the quark-gluon vertex (a) and the gluon-gluon vertex (b).

tion the gluons are represented by curly lines, as opposed to the wavy lines of the photon and W and Z particles.

In chapter 6, we made the point that the group of rotations in three or more dimensions of everyday space is non-Abelian. When you successively apply two rotations to an object, such as the box we used for our prop in chapter 6, the final orientation of the object generally depends on the order in which the two successive rotations are applied. Earlier, we noted that the group SU(2) of rotations in two complex dimensions is also non-Abelian, and here we'll note that, not surprisingly, the group SU(3) of rotations in three complex dimensions is also non-Abelian.

We've also learned in this chapter that the non-Abelian nature of the underlying Lie group of internal symmetries has a very direct and profound consequence on the nature of the corresponding interaction: for any such interaction, the field quanta introduced to ensure local invariance under the Lie group's rotations will interact with themselves. Thus, when you work through the process of introducing the eight gauge fields that you need to ensure invariance under local SU(3) transformations, you will find that the corresponding quanta—the eight gluons—each carry some of the red/blue/green color charge that we originally identified as being a property of the quarks. Thus, you get minimal interaction vertices that involve the field quanta alone (see fig. 8.13b).

Furthermore, the properties of these gluon-only interaction vertices (remember that there are eight gluons, so there are a lot of possibilities) are set by the central characterizing property of the Lie group itself—the Lie algebra of ordering relations between the eight generators of SU(3). Thus, a critical test of quantum chromodynamics is the verification of the existence of the gluon-only, or triple-gluon, minimal interaction vertices of figure 8.13b and the demonstration that their properties are precisely those expected from the Lie algebra of the group.

A particularly interesting diagram, representing a mixture of weak and strong interactions (enabled by the fact that quarks carry both weak and strong [color] charge) is shown in figure 8.14. In this diagram, the interaction is begun by the annihilation of an electron and a positron into a weak nuclear force Z^0 gauge quantum, which subsequently decays into a quark/antiquark ($q\bar{q}$) pair. From there, however, the strong nuclear force takes over. In this diagram, the quark (it could just as well have been the antiquark) employs a quark/gluon minimal interaction vertex to emit (radiate) a gluon. The gluon then exploits the existence of the triple-gluon vertex,

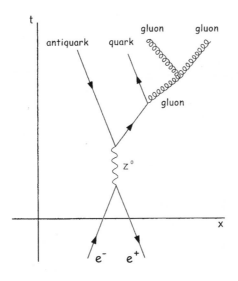

After interaction, what's left is a
quark, antiquark, and two gluons.

Fig. 8.14. In this mixed interaction, a quark-antiquark pair is produced through
the annihilation of an electron and a positron to the neutral weak field quantum (which, as we'll see in chapter 9, is really the Z^0, rather than the W^0 that
we introduced above). Then, the strong interaction takes over, employing both
quark-gluon (fig. 8.13a) and gluon-gluon (fig. 8.13b) vertices to produce two
more gluons. Quite a number of collisions like this, with a total of two quarks
and two gluons coming away from the collision, were observed by the experiments at CERN and SLAC that studied the production of Z^0 bosons produced
through electron-positron annihilation.

our current focus of interest, to produce a total of two gluons flying out from
the collision point in addition to the quark/antiquark pair.

Thus, for the process represented by this diagram, what begins as an electron and positron hurtling toward each other with great energy winds up as
four fundamental "colored" (strong-force-charged) particles moving quickly
outward from the collision point. So, an experiment that looks for such
events, for which exactly four fundamental colored particles emanate from
the point of annihilation of an electron and positron, will be able to confirm
the existence of the triple-gluon vertex and measure its properties (note 8.9).

Such experiments were done in the mid-1990s at the Large Electron
Positron (LEP) collider at CERN. It was found that (1) eight different glu-

ons are needed to explain the results; (2) the results cannot be explained without the process of figure 8.14, that is, that the triple-gluon vertex exists; (3) the characteristics of the triple-gluon vertex of figure 8.14 necessary to explain the LEP four-particle data are precisely those prescribed by the Lie algebra of SU(3). These experiments thus explored, and confirmed, a very specific prediction of the SU(3) gauge theory (quantum chromodynamics) of the strong nuclear force, a prediction that lies at the heart of the remarkable connection between the abstract mathematical properties of Lie groups and the concrete world of matter and its interactions.

You might come away from this discussion with the impression that the consequences of the non-Abelian nature of SU(3) are somewhat incidental, providing a definitive way to test the connection between the gauge principle and the properties of SU(3) to the force that binds atomic nuclei together but, otherwise, having little to do with the essential qualities of that force and the way it shapes the world in which we live.

Nothing could be further from the truth. The existence and precise nature of the by-product of the non-Abelian structure (Lie algebra) of SU(3), the triple-gluon vertices of figure 8.13b, seems to be an essential ingredient in the recipe for making nucleons (neutrons and protons), and so is an indispensible component of what it is that makes atomic nuclei, and thus atoms, and thus molecules—and thus us—possible. Before we can discuss why this is the case, we need to learn a little more about the electroweak force in chapter 9. We'll then be ready to return to this issue in the context of our discussion of "grand unification" in chapter 10.

For now, though, suffice it to say that it is precisely the non-Abelian character of SU(3) and the manifestation of this character in the three-gluon vertices that leads to the essential strong-force characteristic of "asymptotic freedom," as Gross, Wilczek, and Politzer managed to show in their 1973 papers. What is meant by the term "asymptotic freedom" is that quarks, imbued as they are with the powerful color charge, are not observed in nature as independent particles, free of partners that bind with them to form long-lived particles with no net color charge. So, for example, one doesn't observe the quarks within the proton that forms a hydrogen nucleus but, rather, sees the proton as a whole.

Unless one probes the proton with a high enough energy probe. As the energy transferred to the proton by the probe gets large, approaching the mass-energy of the proton itself, the probe can begin to discern the individual quarks within the proton. It's not that the quarks become free, it's just that the energy of the probe begins to exceed the energy with which the quarks are bound into the proton, so they can begin to manifest themselves as individual particles. As the energy of the probe (say, a high-energy photon released from a scattered electron) gets higher and higher, the struck object looks more and more like one of the individual quarks, rather than a bulky proton. At very high probe energies—"asymptotically high," if you will—the quarks manifest themselves as fully individual particles, completely free of the binding effects of the proton in which they reside. Hence, the term "asymptotic freedom."

At everyday energies, however, quarks are not free, asymptotically or otherwise, and the unit that we must concern ourselves with in our physics is that of the nucleon—the proton and neutron. But it's these particles, and not individual quarks, from which we construct the nuclei of atoms. In other words, the other side of the coin of asymptotic freedom is the tight binding of quarks into color-neutral nucleons, without which atoms and molecules, and thus biological systems, would not be possible.

When we return to this discussion in chapter 10, we'll see that the phenomenon of asymptotic freedom is a result of the fact that the overall strength of the strong force, as represented by the size g_s of the strong-force charge, itself varies with the energy of the probe. The precise way in which it does this is intimately connected to the nature of the Lie algebra of SU(3), as it manifests itself through the triple-gluon vertices; the first people to establish this connection, quantitatively, were our three young QCD enthusiasts: Gross, Wilczek, and Politzer.

Parting Thoughts

On one level, particularly to the first-time reader, the principles of gauge theory may seem woefully complex. Once one becomes comfortable with the mathematical language of gauge theory, however, one realizes that the notion of local gauge invariance, with its implications in terms of precise laws of interaction, is tremendously efficient, providing an underlying framework of almost comical physical simplicity, but which is, at the same time, both profoundly general and stunningly precise.

Ponder for just a moment the nature of the world around you, that greets you at every turn through the window of your senses—the tastes, the smells, the feel of sand between your toes, the view of the cosmos permitted by a clear and moonless night. The gamut of sensual experience seems so vast and incomprehensibly diverse that one can begin to wonder how it is that our species ever surmised that it could be comprehended in terms of some finite number of definite and infallible organizing principles.

And yet so we have, and within the last few decades of the most recent century, our success at doing so has begun to seem great indeed. There is no known physical phenomenon that can be held forth as being at odds with the notion that the workings of the universe are influenced by the four forces of nature that were introduced in chapter 2. As complex as any given natural behavior may be, and even though it may be all but impossible to provide a precise and quantitative description of the behavior in terms of its underlying fundamental basis, it is nevertheless invariably consistent with the notion that it is originated and controlled by these underlying principles (note 8.10).

Of these, three (the electromagnetic, the weak nuclear, and the strong nuclear interactions) are understood, from the now well-established outlook presented by gauge theory, to be nothing more than an inevitable consequence of certain redundancy in the natural world.

Quantum mechanical phase, and its generalization in the context of internal symmetry spaces, is an unavoidable component of the wavelike nature of matter—the direct implication of the renowned notion of wave-particle duality originally introduced in the 1920s, and now on about as solid an experimental footing as any conjecture known to science. While the relative phase of the various objects in a system is a crucial determining characteristic of the behavior of the system (via its bearing on the precise way in which the objects' representative waves interfere with each other within the system, as discussed in chapter 3), the overall phase of the system's combination of waves can have, as we have discussed rather extensively, no physical consequence.

Wave functions that differ only in their overall phase provide equivalent, and redundant, representations of any given physical system. A direct exploration of the consequence of this redundancy, consistent with the notion of the locality of space-time—that the behavior of any system at two well-separated space-time points must be to some degree independent—leads inexorably to the conclusion that the objects within the system must be subject to some precisely construed law of interaction. The connection between these two notions, the invariance of the wave equation with respect to changes in phase (or, more generally, rotations in some internal symmetry space) and the corresponding introduction into the wave equation of terms providing a definitive description of a natural mode of causation (interaction), is just what we have been referring to as the *gauge principle*.

From the point of view of gauge theory, there are but two questions that need to be addressed to develop the fundamental picture underlying the rich array of phenomena that comprise the actions and implications of any given law of physical interaction: the specification of the nature of the phaselike invariance obeyed by the wave functions of natural objects (i.e., the underlying symmetry group) and the nature and strength of the charge associated with that invariance.

Given the list of abstract Lie groups that specify the structure of the various internal symmetry spaces and the nature and strength of their associated charges, the fundamental rules of the behavior of matter are completely delineated. This delineation is as precise as it is complete in the sense that it provides, through the framework of quantum field theory, an exact prescription for the calculation of the properties of the interaction between any set of fundamental objects. Furthermore, the approach thus delineated, when one considers the connection between local phase invariance and renormalization, seems to be the only self-consistent approach to modeling fundamental interactions that anyone has dreamt up to date.

But we can pursue this line of thought even a bit further because upon some additional reflection, the gauge principle seems to suggest the ontologically comforting notion that interaction, and thus causation, are inexorable properties of a quantum-mechanical universe. Let's look into this a bit.

Be it the electromagnetic, the weak nuclear, or the strong nuclear force, the gauge principle introduces the corresponding interaction through the "cheating terms" of the form $gA(x)\psi(x)$, where $A(x)$ represents the field (really, potential, for those still keeping track of the distinction) of the interaction, $\psi(x)$ the wave function of the interacting object, and g the amount of the interaction's associated charge possessed by the object. For the moment, let's focus on the latter of these factors: the charge g in this expression for the "cheating" interaction term.

If an object possesses none of the charge associated with the interaction (electric charge, color, weak isospin), then $g = 0$ for that object, and the cheating term is also zero because anything multiplied by zero is zero. This should come as no surprise. If an object doesn't have the charge associated with a given interaction, then it won't influence or be influenced by that interaction, and so it shouldn't form any of the minimal interaction vertices with the field quantum $A(x)$ that are implied by the term $gA(x)\psi(x)$.

In other words, if $g = 0$, there won't be any minimal interaction vertices of the form $gA(x)\psi(x)$ connecting the object, represented by its wave function $\psi(x)$, to the force field of the interaction, represented by its potential function $A(x)$. And this is exactly what we expect because according to the most basic aspects of the description of a force, an object bearing none of the charge associated with that force will not be subject to the influence of that force.

Recall the weak-isospin doublets that generate the characteristic patterns of the fundamental particles shown in table 5.1. These were just the isospin-up and isospin-down pairs of fermions (quarks and leptons), such as the u and d quark, t and b quark, muon (μ) and muon neutron (v_μ), and so forth. Our conjecture that the weak nuclear interaction can be modeled as an SU(2) gauge theory was based on the observation that none of the properties of any given weak interaction process are changed when you rotate, or reorient, the wave function of all participating objects in the complex two-dimensional space of weak isospin, that is, interchange isospin-up particles with isospin-down particles and vice versa.

Recall also that, in the process of modeling the weak nuclear interaction in this way, we learned that the charge associated with the weak interaction is just weak isospin itself. In other words, if an object possesses weak isospin, that is, is doing something that we can think of as spinning about some axis in the abstract internal space of weak isospin, then it possesses a weak-interaction charge g, and it will thus form minimal interaction vertices with the A's, the weak-interaction force-field quanta (which we initially called the W_1, W_2, and W_3 but will soon rename as the W^+, W^-, and Z^0).

If the weak-interaction charge g is zero, then the object possesses no isospin. But, in this case, the notion of reorienting the axis in weak isospin space (about which the object is weak-iso-spinning) is meaningless. If the weak-interaction charge is zero, then the object has no weak isospin, and so it has no axis to reorient, via rotation elements from the group SU(2), in the abstract internal space of weak isospin. A particle with no weak isospin (no weak-interaction charge) simply has no truck whatsoever with the abstract internal space associated with the weak interaction; its wave function simply doesn't extend into weak-isospin space.

Let's turn this argument around. For an object to extend itself into the internal space associated with a given force, it must possess some of the charge associated with that force. You can think of it like an invitation to an exclusive party: possessing the charge associated with a given force is like possessing an invitation to enter the internal symmetry space of that force. With some of the charge of the weak nuclear force in hand, you are cordially invited to enter into and perform your SU(2) gyrations within weak-isospin space; without the charge, you are condemned to lurk forever outside of the party, under the unwaveringly watchful eye of the particularly stern and surly weak-isospin-space bouncer.

But once you are inside, then you have to behave yourself according to one inviolable rule: your physical (experimentally observable) behavior simply cannot depend on how you find your weak isospin oriented in weak-isospin space as you wheel and turn within it, even if you do so locally, that is, turn in isospin space by an amount that depends on where you are in ordinary space-time. This, in turn (as demonstrated by Yang and Mills), means that you must partake in the weak nuclear interaction, with a behavior specifically and definitively governed by the properties of the group SU(2) of possible gyrations on the dance floor of weak-isospin space.

But the content of the universe—what the universe is—is nothing more or less than the set of objects it contains. As far as we can tell, there is no such thing as "pure energy" that exists independently of the objects, or fundamental particles, that populate the universe. When we peer down to the most fundamental level, all the energy we know of is either *kinetic* energy, associated with the motion of objects, or *potential* energy, associated with the mass of the constituent objects and the configuration of force fields. Thanks to the development (and stunning confirmation) of quantum field theory, we now understand that even this field energy only arises in association with fundamental particles: the force-field quanta.

So, a universe without particles is nothing. But, just so, a particle without any charge whatsoever (be it even just the mass-energy charge of the gravitational interaction) is equally nothing. There is no substance, or being, that we can ascribe to a particle that possesses none of the charges associated with the known forces of nature. It is precisely in terms of these charges—mass, electric charge, weak isospin, and so forth—that we identify and differentiate different fundamental particles. So, for the universe to be *something*, we need particles, and for there to be *particles*, we need to have some sort of charge for these particles to carry around with them.

And now, for the point of all this: if a particle possesses the charge of some particular interaction, and it must possess the charge of at least one such interaction if it is to be a particle, then it lives in the internal symmetry space of that interaction, so it must obey the rules associated with that privilege. It must partake in and be influenced by that interaction. Gauge theory suggests that charge is, most fundamentally, an entrée into an internal world — an internal symmetry space within which an object's orientation can have no physical relevance—that bears with it the onus of subjugation to the influence of an interaction.

So, there is no charge without interaction, and equally, there is no universe without charge; thus, we conclude, there is no possibility for a universe without interaction. Any universe must, at its most fundamental level, be interesting. So, perhaps it's not so surprising that ours is so much so.

But even more than this, quantum mechanics demands some sort of internal space of possible changes to the wave function under which the wave equation and its wave function solution are invariant. As we discussed some pages ago, the overall phase of the wave function—the precise time at which the undulating wave function has a value of zero at a given point in space — can have no physical meaning. The size, the wavelength, and the speed of propagation through space of this undulation all have direct consequences on the behavior of the object being described by the wave function, but its phase does not.

Thus, at the very least, the wave equation and its wave function solution must be invariant under changes of phase of the wave function. The set of changes to the wave function that produces these changes in phase can form any of the Lie groups acting within the various internal symmetry spaces that we've considered above—the straightforward $U(1)$ of electromagnetism or the more intricate $SU(2)$ or $SU(3)$ of the nuclear forces, or even another of the many Lie groups we haven't talked about. The group $U(1)$ directly represents the set of possible changes of phase of the wave function. The other Lie groups, while more complicated, nevertheless also count simple changes of phase of the wave function among the possibilities represented by their elements.

But, once we have one or more of these internal symmetry groups at play, which we must if the basic tenets of quantum mechanics are to be satisfied, then the contention of Yang and Mills applies: since well-separated points in space can't be in immediate communication with one another, then the invariance evidenced within the internal symmetry space must be a local one. But, as we've discussed, the imposition of local invariance under the Lie group of symmetry transformations requires the introduction of an interaction into the wave equation.

We can only conclude, then, that gauge theory tells us that any universe with a behavior consistent with the tenets of quantum mechanics must count elements of causation among its fundamental physical laws. Quantum mechanics, completely independently of what particles may or may not have emerged, ruled out the possibility that the universe that evolved just after the big bang would be one devoid of the meddling intertwinings of its fundamental constituents, devoid of the stuff from which organization, structure, and life itself, spring forth. Interaction is as inexorable a quality of the universe as the very space-time that defines its presence.

We conclude this chapter with a philosophical musing that underscores a similar rhapsody from the end of the previous chapter. In that chapter, we reflected on the physical nature of the (global) symmetry spaces that are associated through Noether's theorem with conserved physical quantities.

Invariance under rotations in three-dimensional space, a very real space from any vantage point, led to the conservation of angular momentum. Invariance under rotations in "spin space," a seemingly real space with some unreal twists, was associated with an extension of the notion of angular momentum conservation to include particle spin in addition to the conventional angular momentum a particle possesses because it's in orbit at a distance from some axis. Finally, invariance under rotations in nuclear isospin space, within which neutrons can be turned into protons and vice versa—a thoroughly unreal space, it would seem—was nevertheless associated with the very real phenomenon of the conservation of nuclear isospin. This begged the question of whether this nominally abstract internal isospin symmetry space is in fact concrete and physical.

In this chapter, we deepened this mystery by fathoms, for now our (local) symmetry spaces are not associated with demonstrable but admittedly arcane physical properties of subatomic particles but, rather, with the very essence of the natural world, with the nature and existence of the causative agents that render the universe both interesting and knowable. Yet, still, we have no more substantial a notion than before of what these internal symmetry spaces really represent in terms of the basic nature of the physical universe. The only thing we can say with certainty is that our interest in this issue is piqued all the more. And, since that's all we can say with certainty on this topic, then that's the last we'll hear of it. You are, at last, left to draw your own conclusions about the relation of these internal spaces to the tangible world of physical reality or, perhaps, about the nature of physical reality itself.

9

The Current Paradigm

———∿∿∿∿∿∿∿∿———

Hidden Symmetry, the Standard Model &

the Higgs Boson

In the mid-1960s, as the United States was in the midst of its murky military engagement in Southeast Asia, there was a less heralded struggle underway in the particle physics community. This was not a battle that pitted various camps of physicists against one another so much as it united them against their perennial enemy: ignorance. The battle being waged was that of the formulation of a workable theory of the weak nuclear interaction.

In a terse two-page note in the November 20, 1967, edition of the *Physical Review Letters*, thirty-four-year-old Steven Weinberg of the Massachusetts Institute of Technology and the University of California at Berkeley outlined a somewhat speculative but seemingly self-consistent model of the weak interactions of electrons (e^-) and their corresponding neutrinos v_e. This model incorporated a gauge theory based on local invariance (symmetry) with respect to the Lie group SU(2) of rotations and phase changes in weak-isospin space (the space in which the weak-force identical particles e^- and v_e are swapped between each other).

This theory also incorporated a U(1) gauge symmetry, the same sort of symmetry that reduced Feynman's theory of quantum electrodynamics to a single principle of almost juvenile simplicity.

One telling aspect of this model is that its "photon"—the single gauge quantum associated with the U(1) symmetry—is not the true photon-field quantum of the electromagnetic interaction. The properties of the associated minimal interaction vertex between this U(1) gauge quantum (known as the B^0) and the electron do not quite reproduce those of the well-established theory of quantum electrodynamics. In addition, though two of the three SU(2) field quanta are identifiable as the W^+ and W^- weak field quanta, the third, electrically neutral W^0 quantum was a prediction of the model without an experimental basis. Recall (chapter 8, "An Experimental Triumph") that weak-interaction processes involving the exchange of an electrically neutral weak field quantum were not discovered until 1973.

For this model, the U(1) B^0 and SU(2) W^0 field quanta, both electrically neutral, are really quite similar to each other. They're not precisely similar, but they're similar enough that one could never tell whether, in any given fundamental process, the B^0 or the W^0 was responsible for the interaction. For example, figure 9.1 shows such an interaction between two electrons. In 9.1a, the U(1) B^0 quantum is exchanged, while in 9.1b, the SU(2) W^0 does the honors. Comparing these two diagrams, there's no way to tell, based on what went into and came out of the reaction, which of the two field quanta is responsible. In both cases, all you see are two electrons bouncing off each other.

Since, in any such interaction, there's no way to tell whether the B^0 or W^0 is exchanged, then it's entirely possible that the exchanged quantum could be a combination of the two. That this model needed to be considered seriously hinged, to a large degree, on the fact that it was able to take a piece—a certain well-defined amount—of the B^0, and combine it with a supplementary and similarly well-defined piece of the W^0, to arrive at a combined field quantum with precisely the properties of the photon of quantum electrodynamics.

It's just like forming an alloy of two metals. You have a pure gold coin, which is wonderfully impervious to corrosion but very soft. You also have a copper coin, which is hard but tarnishes easily. If you take 80% of the gold coin and 20% of the copper coin and melt them together, you come up with an alloy. It looks pretty much like gold, but if you make a wedding band out of it, it's not just shiny, it is (for better or worse) also durable. It has just the properties the jeweler wants. Likewise, an alloy of precise amounts of the B^0 and W^0 field quanta (roughly 80% and 20%, respectively) has just the properties the quantum-electrodynamicist wants; it's precisely the photon.

Then there are the leftovers, the 20% of the B^0 (the gold coin) and the 80% of the W^0 (the copper coin). You can melt these together into a

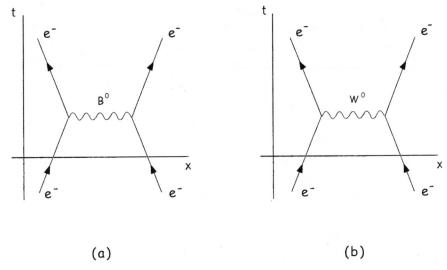

Fig. 9.1. The mutual repulsion of two electrons through the exchange of the U(1) B^0 boson and the neutral SU(2) W^0 boson. There's no way to tell which of the two actually participates in any given interaction, so the exchanged field quantum is thus a quantum-mechanical mixture of the B^0 and W^0.

quantum-mechanical alloy, and the result is a second electrically neutral field quantum that's available to mediate an interaction between matter particles. This leftover piece is the quantum of the neutral interaction of the weak nuclear force; since this carrier of the neutral weak force is not quite a gauge-field-quantum partner of the W^+ and W^-, it needs a somewhat independent name, and for want of a more descriptive term, it has come to be known as the Z^0 boson.

Weinberg's model (really, as we'll see, the Glashow-Salam-Weinberg model) is not a model of independent electromagnetic and weak nuclear forces. Rather, it is a model of single, interconnected electroweak force, with two separate facets: the observed electromagnetic and weak forces. The Glashow-Salam-Weinberg model is based on both a U(1) and an SU(2) gauge symmetry, neither of which is exclusively associated with either the weak or electromagnetic force. Instead, they are together representative of a combined, a *unified*, electroweak interaction.

Because the properties of the photon are so well known, the exact amounts of each of the U(1) B^0 and SU(2) W^0 needed to make a photon are known very precisely. Correspondingly, the complementary amounts of leftover B^0 and W^0 needed to form a Z^0 are inferred with equivalent precision. Because of this, the properties of the Z^0 are a characteristic and precisely predicted mix of the properties of the B^0 and W^0 field quanta.

It's not that you can separately measure the properties of the B^0 and W^0 and then combine these properties to predict those of the Z^0; you can't! It's not the B^0 and the W^0 but, rather, their alloyed states of the photon and Z^0, that are the physically real—the observable—field quanta. Instead, given our precise knowledge of the properties of the photon, and the unequivocal nature of Weinberg's model, we can derive exceedingly sharp predictions of the properties of the Z^0. The experimental confirmation of these precisely predicted properties was, during the decade of the 1990s, a focal point of the effort of the particle physics community and absorbed a large fraction of my own time during that period.

Weinberg appreciated that his model applied equally well to weak interactions involving the heavier μ^- lepton and its neutrino v_μ; with the discovery of quarks soon after, it was quickly appreciated that the model applied to the electroweak interactions of all known fundamental particles.

Thus, with the discovery of Z^0-mediated weak neutral currents in 1973, as the United States called in its troops from the less fruitful conflict in Southeast Asia, the international physics community declared a tentative victory in its battle over the secrets of the weak nuclear force. Not too long afterward, in 1978, exacting experiments masterminded by Charles Prescott of the Stanford Linear Accelerator Center and Lev Barkov and Max Zolotorev of the Budker Institute in Novosibirsk, Russia, were the first to quantitatively confirm the precisely mixed (part B^0, part W^0) properties of the Z^0.

The physics community was compelled to acknowledge that this model was able, at last, to provide a unified and strikingly accurate description of the electromagnetic and weak nuclear interactions. For his work in developing this "Standard Model" of the electroweak interaction, Weinberg was accorded the 1979 Nobel Prize in Physics. He shared the prize with the two other physicists who lend their name to the Glashow-Salam-Weinberg model. Sheldon Glashow (at the Neils Bohr Institute in Copenhagen at the time) had, in 1961, published a model based on U(1) and SU(2) gauge symmetry but which lacked the notion of spontaneously broken symmetry that will soon enter our discussion. The Pakistani theorist Abdus Salam of Imperial College in London had independently developed a model essentially identical to that of Weinberg's, publishing only slightly later, in 1968. The Nobel Committee appropriately recognized that Weinberg and Salam had worked contemporaneously and that both had benefited greatly from Glashow's earlier work. The three were cited by the Committee for "their contributions to the theory of the unified weak and electromagnetic inter-action between elementary particles, including, *inter alia*, the prediction of the weak neutral current."

Were this chapter restricted to a discussion of the theoretical forging and experimental assay of the alloyed Z^0 boson, it would be mercifully short. There is a thorny problem, however, that needed to be addressed on the way to developing the Standard Model. While the W^+, W^-, Z^0 field quanta are all quite massive, each with about 100 times the proton mass, the field quanta generated by gauge theories are always massless. The solution to this problem required the introduction of an entirely new and decidedly subtle notion: that of hidden gauge symmetry. Out of the formalism of hidden symmetry arose, in addition to the masses of the weak-interaction field quanta, the prediction of an exotic new field quantum—the Higgs boson.

Gauge Theory and the Strength of Forces

A standard first-day-of-class lecture demonstration that beginning students of electromagnetism find themselves sitting through is the following. The lecturer rubs two materials together, then divides the static charge accu-mulated on one of the materials between two separate conducting spheres. The spheres are then suspended, close to each other but not touching, from two insulating threads. The point of the exercise is simply to demonstrate that the like charges deposited on the spheres repel each other. The stu-dents are told (and none seem to question it) that the repelling force is that of static electricity, and the lecturer then goes on to probe the qualitative properties of the electrostatic force by varying the amount of charge on the balls, their separation, and so forth.

But is it really true that the electrons added to the spheres interact and repel each other solely by the electric force? We know that there is not one, but four forces through which objects can interact with one another. What about the other three forces?

Since leptons, in general, and electrons, in particular, don't carry color charge, we don't need to concern ourselves with the strong nuclear force. However, electrons do carry the charges, mass and weak isospin, appropri-ate for participation in the gravitational and weak nuclear interactions (note 9.1). There's no reason why the static charges on the spheres shouldn't also interact with each other through these two other forces, and in fact, they do. It's just that gravity and the weak force are so much weaker than the electrostatic force that their effects are too minute to matter. The lecturer's claim that the demonstration probes the properties of the electrostatic force alone is correct, for all practical purposes.

But why is it that the gravitational and weak forces are so weak? In the case of gravity, all we can say is that the relative size of the gravitational and electric charge of an electron is vastly different (note 9.2).

In the case of the weak nuclear force, the weak-isospin charge of the electron is not that different in magnitude from its electric charge; were this not the case, the Standard Model, which provides an intermixed description of the electromagnetic and weak nuclear forces, wouldn't work. The strengths of the U(1) and SU(2) interactions of the Standard Model need to be about the same if we're going to combine the U(1) B^0 field quantum with the SU(2) W^0 quantum to make the physical photon and Z^0.

So if the weak-isospin and electric charges of matter particles are roughly the same, then why is the weak force weak? As mentioned in the section "Particles of Force" in chapter 5, the answer has to do with the large mass, roughly 100 proton masses, of the $W^\pm$ and Z^0 field quanta. Recall the de Broglie relation

$$\lambda = \frac{h}{p};$$

the wavelength λ of an object is inversely proportional to its energy/momentum p. The W and Z field quanta, each possessing substantial mass-energy, thus have associated with them a very short wavelength, about 10^{-18} meters, which is small even in comparison to the 10^{-15}-meter radius of the proton.

So it's not that the W and Z weak field quanta interact with matter any less readily than the photon but, rather, that their sphere of influence simply doesn't extend very far. If two objects, such as the repelling spheres of our lecture demonstration, are not within 10^{-18} meters or so of each other, then they can't influence each other through the weak force. Put another way, it's not that the electron's weak-force charge is too weak, it's just that, as far as the weak force is concerned, they simply miss each other. It's easy to misinterpret this lack of contact as an inherent feebleness in the weak force and that is what was done in the early days and why physicists mistakenly thought of the weak force as being weak when, in fact, it is really just short ranged.

If this picture is correct, then to the extent that particles such as electrons become fast-moving objects in an accelerator, rather than static charges in an introductory physics demonstration, the now sizable energy/momentum that can be possessed by the exchanged photon should be associated with a substantially shorter photon wavelength, again according to the de Broglie relation. Thus, as we go to higher and higher energy in the lab, the sphere of influence of photons that exchange a lot of energy should begin to shrink, and the likelihood of the electrons interacting with each other through the electromagnetic force should get less and less.

This is exactly what happens, and at beam energies of the order of the mass-energy of the W and Z or higher, the electromagnetic force is indeed observed to be roughly the same strength or really, of similar influence, as the weak force. At these energies and above, the unification of the weak nuclear and electromagnetic interactions becomes patently clear.

In any regard, we have a point of absolutely central importance to our development of a unified theory of the electroweak interaction. Our theory is a gauge theory, based on chapter 8's principle of local invariance under a $U(1)$ gauge group of phase changes and a separate $SU(2)$ gauge group of phase/weak-isospin changes. Identifying the gauge group(s) at play is most of the work we need to do to specify any given gauge invariance–based theory, but we also need to specify the nature of the charge associated with each gauge group, and then determine the overall strength of that charge with some appropriate experiment (such as measuring the strength of the repulsive force between two fundamental particles).

The charge associated with the $SU(2)$ gauge group is, as we have discussed at some length, a quality known as weak isospin. The charge associated with the $U(1)$ force, however, is something new, which we haven't yet discussed. Just as the photon acts to exert an influence between particles that carry electric charge and the W^+, W^-, and W^0 act to exert an influence between particles that carry weak isospin charge, the $U(1)$ field quantum (the B^0) acts between particles possessing something known as *weak hypercharge*.

What exactly is weak hypercharge? To be frank, we don't really care because there is no $U(1)$ field quantum; the real field quanta are the photon and Z^0, which are mixtures of the $U(1)$ and neutral $SU(2)$ field quanta.

Since the photon concerns itself solely with electric charge, electric charge is thus some combination of U(1) weak hypercharge and SU(2) weak isospin. Flip it around, and you see that weak hypercharge is just some combination of weak isospin charge and electric charge. So, if you really want to know what weak hypercharge is, there's your answer.

But the essential point is that, whatever the U(1) weak hypercharge and SU(2) weak isospin charges are, they have about the same strength for any given fundamental matter particle, quark or lepton. We haven't discussed how they are determined experimentally, and we won't, because the principle of hidden symmetry obviates the need to do so. But they are of about the same magnitude, and the "weakness" of the weak force—the characteristic of the weak force that first led physicists in the early twentieth century to suspect that it was a new and independent force—is due to the fact that its field quanta happen to be very massive, rather than massless like the photon. It is clear that the issue of field-quanta masses must play a central role in any useful theory of the weak, or electroweak, interaction.

Massive Force Quanta: The Downside

But there's a serious problem. There's no straightforward way to incorporate the mass of force-field quanta into a gauge theory. Everything begins with the quantum-mechanical wave equation—the Schrödinger equation, Dirac equation, or whichever of the several possibilities is appropriate for the task at hand. The requirement of invariance under local transformations belonging to a Lie group of "gauge symmetries" then leads to the introduction of the quanta of the interaction's force field. Recall, though, that the wave equation is just a quantum-mechanical representation of the notion of energy conservation. Since Einstein tells us that there is energy associated with mass, then if we want our gauge quanta to have mass, we have to include their mass-energy in the energy balance accounting represented by the terms in the wave equation.

The precise form of these additional mass-energy terms is beyond our scope, but the procedure for including them is well defined and can be found in any modern textbook on relativistic quantum field theory (note 9.3). What can also be found in these same books is a straightforward, two-line demonstration that such a term, incorporating the mass of the field quanta into the wave equation, destroys the local Lie group (gauge) invariance of the wave equation that we worked so doggedly to preserve. It doesn't matter what Lie group of phase/internal-symmetry transformations we're working with. If we try to incorporate field quanta masses by brute force—by just adding into the wave equation the extra terms that account for the mass-energy of the field quanta—the wave equation will no longer be invariant under local transformations selected from the Lie group.

Is this really something to worry about? We are trying to develop a working theory of the electroweak interaction. Maybe we could arrive at such a theory in two steps. First, we adhere, as Yang and Mills instructed, to the principle of local gauge invariance to introduce the field quanta and develop the rules for their minimal interaction vertices. With that done, perhaps we could just throw that principle out at the last minute by then adding in the terms accounting for the mass of the field quanta. We'd still be making use of the principle of local gauge invariance to develop the theory's rules of interaction, although the final wave equation we end up with, after tacking the mass-energy terms on, would no longer exhibit local gauge invariance. But who cares; as long as the theory works, we should be satisfied.

But it *won't* work! To write down a relativistic quantum field theory for any given force—a set of field quanta and associated minimal interaction vertex rules for their associations with appropriately charged matter—is one thing. To have this theory yield sensible results for the modeling of all possible processes (Feynman diagrams), and for any energy of the participating particles—to be renormalizable—is another. But the only theories we know of that are renormalizable are theories based on local gauge invariance. So any term we wantonly tack onto the wave equation that compromises this invariance necessarily destroys the theory.

So the brute force introduction of the mass-energy terms in the wave equation, with their corresponding destruction of local gauge invariance, does a much more practical damage to our theory than the loss of a mere philosophical tenet. It renders the theory useless as a representation of the interaction we are attempting to model. Without local gauge invariance, there are guaranteed to be processes whose calculation within the theory gives results that are nonsensical. Gauge theory and intrinsically short-ranged forces mediated by the exchange of massive field quanta are simply incompatible, it would seem, and this incompatibility deeply threatens all the progress we have made heretofore.

Screening and Effective Mass

You can't listen to a portable radio if you're sitting, fully enclosed, inside of a conducting metal cage. A metal box—or even a metal mesh—somehow absorbs the radio waves that would otherwise make it to the radio's antenna. However, radio waves seem to make it just fine through most building materials and through stretches of the earth's atmosphere in excess of 100 miles.

The reason for this is that a metal conductor, unlike the insulators that comprise the atmosphere and most building materials, have a high density of electrons that are free to move under the influence of electric and magnetic fields. These electrons, known as *conduction electrons*, are the one or two electrons per atom that exploit the regular pattern of closely spaced atoms in the metal to become the property of the metal as a whole, rather than that of the individual atoms to which they belonged before the metal was forged, and thus are able to move freely about within the conductor.

Radio waves are relatively low frequency waves of electromagnetic radiation, like light, but with a relaxed oscillation frequency of roughly 1 million cycles per second (visible light has a frequency of roughly 10^{16} cycles per second). The medium that oscillates in the case of electromagnetic radiation is the electromagnetic field, the magnitude and direction of the force that would be exerted by the radiation (radio wave) on any electrically charged object placed in its path.

Electrons, of course, are electrically charged, so if an electron that's free to move, such as a conduction electron in a piece of metal, is in the path of a wave of electromagnetic radiation, it will begin to jiggle back and forth at the frequency of whatever radiation is incident on it. But jiggling electrically charged particles create electromagnetic radiation of their own, and this induced radiation acts to cancel the incident radiation as it moves through the metal.

No single electron will respond enough, in its jiggling motion and subsequent canceling radiation, to diminish much of the original radio wave's amplitude. However, if a large number of free electrons are packed densely together, as they are in a conductor, the accumulated effect of each of the large number of individual free electrons lying in the path of the radio wave will lead to the incident wave's total cancellation in the region beyond the conductor. For example, in a good conductor such as aluminum or copper, the density of free electrons is high enough so that a foil just a few tenths of a micron (roughly one hundred-thousandth of an inch) thick will completely cancel any radio wave incident on it.

So, if you're sitting inside a conducting box, even if the walls are relatively thin, no electromagnetic radiation from outside the box will penetrate the box to the antenna of your radio. This property of conducting surfaces—their tendency to block the passage of electromagnetic radiation—is referred to as *screening*.

The relevance of this discussion to that of the massive force-field quanta is direct (note 9.4). From the point of view of quantum electrodynamics, electromagnetic radiation is composed of a stream of electromagnetic force-field quanta (photons) propagating in an organized way through space. These photons have a range of influence that is essentially infinite; you can often pick up an AM radio station hundreds of miles from its transmitter. In fact, the field of radio astronomy is predicated on the detection of radio wave signals from sources located in the farthest reaches of the universe.

However, if you place a conductor—a high density of free electrons—in the path of a radio wave, the radio wave will penetrate only a short distance into the conductor. The range of influence of the photons that comprise the radio wave is no longer infinite but is instead reduced to a mere fraction of a micron.

Now, we've argued that the weak nuclear force is not really weak; it only appears so because it is short ranged, with a range of influence much smaller than even the radius of a proton. The short range of the weak force is due, in turn, to the fact that its field quanta—the W and Z particles—are massive. If the quanta associated with the mediation of a given force are massive, then the range of influence of the force becomes limited.

But when a photon, which we know in truth to be massless, enters a conductor, its range of influence becomes limited. It's as if, on entering the conductor, the photon suddenly becomes a massive field quantum: the cumulative screening of the electrons in the conductor acts to make the photon, in effect, massive. We say that the screening process generates an "effective mass" for the photon.

Formally, then, we can model the behavior of electromagnetic radiation inside a conductor by treating each photon in the radiated wave as a massive field quantum. However, to say that we can formally treat a photon inside a conductor as being massive is not to say that a photon really *does* become massive when it enters the conductor.

In fact, we know that this "mass" is really not a property of the photon itself because it depends on the environment in which the photon happens to find itself at any given time. The more dense the free electrons in the conductor, the greater the screening effect, and the larger the effective mass of the photon. The assignment of the property of "mass" to a photon inside a conductor is really just a fabrication, a modeling tool with no physical basis. To find out what the mass of the photon really is, we have to study its properties in a vacuum, when it's not under the influence of any jiggling charged particles. When we do that, we see clearly that the range of influence of the photon is unrestricted: the photon is massless.

But what if the whole universe were a conductor (technically, a "plasma") filled with a uniform density of free electrons? In this case, no matter where we did our experiment to probe the properties of the photon, it would always behave as if it had a particular mass. This mass would depend on the density of electrons in the universe, but since that density would be uniform (unchanging from point to point in space), the effective mass of the photon would always be the same, no matter where one did the experiment. There would be no place at which one could study the properties of the photon away from the influence of the jiggling electrons, for the jiggling electrons would be everywhere. How then would one know that the photon is, in truth, massless? We wouldn't.

The Solution, Part 1: The Higgs Field

At this point in our discussion, our U(1) and SU(2) gauge theory of the electroweak interaction is under a grave threat from the need to introduce gauge symmetry-destroying masses for the W and Z force-field quanta. In the next couple of sections, we'll rescue the theory, and the general notions that will save it from the scrap bin are those of "background fields" and "hidden gauge symmetry." These ideas and their application to the theory of the electroweak interaction did not spring forth at a discrete point in time from a single imaginative mind. Instead, they were developed steadily over a period of roughly seven years, between 1960 and 1967, through a series of conjectures put forth by a number of physicists, not all of them particle physicists, among whom we can find some of the greatest theoreticians of the time.

In 1960, the Japanese American physicist Yoichiro Nambu of the University of Chicago suggested that it's not necessary to think of the vacuum, the seeming void that forms the bulk of space in the universe, as being empty. Roughly ten years earlier, the development of quantum field theory had introduced the notion of the living vacuum in which virtual particle-antiparticle pairs are continuously popping into and then annihilating themselves out of existence in the void of space. Nambu suggested that we are free to take this one step further. There's no reason, he claimed, to demand that the fluctuations of the living vacuum occur against a backdrop of nothingness. The underlying, unfluctuated state of the vacuum could instead be one in which some fields—be they fields associated with the presence of matter, forces, or even something altogether new and different—provide an enduring background presence, against which the machinations of the universe play out.

The behavior of an electromagnetic field in a conductor, a phenomenon more likely to pique the curiosity of a solid-state physicist interested in the bulk properties of matter than a particle physicist, can be treated formally by ascribing an effective mass to the photon. Accordingly, it was Philip W. Anderson of Bell Laboratories in Murray Hill, New Jersey, one of the most influential figures in the history of solid-state physics, who first suggested that the phenomenon of screening could be responsible for the mass of force-field quanta. Anderson's original suggestion, however, was in the context of the strong nuclear force, an interaction we now believe to be mediated by the *massless* gluon.

Several papers that appeared in 1964, the most prominent of which was due to Peter Higgs of the Institute of Mathematical Physics at the University of Edinburgh in Scotland, enunciated Anderson's connection between screening and mass in the full context of relativistic quantum field theory. This was the point at which the name Higgs became associated with the developing notion of hidden gauge symmetry.

Finally, the 1967 work of Weinberg and Salam applied the notion specifically to the U(1) plus SU(2) gauge theory of the electroweak force, clearing the last major hurdle in the formulation of the Standard Model of the electroweak interaction.

We'll begin by considering Nambu's 1960 conjecture. According to Nambu, there should be nothing wrong in principle with the notion that the vacuum—the nothingness that fills the voids in our predominantly empty universe—contains an appreciable, uniform average value of the free-electron matter field, that is, a uniform density of electrons.

Were this the case, the nature of the electromagnetic interaction—the repulsion, say, of two like electric charges—would be drastically altered.

The electromagnetic field quantum (photon) that the two charges exchange to effect this repulsion would now be screened by the ever-present background matter field. Thus, to model quantitatively the nature of the repulsion, we would ascribe to the photon a mass—effective, but a mass nonetheless—of a magnitude set by the density of the uniform background electron field. As a result of this effective mass, the interaction mediated by the photon would become short ranged, and so any two charged objects separated by an appreciable distance—by a distance greater than the restricted range of influence of the now-massive photon—would have essentially no effect on each other. For example, the two charged spheres in the physics demonstration at the beginning of this chapter would no longer repel each other, and the theory of the electric force that we teach college students would be vastly different from its current form.

But now, consider the question posed at the end of the previous section: If the average value of this background electron field is uniform, pervading every nook and cranny of space-time, how would we ever know that the photon is actually massless? How could we tell that the short-ranged nature of the electromagnetic interaction is due to the fact that the fundamentally massless photon is only *acting* as if it has a mass? We couldn't. Since there would be no unpolluted region of space free of this pervasive background field in which to mount an experiment to determine the true mass of the photon, we would have no way of knowing that the photon is massless. We would conclude, from the result of any experiment done to determine it, that the photon is massive.

Although possible in principle, physically this is not the case; the photon is observed to be massless. Instead, our interest in these ideas stem from the realization that this picture may well be appropriate for the description of the massive W and Z force-field quanta of the weak nuclear interaction. Let's explore this possibility.

If we try to interpret the observed masses of the W and Z quanta in this way, we need to hypothesize that there is a background density of some field with which the apparently massive W and Z quanta (but not the massless photon) interact. For example, neutrinos take part only in the weak interaction, so a background neutrino matter-field density would generate an effective mass only for the field quanta of the weak nuclear force. So, the neutrino matter-field would be a candidate for this background field, although there are many other possibilities.

Since we can't yet say much about this hypothetical background field (other than the fact that it carries weak-isospin charge so that the weak force-field quanta interact with it), it could possibly be a new and, heretofore, unobserved fundamental field. For all we know, at this point in the discussion, it could be the matter field associated with the neutrino, but it also could be some new matter field similar to that of the known quarks and leptons, some new force field similar to that of the known force-field quanta, or something completely unlike any fundamental field that has been observed to date. For the moment, absent any information as to what this field might actually be, we'll give it an arbitrary name. We'll call it the "Higgs field."

The Solution, Part 2: Hidden Symmetry

Intriguing as this suggestion may be, that the mass of the weak nuclear force-field quanta may be a convenient but physically misguided interpretation of the screening effect of some pervasive background field, one can ask what it buys in our attempt to formulate a theory of the weak interaction. Regardless of whether the mass is real or is an effective mass generated by screening effects, if we include the mass-energy term of the force-field quanta into the wave equation, won't our precious gauge symmetry still be violated, leaving us with the same quandary we started with?

The answer is no, not necessarily. However, to introduce force-field quanta masses, even effective masses generated by screening effects, without violating gauge symmetry, we must be very clever. This is where the notion of hidden gauge symmetry enters.

Most of us who are unfortunate enough to remember what life was like when we were thirteen years old will recall that one of the predominant influences that shaped our lives at that age was the unrelenting coercion of peer pressure.

Picture a schoolyard on a crisp December morning, with a delicate layer of frost adorning the otherwise brown grass. Add to the picture several hundred barefoot teenage boys, each one confronted by a choice between one of four pairs of jogging shoes produced by four different manufacturers. Frankly, even in the best of conditions, it doesn't really matter which pair one chooses; they are all comfortable and well constructed. And for the situation at hand, a schoolyard full of young men shuffling around barefoot in the cold, it should matter even less. Any of the four pairs will separate the souls of their feet from the numbing cold.

The situation is *symmetric* with respect to the choice of shoes; the desired effect (saving one's toes from frostbite) is independent of which company's shoes each student chooses. But each student *does* have to make a choice.

Teenage mentality being as it is, none of the boys dares to put on a pair of shoes, for fear of picking the pair that exposes their lack of awareness of fashion. Eventually, some student decides he can't take the cold any longer, and he grabs one of the pairs of shoes and puts them on as quickly as his numbed fingers permit. His neighbors, accepting his cue, quickly follow suit, and in relatively short order, the entire field of adolescents stands in a state of relax, uniformly clad in the same brand of shoe.

At this point, a department store employee on his way to work happens to glance over to the schoolyard and sees the multitude of young men standing there, all wearing the same brand of shoe, surrounded by discarded pairs from the three other manufacturers. His gut reaction is to rush into work to tell the manager to order a large number of the selected brand's shoes, but on further reflection, he thinks better of it. Why?

The reason is that the symmetry, the intrinsic underlying equality of all four brands, still pertains. A student from another school who hadn't been at this particular schoolyard at the time of the *arbitrary* selection of the chosen brand would be no more likely to purchase that brand than any of the other three. In our schoolyard on that frigid December morning, the symmetry was not *broken* by the capricious selection of one of the brands over the others, but it was masked, or *hidden*, by the choice. A careful study, in this case, appropriate interviews by a person well schooled in adolescent psychology, would reveal that the symmetry is still there. Eventually, the school counselor would get the students to admit they couldn't have cared less which of the pairs of shoes ended up on their feet.

This is the essence of the notion of hidden symmetry. One can be presented with a physical system whose appearance is one of broken symmetry (the selection of the same pair of shoes by all the students seems to suggest that pair is somehow preferable) that belies the true underlying symmetry of the system (the fact that there really is no difference whatsoever in the quality of the four choices of shoes, and the choice of a particular pair was completely arbitrary). Some additional external factor, in our example, the discomfort of standing around barefoot, required that some choice, no matter how arbitrary, be made by the system. Once the choice was made, some internal property of the system—the psychological interdependence of the students—required that the entire system adopt the same arbitrary choice, leading to the masking, or hiding, of the true underlying symmetry. But the symmetry is still there. No one really could give a damn about which pair of shoes are on his feet.

Back in the world of particle physics, the question in the early 1960s was whether it is possible to use this hidden symmetry idea to introduce the mass of the W and Z weak force-field quanta in such a way that the corresponding mass-energy terms in the wave equation reflect a *hidden*, as opposed to a *broken*, gauge symmetry. If so, then at its core, the model would still be a U(1) and SU(2) gauge invariant, and the theory might thus be well behaved.

A slightly different point of view may shed more light. If we include the mass-energy terms in the wave equation, gauge invariance, symmetry with respect to local U(1) and SU(2) transformations, is destroyed. The two-line mathematical demonstration that the inclusion of the mass-energy terms into the wave equation ruins the gauge symmetry is clear and incontrovertible. The theory as such will not work.

However, perhaps there are additional, compensating terms that we can add to the wave equation, along with the mass-energy terms, that fix things up for us by restoring the gauge symmetry of the model. Mathematically, it's not possible to do this in a way that leaves the wave equation *apparently* symmetric with respect to local U(1) and SU(2) transformations. It *is* possible, however, to introduce terms that reflect a *hidden* symmetry.

Thus, by introducing the mass terms for the W and Z quanta along with additional, compensating terms, it's possible to avoid the completely broken symmetry that is suffered when the mass-energy terms are inserted by themselves into the wave equation. Might it be that these extra terms, which restore a hidden, rather than apparent, gauge symmetry to the wave equation are enough to rescue the self-consistency of the theory?

The answer is yes, and the key to the procedure is to introduce all of these terms, the mass-energy terms as well as the additional, compensating terms that restore the hidden gauge symmetry, using the introduction of a new, all-pervasive background field: the Higgs field.

The Standard Model

In table 5.1, the table showing the array of currently known fundamental matter particles, the quarks and leptons come in groups of two. This grouping reflects an essential property of the weak force: that the weak nuclear interaction cannot distinguish between leptons and their associated neutrino, or between associated up-type and down-type quarks. This led us to hypothesize that the weak force should be invariant under changes (really, local transformations) that come from the group SU(2) of transformations in weak-isospin space—transformations that swap electrons with neutrinos, down quarks with up quarks, and so forth. The SU(2) invariance of the weak force is telegraphed by the fact that fundamental particles come in doublets whose members can be freely substituted for one another in any given weak process without changing the nature of the process.

When we apply the gauge principle of chapter 8, our desire to preserve invariance with respect to these transformations in weak-isospin space forces us to introduce three massless force-field quanta: W^+, W^-, and W^0. When we mix the neutral W^0 quantum with a little of the B^0 quantum of the U(1) "weak hypercharge" symmetry group, we get the Z^0. Together, the W^+, W^-, and Z^0 act as the three field quanta of the weak nuclear force but with one important shortcoming. Since the weak interaction is short ranged, the W^+, W^-, and Z^0 must all be massive, but introduced in this way, they are massless, so something additional must be done to make this SU(2) and U(1) theory work out. Here's how Weinberg and Salam used the ideas of the previous two sections to finesse this issue in their presentation of what has come to be known as the Standard Model of the electroweak interaction.

The idea is to hypothesize a new doublet of fields that, just like the quark or lepton matter-field doublets, is invariant under SU(2) and U(1) transformations in the spaces of weak isospin and hypercharge (see fig. 9.2). As for

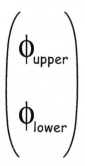

THE HIGGS DOUBLET

Fig. 9.2. Each of the two components (ϕ_{upper} and ϕ_{lower}) of the Higgs doublet is a complex field, that is, a function that assigns a complex number (representing the strength of the field at that point) to each point in space and time.

any of the doublets of matter fields in table 5.1, we ask that this Higgs doublet of fields be invariant with respect to local SU(2) transformations. Just as for all the other doublets, this introduces a minimal interaction vertex between the Higgs doublet fields and the W and Z field quanta: the particles of the Higgs doublet fields can interact with each other by exchanging W and Z bosons.

But that's not what we're after! What we want to do is to have the Higgs field be an all-pervasive background field, so that the interaction of the W^+, W^-, and Z with the Higgs field leads to the screening effect we just learned about (in the section "The Solution, Part 1"), and thus generates the (effective) masses of the W^+, W^-, and Z^0 that we need if the theory is to make sense. We don't need a theory that describes the interaction of the particles of the Higgs doublet; we've never observed such particles, so we don't further our understanding of nature by describing their interactions. What we do need is a Higgs field that, rather than as a discrete particle that exists at a single point in space and time, manifests itself as a uniform presence throughout space-time, providing the ubiquitous background field that we need to generate the effective masses of the W^+, W^-, and Z^0. Recalling Yoichiro Nambu's inspiration, we need to hypothesize that the vacuum of the cosmos, rather than being blank and empty, is uniformly filled by the Higgs field.

And herein lies the beauty of hidden symmetry. Let's take one of the two fields of the Higgs doublet, say, the lower member of the Higgs doublet of figure 9.2, and ask what happens if that field pervades every nook and cranny of the cosmos. Just as we don't perceive the continual fluctuation of virtual particle/antiparticle fields in and out of the living vacuum of quantum field theory, we can't perceive this uniform background presence of the lower field of the Higgs doublet. Yet, we hypothesize that it is indeed there, exerting a sort of cosmic drag force on anything that interacts with it, and thereby giving mass to the W and Z quanta.

What have we done? We've *hidden the symmetry* of the Higgs doublet. By choosing one of the two doublet fields and not the other to provide our all-pervasive background field, we seem to be suggesting that the universe is not symmetric with respect to the SU(2) transformations that swap the two doublet fields. The swapped universe, with the background provided by the upper Higgs doublet field, is a manifestly different universe than the universe we chose. The symmetry appears to be broken.

But the symmetry is not broken. The W and Z particles still interact just as readily with the upper as the lower Higgs doublet field. It's just that the upper Higgs doublet field isn't present everywhere in space, and the lower doublet field is. For some reason (more on this reason in a moment), the universe had to choose one of the two Higgs doublet fields to be the all-pervasive background field, and it chose the lower. It could just as easily have chosen the upper, but it didn't. Just as the barefoot students had to make a choice of tennis shoes to protect their feet from the cold, the universe had to (just after the big bang) make a choice of which Higgs doublet field will develop a uniform presence throughout space. Due to the underlying symmetry, the choice was arbitrary, but the choice had to be made. And once made, the choice led to a physical state that belies the true, underlying symmetry of the Higgs doublet field.

And this is exactly what we need. Hiding the symmetry of the Higgs doublet in this way provides an all-pervasive background field that, through the screening effect we discussed a few pages back, generates the mass of the W^+, W^-, and Z^0. But, in addition—and this is the essential point—it provides that mass through interactions between the Higgs, W, and Z fields for which the SU(2) symmetry is hidden, rather than broken outright. By preserving the underlying gauge symmetry in this way, Weinberg and Salam speculated that this approach would lead to a self-consistent, or renormalizable, theory.

Now, this being physics, and (I hope) not fiction, we need to compel the symmetry to hide itself according to some physical principle—we need to give the universe a concrete reason to hide the SU(2) symmetry by arbitrarily choosing a preference for one of the two members (upper or lower) of the Higgs field doublet. In the model of Weinberg and Salam, this was accomplished by introducing a new and somewhat ad hoc notion into the picture: that of the *Higgs potential*.

To understand the role of the Higgs potential, consider the following analogy. A large number of identical weights, each attached to its own identical spring, lie flat on a lab bench. Lying flat, there's nothing to compel the springs to extend; the value of the spring extension field throughout space (for all springs) is zero. But now, dangle the springs from a clothesline, and suddenly an external property, the pull of gravity, causes each mass to extend its spring by an amount that is the same from spring to spring. Dangling, the value of the spring extension field is nonzero and uniformly so from spring to spring (point to point in space). The Higgs *potential*, in this analogy, is provided by the external pull of gravity; the value Higgs *field* is just the degree of extension of each of the springs.

Or, in terms of our schoolyard metaphor, the external Higgs potential is the sensation of cold on the feet of the students, and their resulting physiological need to get something on their feet, while the magnitude of the Higgs field can be one of two possibilities: zero (no shoes on) or one (one of the four equivalent pairs of shoes on each student's feet).

So, the role of the external Higgs potential is to compel the universe to adopt a particular magnitude, at every point in space, for the value of the Higgs field. Which of the two components of the field—upper or lower—the universe selects to develop this pervasive field is arbitrary. The external property (gravity in our analogy) compels the universe to adopt a nonzero value of some component of the Higgs field (the amount of extension of each hanging spring) uniformly throughout space-time (from spring to spring), but the choice of which component of the Higgs field develops this uniform magnitude is arbitrary.

So the external Higgs potential, whatever it is, compels the universe to adopt a nonzero value of one component of the Higgs field that uniformly pervades all of space and time and, as a by-product, gives mass to the W and Z bosons. The universe, arbitrarily, has chosen this to be the lower of the two Higgs field components. What role, then, does the upper component play?

Consider, say, the upper component of the up/down quark doublet of table 5.1—the up-quark matter field. A nonzero value of this field at any point in space-time indicates that, at that location and at that time, one or more up-flavored quarks is present. At most space-time points, the value of the up-quark matter field is zero, but wherever there happens to be an atomic nucleus that contains up quarks, the value of this field is nonzero. Similarly, a nonzero value of the down-quark matter field indicates the presence of one or more down-flavored quarks, and so on for all the matter- and force-field quanta of tables 5.1 and 5.2.

Now, what I would like to be able to say is the following. Just as a nonzero value of the up-quark matter field indicates the presence of an up quark, a nonzero value of the upper component of the Higgs doublet field indicates the presence of a quantum of the upper-component Higgs field. Thus, the upper component of the Higgs doublet is a field associated with a Higgs particle, a particle that goes by the name of the Higgs boson.

In spirit, this is true, and for the most part we won't get into any trouble thinking of the Higgs boson in this way. However, throughout this book, I've taken some pains to avoid confusing those readers who have some background in this field, so in that spirit, the next few paragraphs will provide a more accurate picture of how the Higgs boson arises from the Higgs doublet field.

The Higgs doublet is in fact a doublet of complex fields; rather than two fields components, it represents four field components: one component for each of the real and imaginary parts of the upper and lower field in the doublet (note 9.5). Also, just as the upper and lower components of the matter fields of table 5.1 differ by one unit of electric charge, so do the upper and lower Higgs field components; the upper component is charged, while the lower component is electrically neutral.

In quantum field theory, the representation of a massive spin-1 particle (such as the W and Z force-field quanta) is qualitatively different from that of a massless spin-1 particle. The massive spin-1 particle has more types of motion available to it, so it takes more field components to describe its motion. So, when the two W and one Z boson get effective mass from the screening effect of the pervasive Higgs field background, they need to incorporate more field components into their description. They do this by co-opting three of the four components of the (complex) Higgs doublet field. The two charged W bosons each take one of the two charged (upper) components, while the neutral Z boson takes one of the two neutral (lower) components. The leftover neutral component is the one that has developed the uniform, pervasive nonzero value (in response to the influence of the external Higgs potential) that causes the screening effect that makes the W and Z bosons behave as if they have mass.

The leftover component of the Higgs field—the one that develops the uniform nonzero value—can deviate, at any individual point in space-time, from this pervasive uniform value. Differences relative to this uniform background value (rather than differences relative to zero) indicate the presence of the quantum of this component of the Higgs field. This quantum, an electrically neutral, spin-0 particle, is the Higgs boson.

Whether you think of it the easy but less precise way presented first or in the more rigorous terms of the preceding three paragraphs, one point stands. The Standard Model of Weinberg and Salam is able to account for the masses of the W and Z bosons through the introduction of a new field, the Higgs doublet field, which is symmetric with respect to the group SU(2) of transformations in the internal space of weak isospin. Imposing an external factor (the Higgs potential) that hides this symmetry, the W and Z bosons get an effective mass but in such a way that the theory remains intrinsically symmetric with respect to SU(2) weak-isospin transformations. When this is done, though, something altogether new pops out of the theory: an as-of-yet undiscovered particle, electrically neutral and with no spin, known as the Higgs boson.

The Higgs boson, if in fact real, is unlike any currently known field quantum; among other things, there are no known fundamental particles with no intrinsic spin (spin-0), as the Higgs is predicted to be. At energies high enough to excite this quantum of the Higgs field, the Higgs boson would mediate a new type of electroweak interaction that lies completely beyond the realm of our current experience. In addition to the effect this would have on our view of the fundamental interactions of nature, determining the properties of the Higgs boson and associated Higgs potential would likely have (although in ways we won't discuss) a profound influence on our formulation of cosmology, the theory of the birth of the universe and of the evolution of its overall structure.

The success of the Standard Model (to be emphasized shortly) compels particle physicists to go out on a limb—to predict the existence of a new and unprecedented fundamental component of nature. Just as Paul Dirac correctly postulated the existence of antimatter to make his relativistic theory of quantum mechanics work, contemporary particle physicists feel confident that the Higgs, or something very much like it, lies within the reach of the next generation of particle accelerators.

One more thing: although we've talked at length about the mass of the W and Z force-field quanta, we've made no mention whatsoever about the masses of the other fundamental particles, the leptons and quarks. In one very important way, the twelve matter-field quanta are no different than the W and Z force-field quanta: the brute force inclusion of their masses in the wave equation destroys the underlying Lie group gauge symmetry that we so covet. Just as we need to be ingenious by introducing the force-field quanta masses in a way that hides, rather than destroys, the underlying gauge symmetry of the theory, we must be similarly clever when we incorporate the matter-field quanta (fermion) masses.

Luckily, the background Higgs field accomplishes this for us with no additional effort. The fundamental fermions all carry weak-isospin charge (they all interact with the W and Z weak force-field quanta), and so they all interact with the Higgs background field. The screening process responsible for providing the quarks and leptons with their (effective) masses is essentially identical to that of the force-field quanta.

So we pose the following question: If this Higgs mechanism approach is correct, then what objects can we point to as being truly and fundamentally massive, rather than having instead an *apparent* mass due to the screening effects of the background Higgs field? Everything we know of is composed of various combinations of the twelve quarks and leptons, bound together by the continuous exchange of force-field quanta. If the Higgs mechanism is conceptually accurate, *all* of these fundamental constituents derive their inertial properties—their appearance of massiveness—from Higgs field screening effects. As we'll see shortly, there is substantial circumstantial evidence (although falling short of direct proof for now) that nature does indeed employ the Higgs field in the generation of effective masses for the fundamental particles.

Additionally, according to Einstein's relation $E = mc^2$, for a system of two or more objects that are bound together by the forces of nature, internal energy associated with the mutual orbital motion or the continual exchange of force-field quanta between the objects will also manifest itself as mass. All but a few percent of the mass of protons and neutrons, and thus of the stuff we think of as ordinary matter, is due to such internal energy.

Thus, one of the most basic and common-sense attributes of a physical object—that of mass—has been removed from the conceptual lexicon by the juggernaut of modern physics, having been exposed as the combination of two illusory effects: those of internal mass-energy and of the Higgs field screening currents. The notion of mass, it would seem, is a sham.

The Reflected Universe

Watching a movie or TV show, you may have seen the following. A scene unfolds, apparently normal in all respects. As the camera pans out, an incongruous border suddenly appears on one edge of the picture. As the camera pans even farther back, you realize that the scene is being reflected into the camera lens by a mirror.

Until the edge of the mirror appears, there is nothing whatsoever to hint at the fact that the scene is being filmed in the reverse image of a mirror reflection. There's nothing evident from the flow of activity, from the sequence of physical events in the scene, that seems out of place. Either view of the scene, the direct or the mirror-reflected, seems to flow perfectly naturally. The physical laws that govern the activity in the mirror-reflected scene are, by all appearances, identical to those of our direct-image, unreflected world.

One is thus drawn to the two following, and equivalent, conclusions. First, the laws of physics are unchanged, or invariant, under the operation mirror reflection. Second, there is no experiment that one can do, nor observation that one can make, that can distinguish between the following two possibilities: (a) that the experiment was performed in our universe or (b) that the experiment was performed in a universe that is a mirror reflection of ours but precisely identical to our universe in every other way.

Mirror reflection, or more formally, *parity inversion* (note 9.6) is an example of what is known as a *discrete symmetry operation*. When we talk about a particular symmetry of a given physical law, we imply that there is a group of operations under which the law is invariant, or symmetric. In the case of parity inversion, the group of operations has but two elements—one for which you do nothing to the system under study, and one for which you mirror-reflect, or parity-invert, the system. If you parity-invert the same system twice, you get back to where you started from, the uninverted system. (Given this, can you convince yourself that this two-element set satisfies all the requirements of a mathematical group that were laid out in chapter 6?)

The label "discrete" arises because parity inversion is all-or-nothing—either the system is mirror reflected or it isn't. For rotations, one has a full range of possible angular changes to the system's orientation; there is a continuum of possibilities, corresponding to the continuum of possible angles of rotation. The invariance of physical laws under such changes of orientation establishes rotation as a continuous symmetry of nature. For all-or-nothing parity inversion, there is no corresponding continuum of possibilities that smoothly ease the system from a direct to a mirror-reflected orientation. Instead, the transition is discontinuous, or discrete, with but two elements: mirror-reflect or do nothing. Parity inversion is a discrete symmetry of nature.

Since the two-element (mirror-reflect or do nothing) parity group represents a symmetry respected by natural laws, there ought to be, in the spirit of Noether's theorem, an associated conserved physical quantity. This physical quantity is known, fittingly enough, as "parity," and it has only two possible values, usually referred to as "even" and "odd" (note 9.7). Thus, any system with, say, odd parity, and governed by laws that are parity-inversion invariant, as it seems our familiar physical laws are) will remain with odd parity in perpetuity, unless acted on by some external agent.

Now, in the mid-1950s, among the confounding zoo of elementary particles were two near twins: the θ and τ mesons (the latter not to be confused with the τ lepton, discovered twenty years later). In almost every respect, the θ and τ were alike. They had the same mass and lifetime, and they decayed into similar sets of particles. The only difference was that when the θ meson decayed into a collection of lighter particles at the end of its lifetime, it decayed into a system with even parity, while the τ meson decayed into a system with odd parity.

Now, under the assumption that physical laws are parity-inversion invariant, the process of particle decay, which is governed by those laws, cannot change the parity of the system. So, the θ meson, decaying in isolation, must possess an intrinsic parity that is even, while the τ meson must have odd intrinsic parity. The θ and τ, even if differentiated only by their intrinsic parity, thus have different properties and so must be different particles.

Physicists of the day were uncomfortable with the notion that different particles should somehow have identical masses and lifetimes, so much so that they referred to the situation as the "τ/θ paradox." They also knew that there was an easy way to resolve the paradox. If the physical law governing the decays of the τ and θ is not invariant under mirror reflection, in which case parity is not conserved, then the τ and θ could in fact be one and the same particle, decaying sometimes to a system with even parity and sometimes to a system with odd parity.

The problem with this possibility is that it flies so directly in the face of physical intuition that it just can't be the case. The notion that space is rotationally invariant—that the orientation of an experiment performed in outer space should have no affect on its outcome—seems manifestly obvious. Similarly obvious is the notion that the experimental outcome should not be altered in any way if you rebuild the experiment to look like its mirror image. Electrical charges should repel one another as strongly, radiate light as readily, and so forth. What difference could mirror reflection possibly make?

There was one thing, however, that everyone agreed on: the relatively long (by particle physics standards) 10^{-8} second common lifetime of the τ and θ mesons clearly established that the decay was being governed by an interaction that was not particularly strong. The decay of the τ and θ had to be mediated by the weak nuclear force.

In 1956, T. D. Lee (Columbia University) and C. N. Yang (Princeton University [note 9.8]) recognized, in the spirit of Einstein, that intuition, while often guiding us onto the golden pathway to the truth, can at other times be our biggest impediment toward finding that path. Lee and Yang decided to look carefully at the full corpus of experimental data on weak-interaction processes in an attempt to find supporting evidence for the obvious notion that the weak interaction is invariant under parity inversion. To no one's greater surprise than their own, they were able to find none. Their exhaustive analysis revealed that the experimental data available at the time shed no light whatsoever on the issue.

Thus, as far as the experimental evidence of the day was concerned, the issue of the parity-inversion invariance of the weak interaction, the most obscure of all the physical forces, was a completely open question. Could it be that the weak nuclear interaction, the instigator of τ and θ meson decay, is not parity-inversion invariant? Does the weak interaction not conserve parity?

Performing and comparing the outcome of experiments that are set up as mirror reflections of each other is well within the capabilities of the contemporary particle physicist. And the key to the execution of such comparisons is the manipulation and measurement of the orientation of the intrinsic spin of elementary particles.

Spin and Parity

Consider a particle, say an electron, moving through space. The line along which the electron moves forms an axis through the electron; let's assume that the electron's $\frac{1}{2}\hbar$ spin angular momentum is oriented along this axis, as shown in figure 9.3. If you view the electron as it moves directly away from your eye, the axis about which it spins extends straight from your eye to the electron. The electron can spin about that axis in one of two senses, clockwise or counterclockwise. Now, notice that if you take your right thumb and point it in the direction of the electron, the fingers of your right hand will naturally curl in a clockwise direction. For this reason, we say that a particle is right handed if its spin is seen to be clockwise as it moves away from the viewer. Likewise, if its spin is counterclockwise, we say that it is left-handed.

There's a direct connection between this handedness of the electron's spin and the operation of parity inversion. Consider an initially right-

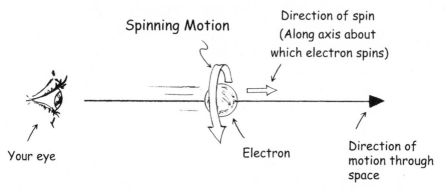

Fig. 9.3. This is a right-handed electron. If the thumb of your right hand points in the direction the electron's moving, then the fingers will curl in the same direction that the electron spins. For right-handed electrons, the spin is seen to be clockwise as the electron travels away from your view. For left-handed electrons, the direction of spin is counterclockwise, or in the direction that the fingers of your left hand would curl.

handed electron after it has been reflected back into your eye by some sort of "electron mirror," in just the way a ray of light is reflected by an optical mirror. After the mirror reflection — the *parity inversion* — you'll still see the electron spinning clockwise, as before, but now the electron is coming back toward your eye. But the handedness of the electron is determined by the sense of its spinning as it goes *away* from your eye, so to determine the handedness of the mirror-reflected electron, you need to turn your head around, letting the electron pass through from the back to the front of your head, and then study it as it recedes.

You can simulate this as shown in figure 9.4, using a pencil in place of the electron. Take the pencil and spin it clockwise as you move it away from your eye; now continuing to roll the pencil in the same direction, pretend that the pencil bounces off a wall back toward your eye. As you keep rolling the pencil in your fingers, direct it past your face so that it begins to travel away from the back of your head. Turn your head by 180 degrees and observe. You'll see that the spin is now perceived to be counterclockwise as the pencil moves away from your eye. What you have just shown is the following essential point: mirror reflection (parity inversion) changes the handedness of spinning objects.

Now *this* is something that Lee and Yang could sink their teeth into! Consider some set of particles taking part in a process mediated by an interaction that is invariant with respect to parity inversion. Since parity in-

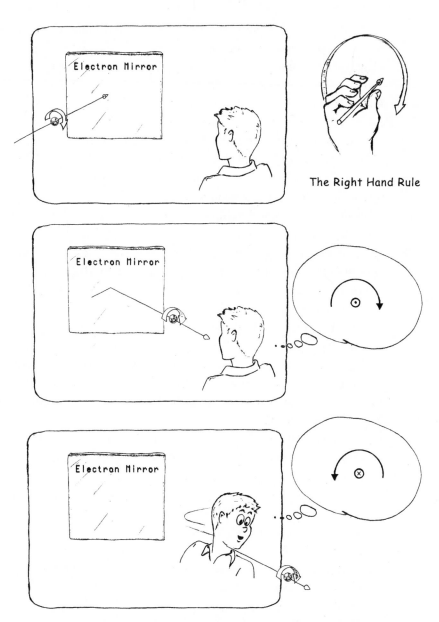

The Right Hand Rule

Fig. 9.4. Mirror reflection changes the handedness of spinning objects.

version flips the handedness of spinning particles, if the interaction is to be invariant with respect to parity inversion, then the process must proceed identically regardless of the handedness of the spin of the particles taking part in it. Or, conversely, if the characteristics of the process change in any way when the spins of all the participating particles are flipped, then the interaction that mediates the process does not respect parity invariance — the interaction violates parity. Exploring how various weak-interaction processes depend on the handedness of the participating particles, Lee and Yang argued, would be an iron-clad way to ascertain whether the weak interaction violates parity-inversion symmetry.

In late 1956, prompted by the work of Lee and Yang, C. S. ("Madame") Wu, a nuclear physicist at Columbia University, teamed up with a small group of low-temperature physicists led by Ernest Ambler at the National Bureau of Standards in Washington, D.C. The $\frac{1}{2}\hbar$ spins of the collection of twenty-seven protons and thirty-three neutrons in the isotope cobalt-60 lineup in such a way the cobalt-60 nucleus possesses a net spin of $5\hbar$. If a sample of cobalt-60 is made very cold (close to absolute zero; hence, the need for the expertise of the Ambler group) in a magnetic field, it will become polarized, which is to say that the axes of rotation of the majority of the $5\hbar$ cobalt-60 nuclear spins will align themselves with the direction of the magnetic field, like compass needles seeking magnetic north in the earth's magnetic field.

Now, cobalt-60 is unstable, disintegrating via beta decay to nickel-60 by switching one neutron to a proton, with the subsequent emission of an electron and an electron-type antineutrino. Using a straightforward argument based on the well-established notion that angular momentum is conserved (note 9.9), it can be shown that right-handed electrons will tend to be emitted in the direction of the cobalt-60 nuclear spin (northerly in our compass analogy), while left-handed electrons will tend to be emitted in the opposite (southerly) direction. Thus, if nuclear beta decay — the archetypical weak-interaction process — exhibits parity violation by preferring to emit right-handed over left-handed electrons, one will simply observe more northward-traveling electrons than southward-traveling electrons; if the weak interaction prefers left-handed over right-handed particles, one would observe more southward-traveling electrons. For parity to be conserved in this process, the number of northward- and southward-traveling electrons would have to be identical.

Using an elegant system of controls and cross-checks, the Wu/Ambler experiment demonstrated a clear preference for the emission of southward-traveling electrons. Parity-inversion symmetry is violated by the weak interaction! This resolved the nettling τ/θ paradox: τ and θ mesons are one and the same thing, which just turns out to be the kaon (in the language of the quark model of the eightfold way, the kaon is the state formed by combining an $\bar{s}$ antiquark with a u quark). In doing so, the Wu/Ambler experiment opened up a fertile new avenue for the exploration of the weak nuclear force: that of parity-violation experiments. The Wu/Ambler experiment is an example—and one of the few (the precise measurement of the electron's magnetic properties, discussed in chapter 4, being another)—of a major experimental advance in particle physics made on a tabletop, without the use of high-energy particles from a particle accelerator or from natural cosmic radiation.

Since the electrons in the Wu/Ambler experiment were emitted preferentially in the southerly direction, the weak interaction manifests its parity violation via a penchant for left-handed particles (note 9.10). With that demonstrated, it's natural to wonder just how strong this prejudice is.

The cobalt-60 nucleus, with its sixty nucleons continually bouncing off one another, is too messy a system to obtain a precise quantitative answer to this question. However, the properties of muons, fundamental particles that are the heavier cousin of the electron, are understood quite well, so electrons emanating from the decays of muons, which decay through the weak nuclear interaction, can be used quite effectively to explore this question.

The Feynman diagram associated with the decay of the muon, which we've come across before, is shown again in figure 9.5. The negatively charged muon emits a negatively charged W^- weak-interaction field quantum, turning into a neutral muon-type neutrino in the process. The W^- subsequently converts into an electron and an electron-type antineutrino.

Experimentally, it's possible to measure the angle at which the electron comes out relative to the direction the muon was traveling just before it decayed. As with the beta decay of cobalt-60, we can use angular momentum conservation to calculate the relative rate of forward decays (electron emitted in the general direction of the muon's original flight direction) versus backward decays (electron emitted opposite to the muon's flight direction).

The result of this calculation is as follows. If the W^- forms minimal interaction vertices solely with right-handed particles, the forward-to-backward ratio will be two to one. If only left-handed particles connect with

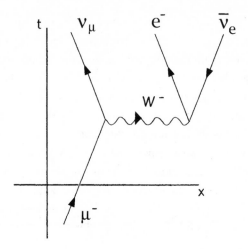

Fig. 9.5. The Feynman diagram representing the process of muon decay; a repeat of figure 8.8*b*.

the W^-, the ratio is one to two. If the W^- shows no preference one way or the other, the ratio will be one to one.

Based on the cobalt-60 studies, which definitively established a preference for left-handed electrons on the part of the weak interaction, we expect a forward-to-backward ratio of between one to one and one to two—more backward decays than forward decays. The bigger the predominance of backward decays, the stronger the preference for left-handed particles.

In the late 1950s, there were many smallish (by today's standards) particle accelerators, known as *cyclotrons*, scattered throughout the world, and perhaps the most productive of all of these was the Nevis cyclotron in the New York City suburbs, operated by Columbia University. Aware of the Wu/Ambler result before its formal publication (by dint of the fact that they shared hall space with Wu at Columbia), a team of three Columbia physicists led by Richard Garwin (note 9.11) quickly assembled an experiment that could analyze the decays of muons that were brought to a halt in a carbon absorber after being produced by the cyclotron beam.

Their result, if not shocking, was quite interesting. To within experimental accuracy (about 10% on the measured ratio), they found the forward-to-backward ratio to be exactly one to two. The weak interaction consorts exclusively with left-handed particles; it will have nothing whatsoever to do with right-handed particles (note 9.12).

We often say that, for the weak interaction, parity violation is "maximal"—you can't imagine any greater prejudice toward left-handedness than the complete preference it exhibits.

These two experiments, those of the Wu/Ambler and Garwin groups, were published back-to-back in early 1957 in the 105th volume of the *Physical Review*. They provided the first hint that, although the most obscure of the interactions from the point of view of everyday experience, there is something entirely special and out of the ordinary about the weak interaction. It marches to the beat of its own drummer, so to speak, and in doing so is responsible for a number of phenomena, such as nuclear stability patterns and the fundamental differentiation between the behavior of matter and antimatter, that we now appreciate as being absolutely essential for the development of life. The discovery of parity violation in the weak interaction was clearly of Nobel Prize caliber, and this time it was the theorists, Lee and Yang, who found themselves draped with gold medallions as the co-recipients of the 1957 Nobel Prize in Physics.

Parity Violation and the Electroweak Force

In the preceding discussion, we've ignored one essential qualification. The processes we've discussed, beta decay of cobalt-60 and muon decay, involve the exchange of a charged W^- weak field quantum but neither involves the exchange of a neutral Z^0 weak field quantum. Thus, the 1957 experiments, while demonstrating that the weak force exhibits maximal violation of parity-inversion symmetry, did so only for the charged-current weak force mediated by $W^\pm$ field quanta. In fact, the first suggestion (due to Glashow) of a neutral weak force was still four years away, and its experimental confirmation (the CERN/Gargamelle weak neutral current experiment) lay fully sixteen years in the future.

The Standard Model, published ten years after the discovery of parity violation in the weak interaction, predicted the existence of the Z^0, which is formed mostly from the neutral W^0 quantum of the SU(2) interaction, with a little bit of the B^0 of the U(1) interaction thrown in for good measure. What does the Standard Model have to say about the degree of parity (mirror-symmetry) violation in the interactions of the Z^0 with matter particles?

The charged weak-interaction field quanta are just the basic, unadulterated $W^\pm$ particles from the SU(2) gauge symmetry Lie group. The 1957 parity-violation experiments showed that these two quanta, the W^+ and W^-, will mediate an interaction if and only if the matter particles (quarks or leptons) involved in the interaction are left-handed.

Thus, in the Glashow-Salam-Weinberg model, the third SU(2) field quantum, the W^0, must also participate only in interactions with left-handed particles, just like the W^+ and W^-, its SU(2) field-quanta partners. This neutral SU(2) quantum, though, is "unphysical"—it's a component of both the photon and Z^0 boson, which mediate the electromagnetic and neutral weak interactions, but it does not itself mediate interactions.

Now, although it's been looked for quite carefully, the electromagnetic force has not been observed to violate parity; it does respect parity-inversion symmetry. In quantitative terms, the photon exhibits precisely zero preference between left- and right-handed matter particles.

But the photon is a combination of a lot of the B^0 and a bit of the W^0. So, since the SU(2) W^0 quantum interacts only with left-handed particles, then the B^0 must show some (partial) preference for right-handed particles, just enough so that when combined with the small amount of the W^0 appropriate to form a photon, the partial right-handed preference of the B^0 exactly compensates for the complete left-handed preference of the W^0, leaving the photon with, as observed, no preference whatsoever for left- or right-handed particles.

We can now answer the question at hand: that of the degree of handedness preference (parity violation) of interactions mediated by the Z^0 boson, the neutral carrier of the weak nuclear force. The Z^0 is composed of a large amount of the SU(2) W^0 quantum and a complementary small amount of the U(1) B^0 quantum. As we've just seen, the large piece of the W^0 interacts solely with left-handed particles, while the small bit of B^0 interacts with a mild preference for right-handed particles. Alloyed in this way from the W^0 and B^0, the Z^0 quantum interacts with a *partial preference for left-handed particles*—the dominant W^0 component of the Z^0 exclusively prefers left-handed particles, although this preference is watered down (but not eliminated) by the mild preference for right-handed particles of the less predominant B^0 component.

So, neutral-current weak interactions, mediated as they are by the Z^0 boson, take place more readily for left-handed than for right-handed particles, although they still occur at some level for right-handed particles. This statement lies at the heart of recent exacting tests of the Glashow-Salam-Weinberg Standard Model of the electroweak force.

Precision in Collision

The model put forth by Weinberg and Salam in 1967 was encouraging in that it laid out a unified theory of the electromagnetic and weak nuclear interactions that was consistent with the experimental data of the day. Yet, it was rather speculative. The incorporation of the W and Z force-field quanta masses through the introduction of a Higgs field with hidden gauge symmetry brought with it the hope that the theory was self-consistent (renormalizable), but the work of Weinberg and Salam fell short of actually demonstrating this essential property.

Thus, the first substantial confirmation of the Standard Model came from the theorists, who have merely to lock themselves in a room with a blackboard and coffee maker to conduct their business. In 1971, Martinus J. G. Veltman and his precocious student Gerardus 't Hooft, of the State University of Utrecht in the Netherlands, demonstrated that the effort expended in introducing mass via the Higgs mechanism had paid off. Veltman and 't Hooft succeeded in proving, with mathematical certainty, that the Standard Model is renormalizable. We need delve no deeper into the detail and difficulty of this task than to say that, for their troubles, Veltman and 't Hooft were awarded the 1999 Nobel Prize in Physics. With this critical issue resolved, the task of confirming the Standard Model fell into the hands of the experimentalists.

Were you to read back through the previous two chapters, highlighting the points at which quantities that need to be determined by experiment were introduced into the Glashow-Salam-Weinberg theory, you would end up with two highlighted passages. These two passages would be associated with the determination of the two overall coupling strengths of the SU(2) and U(1) gauge groups—the magnitude of the weak isospin and weak hypercharge (or equivalently, the weak-isospin and electric charges) of the electron. There's no way to deduce the size of these two charges; they must be measured.

In fact, there really should be a third highlighted section: the discussion of the value of the background Higgs field dictated by the Higgs potential. The stronger this background Higgs field, the more massive the W and Z force-field quanta. So, reversing the argument, a measurement of the W or Z mass yields a determination of the strength of the background Higgs field. Again, the value of this parameter is not predicted by the theory; it must be inferred from a measurement of either the W or Z boson mass (the W^+ and W^-, being a particle/antiparticle pair, have the same mass).

Once these three quantities are precisely measured, everything else about the theory—the enumeration of all its minimal interaction vertices and the delineation of the prescriptions for their application to the calculation of any resulting electroweak interaction process (i.e., of any Feynman diagram one can construct from these vertices)—is specified with corresponding precision. The behavior of any fundamental process governed by the electroweak interaction is definitively predicted once these three quantities are known.

One thing, in particular, that is thus predicted by the SU(2)-plus-U(1) Standard Model is the precise amount of each of the (unphysical) neutral B^0 and W^0 gauge quanta that go into making a Z^0 force-field quantum. This quantity, the small amount of B^0 in the Z^0 (or, correspondingly, the small amount of W^0 in the photon), is referred to by particle physicists as $\sin^2\theta_W$ (sine-squared theta-W), with θ_W known in turn as the *weak mixing angle*. Don't worry about the trigonometry in this expression; think of $\sin^2\theta_W$ as a single number: the admixture of the unphysical B^0 quantum of the U(1) interaction that one mixes in with the equally unphysical W^0 gauge quantum of the SU(2) interaction to form the physical Z^0 quantum of the neutral weak interaction.

Here's the explicit expression relating the three measured inputs to the Standard Model to the value of $\sin^2\theta_W$:

$$\sin^2\theta_w = \frac{1}{2} - \frac{1}{2}\sqrt{1 - \frac{4\pi\alpha}{\sqrt{2}G_F M_Z^2}}.$$

The symbols α and G_F represent, respectively, the magnitude of the electric and weak-isospin charges of the electron, while M_Z is the mass of the Z^0 field quantum. The value of π is the usual ratio of the circumference to the diameter of a circle.

The point of showing this formula is to underscore the statement that once the three experimental inputs to the Standard Model are determined—the mass of the Z^0 boson and the magnitudes of the electric and weak-isospin charges—the admixture of B^0 in the Z^0 is completely specified. Armed with precisely measured values of α, G_F, and M_Z (note 9.13), and a calculator, it's straightforward to calculate the value of $\sin^2\theta_W$. Doing so, one finds that $\sin^2\theta_W$ is precisely 0.21215.

Thus, the recipe for the Z^0 is the following. Take precisely 21.215 percent of a U(1) B^0 neutral gauge quantum and 78.785 percent of an SU(2) W^0 neutral gauge quantum. Mix together until well blended. Yield: Enough to mediate one (1) neutral weak interaction. Note: You must follow this recipe *exactly*, or you'll screw up the universe.

Now, recall the discussion from the end of the previous section. The Z^0 boson, in its interactions with electrons, exhibits a partial preference for left-handed matter particles. The extent of this preference is dictated by the amount of B^0 and W^0 that are mixed together when a Z^0 is formed. If the Z^0 were made purely from the W^0, with no admixture of B^0, then the preference for left-handed electrons would be absolute. If the Z^0 were pure B^0, then its preference would be for right-handed electrons, not to the exclusion of interactions with left-handed electrons but partial to right-handed electrons nonetheless (note 9.14). To the extent that the Z^0 is composed of a mixture of these two, its preference for particles of a particular handedness lies somewhere between these two poles.

But we know, if the Standard Model is correct, that the Z^0 is composed of 78.785% W^0 and 21.215% B^0. With the composition of the Z^0 so precisely specified, the exact extent to which the Z^0 prefers to interact with left-handed over right-handed electrons is thus predicted with corresponding precision. The degree of this preference, the "quantitative extent of parity violation" in the interaction of the Z^0 quantum with matter particles, is the most stringent prediction of the Standard Model accessible to particle physics experimentation. If the Glashow-Salam-Weinberg model of the electroweak interaction is to be accepted, this prediction must be found to be true within the limits of experimental accuracy.

With the advent in 1988 of electron-positron accelerators of energy high enough to produce the Z^0 boson, this issue—the degree of parity violation in the interaction of the Z^0 with electrons—could be explored head on. The process employed in this study is shown in the Feynman diagram of figure 9.6. An electron and a positron come in from opposite directions and annihilate, producing a Z^0 force-field quantum. This quantum lives, on average, to the ripe old age of 3×10^{-25} seconds after which it disintegrates into any matter/antimatter pair (quark or lepton) from table 5.1, whose presence is then sensed by detectors in the experiment mounted around the collision point. By comparing the annihilation probability for a beam of left-handed (counterclockwise-spinning) electrons with that for a beam of right-handed (clockwise-spinning) electrons, one directly and unambiguously measures

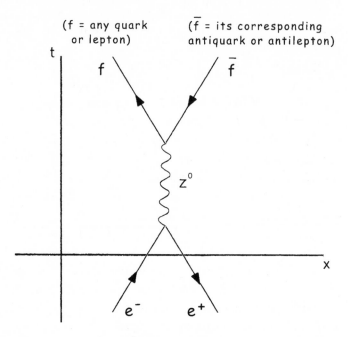

Fig. 9.6. The basic process of Z^0 field quantum production through the mutual annihilation of an electron and positron. The Z^0 is exceedingly short lived and decays almost immediately into any fermion-antifermion (quark-antiquark or lepton-antilepton) pair whose combined mass is less than that of the Z^0 (i.e., anything except a top-antitop pair).

the extent of parity violation in the interaction of the Z^0 boson with electrons.

In 1988, with a freshly minted Ph.D. from the University of Chicago, I took a fellowship at the University of California's Lawrence Berkeley National Laboratory to pursue this measurement with a team of physicists based at the Stanford Linear Accelerator Center. Although CERN's pan-European LEP electron-positron collider was also soon to begin running at energies high enough to produce the Z^0 boson, only the Stanford Linear Accelerator offered the possibility of running the experiment with *polarized* beams—with separate beams of left- and right-handed electrons. The LEP experiments, running with an electron beam composed of equal numbers of left- and right-handed electrons, were still able to measure Z^0 parity violation, but their approach without separately polarized beams was necessarily less direct. However, many other essential studies, including the precise measurement of the Z^0 mass, could best be performed at the European LEP machine.

By 1998, after ten years of steadily improving running at Stanford, the result for the left-handed preference of the Z^0 boson was found to be 15.06%, with an experimental uncertainty of 0.24%: to within an uncertainty of 0.25% or so, a left-handed electron is 15.06% more likely to annihilate with a positron to form a Z^0 than is an electron from an unpolarized beam with equal numbers of left- and right-handed electrons. Working backward in the Glashow-Salam-Weinberg theory, this corresponds to a value of $\sin^2\theta_W$, the small amount of the unphysical B^0 in the physical Z^0, of 23.097%, with an experimental uncertainty of 0.027%. This result could be combined with a corresponding (albeit less direct) value from the LEP experiments, leading to a combined measurement of $\sin^2\theta_W$ of 23.156%, with an uncertainty of 0.017% (note 9.15).

Now, as we've seen, based on the strength of the electric and weak charges and the mass of the Z^0 boson, the Standard Model predicts a Z^0 ingredient fraction of 21.215% B^0; this is what we got when we plugged the measured numbers into the somewhat complicated formula for the ingredient fraction $\sin^2\theta_W$ from a couple of pages back. Compare this with the value 23.156% measured in the parity violation studies. While the difference is only about 2%, it's many times greater than the experimental uncertainty of 0.017%. In this business, close doesn't really count. A prediction must agree with its experimental verification to within roughly the experimental uncertainty, or the prediction is wrong.

So, after all this, is the Standard Model wrong? No. The reason the predicted and measured values of the Z^0 ingredient fraction fail to agree with each other lies at the heart of the framework—quantum field theory—that forms the basis of the Standard Model.

The shortcoming in the way we executed our parity-violation Z^0 ingredient-fraction test is the following: In formulating our theoretical prediction (the prediction represented by the somewhat complicated formula for $\sin^2\theta_W$), we ignored all but the simplest process by which an electron and positron can annihilate to form a Z^0. The Feynman diagram of figure 9.6, which was used to derive the prediction, is the most basic diagram that can be drawn that connects an electron-positron initial state to a fermion-antifermion (lepton-antilepton or quark-antiquark) final state through the production of a Z^0. In actuality, there are many more (infinitely more) ways,

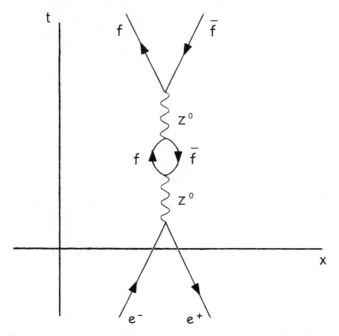

Fig. 9.7. Another way to produce a fermion-antifermion pair in an electron-positron annihilation experiment through the production of a Z^0 boson. In this case, the Z^0 interacts with an independent fermion-antifermion vacuum fluctuation before decaying into the final fermion-antifermion pair.

all more complicated, by which the electroweak interaction can mediate this process. If we want to make the prediction accurate enough to compare with the demanding precision of the measurement, we need to include the effects of at least the most simple of these more complicated processes.

An example of another such process is shown in figure 9.7. In this figure, the Z^0 involves itself with a "virtual loop" from the vacuum, temporarily fluctuating into a virtual fermion-antifermion pair (pick your favorite fermion type; any one of the twelve will do) before decaying into a real fermion-antifermion pair.

Recall chapter 4: the greater the number of minimal interaction vertices employed by the Feynman diagram, the less likely it is that the process will contribute to the interaction. Diagrams like that of figure 9.7, with an extra two vertices relative to the diagram of figure 9.6, won't contribute very much to the overall probability of electron-positron annihilation to a Z^0. Nonetheless, the experimental data are precise enough that these processes need to be considered. These more complicated diagrams slightly alter the prediction for the composition of the Z^0, so one might hope that the addition of these more complicated diagrams might fix up the disagreement between the predicted and measured ingredient fractions.

With a combination of hope and anxiety, we can draw the twelve diagrams represented generically by figure 9.7 (one for each of the six lepton and six quark flavors) and calculate how each of these diagrams acts to slightly change the predicted ingredient fraction of the Z^0. Taking all these small changes into account, we might hope that the theoretical prediction is brought into agreement with experiment.

There was a hitch, however. As the precise experimental data began to accumulate in the early 1990s, the top quark had yet to be discovered. Figure 9.8 is the specific case of figure 9.7 for which the virtual loop is formed

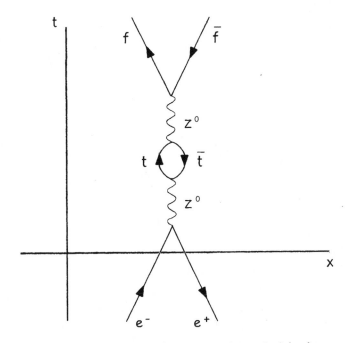

Fig. 9.8. According to the Heisenberg uncertainty principle, because the fermion-antifermion vacuum fluctuation is short lived, its constituents can have a mass-energy greater than that of the Z^0 boson from which it springs. Thus, it is entirely possible, necessary even, that the vacuum fluctuation will sometimes be to a top-antitop pair. Because of this, the process of electron-positron annihilation to a fermion-antifermion pair depends on the properties of the top quark, even though the top quark itself cannot be produced by the interaction.

by the top quark. To calculate the effect of figure 9.8, theorists needed to know the top quark's mass. Without having observed the top quark, there was no way to know how much it weighed and thus no way to predict, with a precision commensurate with that of the new experimental data, the ingredient fraction of the Z^0.

Instead, theorists took the opposite tack and calculated the value of the top quark mass that would be required to make the Standard Model prediction agree with the data. The answer they came up with was shockingly large. The top quark would have to have mass energy nearly 200 times that of the proton if the Standard Model prediction were to be correct. The next heaviest quark, the bottom (b) quark, has a mass-energy of about five times that of the proton, heavy but nothing compared to this. With a mass-energy approaching 170 GeV, the top quark would be the heaviest known particle, weighing in at almost twice the mass of the 91 GeV Z^0 boson.

In short order, the top quark was found. Its discovery at the Fermi National Accelerator Laboratory was announced in 1995, along with the first measurement of its mass. In another triumph for the standard model, the mass-energy of the top quark was found to be 174 GeV, with an experimental uncertainty of about 5 GeV.

In addition to providing a stunning confirmation of the Standard Model, this work also compellingly demonstrated the indispensable complementarity of the two primary approaches to particle physics experimentation. The parity violation measurements that led to the prediction of the top quark mass were performed at the SLAC and CERN electron-positron colliders, while the top quark itself was found at the Fermilab proton-antiproton (Tevatron) collider. Without both of these accelerators operating contemporaneously and with commensurate energies, particle physicists would not have enjoyed this critical and timely advance. As we contemplate the next generation of powerful accelerators, this complementarity between electron and proton machines remains a primary consideration.

With the top quark mass accurately measured at Fermilab, the process of figure 9.8 can be included in the prediction for the ingredient fraction of the Z^0. When this is done, the prediction is found to disagree with experiment by about 0.1%. This is much better than the 2% discrepancy we started with, but it's still not good enough since the measurement is accurate to about 0.02%, a margin of error that remains substantially less than the level of disagreement.

There is, however, another diagram that we have yet to consider — figure

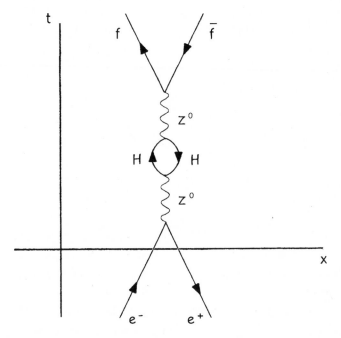

Fig. 9.9. It's also possible for the vacuum fluctuation to involve the Higgs boson, making this process sensitive to the properties of the Higgs (should such a thing actually exist!).

9.9. Just as with figures 9.7 and 9.8, the Z^0 produced by the annihilation of the electron and positron involves itself with a virtual loop of the fluctuating vacuum. In this case, instead of being comprised of a matter particle/antiparticle fluctuation, the loop is due to a spontaneous fluctuation into an entirely different type of field quantum: that of the Higgs boson.

   This diagram is every bit as central to the workings of the Standard Model as those of figures 9.7 and 9.8, so it must also be included in our attempts to predict the ingredient fraction of the Z^0. While the effect of this virtual Higgs boson loop is expected to be substantially smaller than that of the top-quark loop of figure 9.8, the precision of the Z^0 parity violation measurements is fine enough that this process needs to be considered.

   As for the top quark before 1995, the Higgs boson has yet to be discovered. So again, as we did with the top-quark loop process before the discovery of top, we turn the question around and ask the theorists to calculate the value of the Higgs boson mass-energy required to make the prediction of the Z^0 ingredient fraction agree with the measurement.

The answer is enticing: the Higgs boson mass-energy needs to be relatively small, about 85 GeV; roughly the same as the Z^0 mass-energy. However, the effects of the diagram of figure 9.9 are minute enough that the precision of the Z^0 ingredient fraction measurement, impressive as it is, admits a fairly wide range of possible Higgs boson mass-energy values. Nevertheless, given the accuracy of the ingredient fraction measurement, we can say with better than 95% certainty that the mass-energy of the Higgs boson should be less than 200 GeV. This will be well within reach of our next generation of particle accelerators.

Were the measured value of the Z^0 ingredient fraction slightly different—by as little as 0.1%—it would be at odds with the prediction for any value of the Higgs boson mass-energy. Remarkably, this is not the case. Our measurement is indeed consistent with the narrow range of possibility admitted by the Standard Model, given that the Higgs mass-energy is not known. But even beyond this, the measurement is precise enough to pin down, to a substantial degree, where within the range of possibilities nature happens to lie. If the Standard Model, as formulated in 1967 by Weinberg and Salam, is correct, then in all certainty, we will find the Higgs boson with a mass-energy between 35 and 200 GeV.

On the other hand, we've already observed several particles in this mass range. The top quark, with a mass-energy of about 175 GeV, has been clearly observed, while for the Z^0 boson, at 91 GeV, tens of millions were produced at the CERN and SLAC machines. Why has the Higgs not been discovered?

The answer is that the Higgs boson, related as it is to the background field that generates the (effective) mass of the fundamental particles, doesn't really like to interact with the light forms of matter—up and down quarks and electrons—that make up particle beams. The mass of a particle is directly related to the readiness with which it interacts with the Higgs field, so the light, ordinary forms of matter found in particle beams are not overly disposed to forming Higgs bosons when they collide.

The process by which one would produce a Higgs boson in an experiment based on colliding beams of electrons and positrons is shown in figure 9.10. The Higgs boson is happy to interact with the massive Z^0, but there's a price to pay: this process requires the production of *both* a Z^0 and a Higgs boson. Thus, the energy of the colliding electron and positron beams must be high enough to produce both of these field quanta at the same time. In other words, if the Higgs boson were to happen to have a mass-

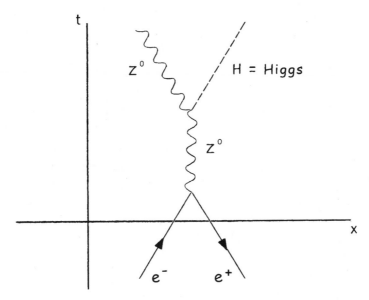

Fig. 9.10. If the combined energy of the electron plus positron is greater than the combined mass-energy of the Z^0 plus Higgs, the Higgs boson will be produced by the process represented by this Feynman diagram.

energy identical to that of the Z^0, the electron-positron collision would have to have twice as much energy to produce a Higgs boson in combination with a Z^0 than it would to produce a Z^0 boson alone.

In 1995, after accumulating a sample of about 20 million Z^0 particles, physicists at the LEP electron-positron collider began a program of steadily increasing its electron and positron beam energies above the 91.187 GeV required to produce the Z^0 boson in isolation. In mid-2000, in a devil-take-all push, the LEP electron-positron collision energy was inched up to a record 207 GeV. Taking into account the Z^0 mass-energy, this allowed the experiments at LEP to search for a Higgs boson with a mass-energy of 116 GeV or less, not all the way up to 200 GeV at the top of the expected range, but well within the region of interest. While some intriguing collisions were observed, with properties consistent with the production and decay of a Higgs boson of 115 GeV, clear evidence was lacking. We do know from this, however, that the Higgs boson has a mass-energy greater than 110 GeV; were it less than this, the LEP physicists certainly would have seen it.

Unable to push the LEP machine any higher in energy, CERN physicists have dismantled it and are assembling in its 27-kilometer-circumference tunnel a proton-proton collider to be known as the LHC (Large Hadron Collider). When completed in 2007 or so, the LHC will operate with a collision energy of 14 TeV, an amount of energy equivalent to that obtained after acceleration through 14 trillion volts of electrical potential. Now, proton-proton (or proton-antiproton; at these high energies, it doesn't make much difference) colliders don't get as much physics per volt of acceleration as electron-positron colliders do; nevertheless, with a collision energy seventy times greater than that of the LEP electron-positron collider, the LHC will be the uncontested atom-smashing champion of the world when it begins to collect data toward the end of the decade. By comparison, the currently reigning champion proton-antiproton collider, Fermi National Accelerator Laboratory's Tevatron (the machine at which the top quark was discovered) runs with a collision energy of 2 TeV. If the Higgs boson is out there, the LHC will find it. If the Higgs boson has a mass-energy close to 115 GeV, as the LEP experiments hinted, the Tevatron may show evidence for it even sooner.

Final Musings

We have laid out at last the conceptual basis of the Standard Model of the electroweak interaction and discussed its compelling triumphs. Yet it was the previous chapter on gauge theory that was presented as the intellectual cornerstone of the Standard Model.

Born into this world, tabula rasa, we are overwhelmed by sensory input coming at us with little apparent purpose or connection. As we begin to develop, we become aware of order, of a relation between cause and effect in our surroundings. Food satisfies our hunger, clothing our desire for warmth, and a parent's soothing embrace our need for companionship and protection. As time progresses and our experience mounts, we begin to perceive a deeper connection between a relatively modest number of underlying causes and the diverse phenomena they are responsible for. We take up the systematic study of the guiding principles of these root causes, allowing us to satisfy our curiosity about the nature of our surroundings, while at the same time permitting us to advance technologically, offering the possibility of increasing the degree of comfort and interest afforded by our lives.

Gauge theory is, at this point in time, the endpoint of this intellectual trajectory. It condenses our understanding of all forms of physical causation (except gravity, for now) to the selection of three underlying internal symmetry groups—U(1) and SU(2) for the electroweak interaction and SU(3) for the strong nuclear interaction—and the empirical determination of the three coupling parameters, or charge strengths (electromagnetic, weak isospin, and strong color) associated with each of the three symmetry groups. Were there no mass (or, more precisely, as we have seen, no physical behavior that we describe via the notion of mass), then this would be it. The majority of all physical effect, the only exception being gravitational pull, would be comprehensively described in terms of this exceedingly simple set of underlying assumptions, in concert with the application of the gauge principle of invariance under local transformations in the internal spaces of these symmetry groups.

Such, however, is not the case, and so we cannot get by with the elegance and simplicity of gauge theory alone. To account for mass without destroying the self-consistency of our theory, we need to introduce another field, whose sole purpose is to provide the all-pervasive cosmic screening effect that causes particles that interact with this field to behave in every measurable way as if they have mass. This Higgs field serves no other purpose; it provides the source of no other known physical phenomenon. It is an additional component that, while necessary for the health of our theory, compromises the wonderful economy and clarity of gauge theory. In this sense, the Higgs field seems to be excess intellectual baggage. It is ad hoc.

Even more disconcerting, the existence of the Higgs field alone is not enough to cause the various matter- and force-field quanta to behave as if they have mass. We also require that this Higgs field have a nonzero average "vacuum expectation" value that uniformly pervades all space. We fulfilled this need by introducing another arcane and ad hoc notion: the Higgs potential.

This Higgs potential, the agent responsible for compelling the Higgs field to adopt the specific, nonzero background strength that we require, is again excess intellectual baggage, perhaps an even worse sort than the Higgs field itself. There is no tenet of the Standard Model that demands that there be precisely three generations of matter fields; there just are, and that's that. So, to add yet another field (albeit a Higgs field unlike any other field previously postulated) is perhaps not too dissatisfying. But to be forced to introduce this arbitrary and unfounded potential, inexplicable in terms of any known physical principle or property, is downright demoralizing. In this light, one might even say that the discovery of the Higgs boson would represent a bold step backward. It's just the smallest step backward that we know how to take.

What is the *physical* origin of the external Higgs potential? What is the *physical* basis of the peer pressure–like internal organizing principle that causes the Higgs field symmetry to be hidden in a uniform manner throughout space-time? These questions, unanswerable within the Standard Model, suggest that, beyond the unearthing of the Higgs boson, there may well be some radically new facet of the natural world awaiting our discovery. To find the Higgs boson and confirm that its mass is consistent with the exacting parity-violation studies would be a wonderful confirmation of the Standard Model. However, with further exacting studies of the Higgs boson's properties and behavior, we may begin to get some insight into these open questions and be led onto the path of some altogether new and revolutionary view of the way in which the universe is pieced together.

The call to arms is clear. From the perspective of the particle physicist, there is no more compelling a goal than the discovery and delineation, to the fullest extent permitted by ingenuity and resources, of the properties of the Higgs boson.

The technology needed to address the issue is in hand. The LHC is essentially guaranteed to see the Higgs boson of the Standard Model if it exists. And if the Higgs is observed at CERN's LHC, a number of its properties, including its mass and some critical aspects of its disintegration patterns, will be measured.

Further, and even more exacting, studies could be performed at a high-energy electron-positron collider, the design of which is currently being finalized by a cooperative worldwide effort. The world's community of particle physicists has reached a consensus that the construction of such a machine should be its top priority after the completion of the LHC. It seems likely that such a machine lies in our future.

But what if the Higgs field approach, the Higgs mechanism, is *not* the route chosen by nature to manifest the properties of mass? Would we then be in the unfortunate position, come 2010 or so, of having learned nothing further about the fundamental workings of nature?

The possibility that we will learn nothing from our next stage of experimentation, while not ruled out, seems to be unlikely. We can't predict what we'll find with absolute certainty—such is the nature of experimentation at the frontier of knowledge—but we believe that something new, Higgs boson or otherwise, must lie just beyond the reach of the experiments we've done to date.

If a theory is not self-consistent, not renormalizable, then there will be interaction processes, usually associated with very high energy for the incoming participants, that are predicted by the theory to have a likelihood of greater than 100%. This is a clear failing of the given theory and suggests that, if not dead wrong, the theory is at best incomplete.

Chapter 8's pure SU(2) and U(1) gauge theory of electroweak interactions (absent the introduction of the Higgs field), with all its successes, is certainly not dead wrong. However, without some serious thought, the SU(2) and U(1) gauge theory of the Standard Model is not renormalizable when the mass of its W and Z force-field quanta is taken into account. It was precisely this issue that forced us to introduce the Higgs mechanism, with all its trappings. The pure SU(2) and U(1) gauge theory of the electroweak interaction, in and of itself and without the Higgs mechanism to incorporate the masses of the fundamental particles, is incomplete.

This incompleteness manifests itself as the energy of the objects participating in the electroweak interaction gets large. For large enough energy, the pure SU(2) and U(1) theory, without the Higgs field, falls apart just as described; it predicts collision processes with probabilities of occurrence greater than one. Something has to give. There must be some new interaction, particularly relevant at high energy, that curbs the growth of the collision probability with the increasing energy of the incoming particles, keeping it from becoming unphysical as the beam energy is increased. Physical systems always behave physically, even if our theories that model them don't.

But the Standard Model—the SU(2) and U(1) gauge theory plus the Higgs field with its hidden symmetry—is renormalizable; it does provide sensible predictions for all interactions at all possible collision energies. So what exactly is it within the Standard Model that protects the calculation of these collision probabilities from becoming unphysically large (greater than one) at high energy? The answer is, perhaps not surprisingly, the Higgs boson. When its contribution to the collision process is added as a possibility, the calculation of the collision probability remains less than one (note 9.16), no matter how high the collision energy.

While the collision energy at which the pure SU(2) and U(1) electroweak theory begins to fall apart is very high, so also will be that of the LHC and the proposed new electron-positron Linear Collider. Even if there is no Higgs, if the Higgs is nothing more than a figment of particle physicists' overactive imaginations, then *something else* must be there that keeps the interaction probability from exceeding its maximum possible value of 100%. Since the energies of the LHC and the high-energy Linear Collider will be high enough to probe the point at which the pure SU(2) and U(1) theory begins to fall apart, they will necessarily be able to explore whatever this "something else" is that nature has chosen instead.

As the data from the LHC and high-energy Linear Collider begin to roll in, they *must* contain within them clues to the resolution of the mystery of the origin of mass in the universe. The precise way in which the collision probability behaves as we reach this energy is, with mathematical certainty, guaranteed to reveal something new and unprecedented, be it the Higgs mechanism or some other less-anticipated scenario.

The question, then, is not one of whether there is something new to be found at the enhanced energies available to the upcoming generation of particle accelerators but, rather, whether our experiments will be sensitive enough to dissect this behavior and figure out what's going on. If what's going on is the Higgs mechanism, then we stand to learn a tremendous amount. If it's something else, who knows? This is where the excitement and uncertainty of experimentation enters. The clues that will help us unravel this mystery may hit us in the face even more abruptly than those expected from the Standard Model's Higgs mechanism, or they may only be vague hints that raise more questions than they answer.

Theorists are a tireless bunch, and a number of possible alternative scenarios have been dreamt up, some of which would yield copious and revealing signals, while others would tax the limits of our experimental capabilities. But one thing is clear: we won't know what's out there until we build the machines and do the experiments. Our experimentation over the next decade or two will take us deeply into the world of the unknown. The only thing that we can say with certainty is that we *know* that there is indeed something unknown out there!

10

Into the Unknown

———∿∿∿∿∿∿∿∿∿———

What Lies Ahead

No single book can aspire to a comprehensive presentation of the field of particle physics. This book is no exception. It is as conspicuous for what it leaves out as it is for what it includes.

We have talked little about quantum chromodynamics (QCD), the SU(3)-based gauge theory of the strong nuclear force, that, along with the SU(2) and U(1) theory of electroweak interactions, forms the basis of the Standard Model of particle physics. Entire avenues of experimentation, including the properties of matter containing heavy quarks (c, b, and t) and the exploration of the inside of the nucleon, have been entirely ignored. We have enjoyed but scant discussion of neutrino physics, and of "CP violation," the supposed source of the slight difference in the behavior of matter and antimatter that led to the elimination of the antimatter component of the universe as it evolved away from its nonpartisan beginning at the big bang. We have had no discussion whatsoever of the immediate and far-reaching effect that developments in the field of particle physics have had on our theory of cosmology—on our understanding of the origin and evolution of the universe and its space-time framework (note 10.1).

Similarly, little mention has been made of the design and operation of the world's great particle accelerators. Nor has there been any discussion of the principles underlying the detection and precise measurement of the trajectories of the subatomic particles that burst forth from the powerful collisions produced by these machines. Either of these would fill up several additional chapters, at a minimum.

However, we have been able to present the essential elements of our current view of the fundamental nature of matter and its interactions. We have seen how this effort has, over the past century, continually clarified, simplified, and reshaped our notions of what lies at the heart of the tremendously broad array of natural phenomena.

Considering those directly related to particle physics alone, in excess of twenty years' worth of Nobel Prizes have been delivered to the forty or so most prominent foot soldiers in the march toward today's elegant paradigm of natural law. Add to this the forefathers—the generation of Einstein, Bohr, de Broglie, Schrödinger, Heisenberg, Dirac, and others—and the list grows substantially longer. To whatever extent the receipt of a Nobel Prize designates a research accomplishment as a triumph of the human intellect, the pursuit of the fundamental laws of nature stands fully decorated.

Despite these successes, the exploration is far from over. Consider the developments that occurred during the writing of this book alone. The last of the twelve suspected fundamental fermions, the τ-lepton neutrino v_τ was discovered. Evidence for neutrino oscillations, spontaneous identity swapping between the different species of neutrinos, became so compelling as to merit the award of a Nobel Prize. A strong hint of the Higgs boson, perhaps spurious but certainly suggestive, was observed at the now defunct LEP collider at CERN. As we enter the twenty-first century, the pace of discovery continues unabated.

As for the near future, the most basic tenet of physical common sense, that the probability of a given event occurring can never be greater than unity, tells us that there must be something else out there and that this something else will, in all likelihood, manifest itself clearly in the next generation of accelerator-based experiments. Of course, it's remotely possible that this most basic tenet of physical common sense may turn out to be nothing more than common nonsense, rendering the projections for our future program all but meaningless. This, however, would be as profound a discovery as any scenario we may have envisioned ahead of time.

In many ways, the mandate for the progression to the next energy scale— the next step in accelerator size and cost—has never been stronger. More so than ever before, the questions we need to answer to peel the next layer off the onion are well focused and clearly formulated, while the technology needed to address these questions lies fully in hand. Existing theory, endowed with the weight of authority afforded by its dramatic experimental confirmation, suggests in no uncertain terms that the exploration of the energy scale between roughly 100 and 1,000 times the proton mass-energy, the electroweak scale, will be a rich and rewarding undertaking.

Yet, despite this, at the turn of the twenty-first century, the future of experimental particle physics lies somewhat in the balance. At the beginning of the previous decade, the design of a proton-proton collider with a circumference of nearly 100 kilometers and a collision energy of 40 TeV (or about 50,000 times the proton's mass-energy and twenty times greater than anything available today) was approved by the U.S. Congress, and construction was begun at a site outside of Dallas, Texas. This machine—the Superconducting Supercollider (SSC)—would have decisively explored the electroweak scale. However, political support for this machine began to falter in the face of cost overruns and a wilting economy, and in 1993, its construction was abruptly terminated (note 10.2).

A competing effort in Europe, the LHC (Large Hadron Collider), has continued to move forward since the demise of the SSC. This European proton-proton collider has the advantage of smaller cost because it is being built in the 27-kilometer tunnel formerly occupied by the LEP electron-positron collider and will run at a substantially lower collision energy (14 rather than 40 TeV). What the LHC lacks in energy relative to the ill-fated SSC, it makes up for, at least in part, with its ten-times-higher event rate, an approach that, while less expensive, was thought to be a bit riskier due to the problems associated with collecting data at such a high rate. Nearly ten years later, technological development seems to have rendered this data rate manageable, and the LHC promises an exciting ride beginning in the latter part of this decade.

At this time, early in the first decade of the twenty-first century, the international physics community finds itself engaged in a debate about what should follow the LHC. Given the decade-long lead time between proposal and activation of these mammoth scientific instruments, the time to settle on the next great initiative is drawing near.

A leading candidate has emerged from this debate: an electron-positron collider with a maximum collision energy of roughly 1 TeV (1,000 proton masses), whose various existing designs are known generically as the Linear Collider (LC) (note 10.3). This machine would cost roughly the same as the LHC; combined, the LHC and LC would cost the international particle physics community less than the United States alone would have spent on the SSC, with a potentially richer and more diverse program than that of the SSC. Yet, there is a strong sense within the community, as well as with the government officers who oversee the field, that even at this level the Linear Collider will be difficult to sell to world legislative bodies.

There is little question that the cost of the Linear Collider is great—about 5 billion U.S. dollars. One can think of any number of worthy causes that would benefit vastly from such a sum. But one must evaluate this cost in view of the promise this research holds to advance one of our most persistent and ineluctable intellectual pursuits. This research represents our epoch's contribution to the millennia-long struggle to comprehend the fundamental laws of nature.

The human capital required to execute this program resides with us now; no amount of careful documentation of methodology can fully capture the art of designing and operating the machines and detection apparatus that are necessary to carry it out. If we are to do so, the thousand or so men and women, sprinkled throughout the nations of the world, who have developed this art will be the effort's most essential resource. The time to do it, it seems, is now.

This research is solely the domain of the science programs of the governments of the world. The direct benefits of the knowledge we hope to gain, in terms of technological advancement, are so far off in the future (maybe infinitely so) that it would be folly for any private corporation to invest in this sort of research. What drives us forward with the energy and motivation necessary to carry it off is not the hope of developing fancy new technologies but, rather, our innate sense of curiosity about our natural surroundings. We are driven forward, first and foremost, by the pursuit of knowledge in its pure form, coupled with the well-grounded expectation that our knowledge stands to increase measurably with this effort. The only agency that can appropriately further this demanding program is the world's system of government, operating with the mandate of its constituency.

This, however, is *not* to say, and I emphasize this strongly, that resources invested in fundamental research have not provided a significant benefit beyond the appeasement of our curiosity. By some estimates (admittedly done by scientists), as much as half of the economic output of industrialized nations is a direct outgrowth of this sort of fundamental scientific research.

In the case of particle physics, these contributions have been spin-offs, incidental developments that were not part of the driving force behind the work. Nonetheless, the contribution that particle physics research has made to our everyday lives is substantial.

The particle accelerator—the plaything whose development was driven by physicists' desire to peer more deeply into the nature of matter—has become as widespread a tool outside the field of particle physics as it is within. Particle beam therapy is the treatment of choice for a number of forms of cancer and other malignancies. Synchrotron radiation, intense and highly focused beams of high-energy photons that are a natural by-product of the acceleration of charged particles, has become an indispensable tool in numerous other scientific pursuits. Synchrotron facilities (a number of them having been developed by co-opting outmoded particle physics accelerators) provide a rich program of study in chemical and materials science, biophysics, and medical physics. The need of the electronics and computer industries to develop microchip circuitry with ever smaller features is spurring the development of particle beam lithography, exploiting Heisenberg's recognition that a particle's sphere of influence becomes finer and finer as the particle's energy is increased.

From the particle physicist's perspective, there's no reason to build more powerful accelerators if the products of their collisions cannot be accurately and comprehensively studied. As the energy and intensity of accelerator beams has grown geometrically over the past few decades, the sensitivity and the resilience of the corresponding particle physics detectors, and the sophistication of software tools necessary to process the information they provide, has kept pace. The apparatus that the particle physicist requires to conduct ever more exacting studies of high-energy particle collisions cannot be purchased from a catalog; it must be continually reinvented through a program of advanced research and development, carried on at national laboratories and within university physics departments throughout the world. The resulting detection and representation technology has been pressed into widespread service in fields such as medical diagnosis, microscopy, and microelectronics manufacturing, as well as "pure science" fields such as astronomy, chemistry, and biophysics.

The most visible of all these contributions, and perhaps the most unforeseen, sprang forth from the need of particle physicists to disseminate developments quickly and freely throughout their global community. In 1991, the Briton Tim Berners-Lee of the pan-European CERN laboratory began to develop an information-sharing protocol and associated programming language (HTML) to take advantage of the growing information super-highway, facilitating the search and attainment of electronic information located at particle physics institutions throughout the world (note 10.4). The value of such a system was immediately recognized by many outside of particle physics, and the World Wide Web's ability to link together networks of computers at particle physics institutions rapidly expanded into the all-pervasive system that delights and informs us today.

Of its avenues of return to the society that patronizes it, particle physics' least tangible benefit is perhaps its greatest. In the United States, about 100 students are granted doctorates in particle physics each year. While many of these emerge as the field's future leaders, roughly an equal number leave particle physics behind, either immediately or after several years as post-doctoral fellows. While they're in the particle physics program, these men and women immerse themselves for five or ten years in an intense profes-sional community for which both creativity and precision are essential to success. Many of them see firsthand how all but insurmountable problems are attacked and solved. Consciously or not, many of these young scientists come away with a greatly expanded sense of the limits of human and, in par-ticular, their own, ingenuity. Most of the students who leave particle physics go on to rewarding careers in a broad range of fields, such as electronics and high-technology industries, computer design and programming, informa-tion technology, finance, and education. In a world driven increasingly by the forward march of technology, these uniquely trained individuals repre-sent a substantial resource.

Despite all this, though, the economic return on our investment in par-ticle physics research does not stand as its primary justification. The study of fundamental behavior is not driven by the goal of bringing new tech-nologies to market. To the extent that it does, the benefit is almost always fortuitous, having been foreseen neither by the protagonists of the research nor the leaders of its funding agencies. The motivation, and expected re-turn, of this sort of research is nothing more or less than the deep but be-nign insight that it provides into the nature of the universe and our existence within it.

What Are We Still Looking For?

The study of physics has a way of sprouting forth vital new shoots just when it seems that all the fertile ground has been fully explored. We've seen how the modern physics revolution grew out of the ashes of the complacency induced by James Clerk Maxwell's profoundly successful classical theory of electromagnetism. We've also seen how the frustration and despair associated with the ever more perplexing array of elementary particles discovered in the 1950s eventually yielded to the inspiration of Murray Gell-Mann's eightfold way, initiating the appreciation of the deep connection between Lie groups and physical law. On these and other occasions, pronouncements of the impending demise of branches of physics, or even the entire field itself, have been laughably premature.

Yet, make note of the fact that even Lazarus, having himself risen from the dead to live life anew, is no longer with us. The few centuries that have passed since Newton, or even the millennia that have elapsed since the time of the philosophers of antiquity, seem fleeting when compared with the vast reaches of time that stretch behind and before us. Our success at probing deeply into the laws of nature will almost certainly reach an end, regardless of how many times we mistakenly foresee that end. From the wizened perspective of the universe whose fundamental behavior these studies purport to unveil, the culmination of this work must be very near at hand.

At the dawn of the third millennium, however, we hold a strong conviction that our odyssey is far from over. The Standard Model, our current theory of all known interactions except gravitation, does a remarkable job of describing, and in some cases predicting, the complete set of natural phenomena that have confronted us so far. Yet, at interaction energies not too far above those we currently have access to, and restricted to operate with the ingredients that have been observed to date, the Standard Model begins to fall apart. As the next generation of particle accelerators comes online, we are all but guaranteed to uncover some resounding new characteristic of the natural world, providing insights that run deeply into the heart of both particle physics and cosmology. Our current best guess for this new component of nature is the Higgs field, although there's no way to know with certainty what will confront us until we actually create it in our laboratories. And even if it is the Higgs field that awaits us, its discovery will raise a host of new, interesting, and experimentally addressable questions.

Independent of the question of the origin of the phenomenon of mass, there are other compelling suggestions that even greater discoveries lie ahead.

In the 1960s, when the eightfold way was formulated, no subatomic particles containing quarks other than the lightest three flavors—up, down, and strange—had been observed. As time passed, however, a total of six quark flavors were found to exist. In addition, experimentation revealed the number of lepton flavors also to be six. As shown in table 5.1, these twelve matter-field quanta fall into a very clear pattern. Quarks and leptons each fall into sets of two, with the two quanta in each doublet being distinguished by a difference of one unit of electric charge. Each quark doublet has an associated lepton doublet; together these four particles form a "generation." As discussed in chapter 8, it was this pattern of repeating doublets of matter-field quanta that led to the hypothesis of an underlying SU(2) and U(1) gauge symmetry, and the subsequent formulation of the Standard Model of electroweak interactions.

But there is nothing within our current understanding of nature that explains why, as shown in table 5.1, there are three generations of matter-field quanta. That these quanta appear as doublets separated by one unit of electric charge is understood, at a satisfyingly deep level, to be due to the underlying SU(2) and U(1) gauge symmetry. That symmetry, however, would be perfectly well reflected were there any number of quark and lepton doublets and regardless of whether the number of quark doublets is the same as the number of lepton doublets.

To be sure, the phenomenon of CP violation (the origin of the preference for matter over antimatter in the constitution of the cosmos) requires at least three generations. In addition, due to the subtle influence of the quantum-mechanical vacuum, the Standard Model's predictions for the properties of fundamental interactions only make sense if the number of quark and lepton generations is identical. But this falls short of being a satisfying explanation of why it is that the quark and lepton doublets fall into a mutually repeating pattern of precisely three generations. Given the current state of our knowledge, all we can say is that, thankfully, that's the way the chips fell when the universe came into being.

To a particle physicist, no statement could be more dissatisfying. In the 1950s, the burgeoning list of subatomic particles, whose constitution bore no apparent relation to the fundamental nature of matter and its interactions, compelled Willis Lamb to proclaim that anyone who discovered another such particle should be fined. What's the point of looking for all these new particles if they provide no insight into the nature of things? But in the end, with the help of Gell-Mann, they *were* found to provide a critical insight into a deeper and much more elegant natural order.

Similarly, particle physicists are convinced that the generational pattern of quarks and leptons is suggestive of an even deeper and more concise underlying organizing principle, just as the notion of quarks and quark flavor underlies the complex patterns of the 1950s-era subatomic particles. Our experience with patterns in nature, from those of biological systems down to those of the fundamental matter-field quanta, gives us the conviction that the generational pattern of quarks and leptons is unlikely to be a mere accident of nature. *Something* has to be responsible for it, and we will not consider our work done until we uncover it.

Another experimental result that gives us substantial enthusiasm for the future relates to the discussion of renormalization in the "The Living Vacuum" of chapter 4. To make sense of quantum field theory, the stunningly successful framework in which the fundamental forces are represented in terms of minimal interaction vertices and the Feynman diagrams that can be constructed from them, we had to recognize that the object whose properties we measure in an experiment on electrons is not just a "bare" electron. Instead, we probe en masse the assemblage of the electron and its seamy cloud of virtual hangers-on, the photons and matter/antimatter particle pairs of the living vacuum surrounding the electron. This persistent cloud of virtual field quanta acts to screen the charge of the bare electron, reducing its apparent strength just as a shade of smoked glass will reduce the brightness of a light bulb that it surrounds.

As the properties of an electron (really, electron plus virtual cloud) are explored with higher and higher energy probes, the definition of the probe becomes successively finer (due to the uncertainty principle). In interacting with the electron, the probe penetrates deeper and deeper into the electron's virtual cloud, and as a result, the charge of the underlying bare electron is screened less and less, so the apparent charge sensed by the probe grows successively larger. Exploiting our light bulb analogy, we could liken this to penetrating the glass shade; the further into the shade we press, the brighter the light source at its center will appear.

Now recall chapter 8 and its discussion of gauge theory. When we reinterpret the relativistic quantum theory of electromagnetism in terms of a gauge theory based on invariance under the group U(1) of changes of the wave function's phase, the strength of the electron's charge becomes the single aspect of the theory of electromagnetic interactions that cannot be derived. It must be measured, and once measured with a probe of a given energy, its magnitude establishes the overall strength of electromagnetic interactions. All the other rich and varied aspects of that interaction follow directly from the formalism of gauge theory. In particular, once we determine the strength of the electromagnetic interaction at some energy, we know exactly what it should be at any energy, including energies far beyond those that we can reach in the laboratory.

Today, we know that the electromagnetic force is really just one facet of a more encompassing electroweak force, described by a gauge theory based on both the Lie groups $U(1)$ and $SU(2)$. In addition, the strong nuclear force is now understood, with substantial experimental backing, to be a gauge interaction based on the Lie group $SU(3)$. The more complicated gauge groups $SU(2)$ and $SU(3)$ are non-Abelian: when you apply two successive transformations from either of these groups to the wave equation and its wave function solution, the identity of the resulting combined transformation depends on the order in which you apply them.

In the section "Beyond Mere Phase 1" of chapter 8, we saw that for non-Abelian gauge groups such as $SU(2)$ and $SU(3)$, the application of the gauge principle forced us to include additional, and wholly unusual, minimal interaction vertices. Rather than connecting together the matter-field quanta with the force-field quanta, this additional set of vertices connects the force-field quanta *to themselves*, allowing, for example, a gluon to mediate a strong-force interaction between two other gluons. In such interactions, there need be no matter-field quanta involved at all—matter doesn't matter, if you will. But, in quantum field theory, force-field quanta associated with a given interaction will only form minimal interaction vertices with objects that carry the charge appropriate to that interaction. For interactions arising from non-Abelian gauge groups, then, the force-field quanta themselves carry the charge associated with the interaction.

Going back to our electron and its attendant cloud of virtual particles, let's replace the electron with a quark (any flavor will do). Bear in mind that, unlike electrons, quarks possess color (strong-force) charge, and so partake of all three of the strong, weak, and electromagnetic interactions.

Just as for the electron, if we probe this quark electromagnetically by bouncing a photon off it, the magnitude of the quark's electric charge seen by that photon will depend on the energy of the photon. The greater the photon energy, the more effectively it slices through the cloud of virtual particles surrounding the quark, and the more directly it probes the quark's charge. The magnitude of the quark's charge—the strength of the quark's electromagnetic interaction—increases with the greater definition provided by a higher-energy probe.

What if, rather than electromagnetically, we were to probe this quark via the *strong* interaction by firing a gluon at it? Quantum mechanics is still at play; a higher-energy strong-interaction probe will still cut through the virtual cloud surrounding the quark more effectively than a lower-energy probe. As for the electron, this virtual cloud contains both matter particles (quarks) as well as force-field quanta (gluons). In this case, however, the force-field quanta themselves carry the charge of strong interaction being used to probe the quark.

The non-Abelian nature of the strong interaction's SU(3) gauge group—the fact that the order in which SU(3) transformations are applied makes a difference—leads to a qualitative change in the character of the virtual cloud surrounding the quark relative to that surrounding the electron. For the quark, both the matter quanta *and* the force-field quanta in the cloud contain the charge that the probe is looking for, and both will affect the result you get when you probe the quark with the gluon. But force-field quanta (gluons) in the virtual cloud interact with the incoming gluon probe in a much different way than its matter-field quanta (quarks); the minimal interaction vertex connecting an incoming force-field quantum to another force-field quantum has quite different properties than that of the vertex connecting the incoming field quantum to a matter-field quantum. It's no longer obvious that probing deeper and deeper into the virtual cloud will lead to an *increase* in the associated charge seen by the probe.

In fact, for non-Abelian gauge interactions, it can go either way. Whether the overall charge seen by the probe, *and thus the strength of the interaction itself*, gets stronger or weaker for higher-energy probes depends on the specifics of the theory. It depends on the precise nature of the way in which the ordering of the underlying gauge group transformations matters (the group's Lie algebra). It also depends on the number and nature of particles that can fleetingly appear as particle/antiparticle pairs in the virtual cloud. In the language of our light bulb and smoked glass shade analogy, it's as if the smoked glass shade is itself emitting some light. Whether the light appears brighter or dimmer as you get rid of more and more of the shade depends on whether the shade emits more light than it absorbs or vice versa.

But what *is* true, regardless of the character of the underlying gauge group and of the specific nature of the virtual cloud, is that the strengths of the three interactions are not fixed. All three of their strengths are expected to depend on, in one way or another, the energy of the object that is probing that strength.

In the late 1980s, the first precise studies of large samples of Z^0 bosons allowed physicists to accurately determine the strength of the SU(3) (strong nuclear) interaction for probes with the energy of the Z^0 mass-energy (about 100 GeV). Since the strength of the strong nuclear interaction had already been measured for lower probe energies, it was possible to compare expectation against hard data. In this way, this curious aspect of quantum field theory—that the strengths of the fundamental interactions depend on the amount of energy at play—was convincingly demonstrated. In fact, given the Lie algebra of SU(3) and the particles available (tables 5.1 and 5.2) to populate the virtual cloud surrounding quarks, the strong nuclear interaction was expected to *decrease* in strength with increasing probe energy, and this is what was observed.

However, it is even more interesting to use the properties of the Standard Model to predict what will happen to the three interaction strengths as the energy of the probe increases above that of the Z^0 boson mass-energy. Since the calculation of this prediction requires nothing more than a little scratch paper, there's nothing to keep us from making the prediction for probe energies far above the reach of any conceivable future accelerator.

Doing this, one discovers something very tantalizing. The SU(3) interaction, already observed to be decreasing in strength with increasing probe energy, continues to do so. The next strongest of the three interactions, the SU(2) interaction of the electroweak force (note 10.5), is also predicted to decrease in strength with increasing energy; however, its rate of decrease with energy is not quite as fast as that of the SU(3) (strong nuclear) interaction. Both strengths are decreasing, but the stronger of the two is weakening faster than the weaker of the two; as a result, at a point sufficiently high in energy, the strengths of the SU(3) and SU(2) interactions should become identical. This energy is very high, about 10^{17} GeV, or the equivalent of about 10^{17} times the proton mass-energy. Since the highest collision energy of any accelerator, current or planned, is only about 10,000 (10^4) GeV, this energy at which the SU(2) electroweak and SU(3) strong-interaction strengths are predicted to become identical lies far beyond the realm of direct experimental confirmation.

The situation becomes more interesting when the strength of the U(1) electroweak interaction, the weakest of the three, is calculated for energies above the Z^0 mass-energy. Recall that the U(1) Lie group, the most rudimentary of the three Standard Model gauge groups, is Abelian. No matter what, its interaction strength will increase with energy. With the stronger SU(2) and SU(3) interaction strengths weakening with higher interaction energy, there will necessarily be an interaction energy for which the U(1) and SU(2) strengths are the same and another such energy for which the U(1) and SU(3) interaction strengths are the same.

Given what we know about quantum field theory, gauge theory, and the array of fundamental particles that contribute to the virtual cloud surrounding an object, we expect there to be interaction energies for which each pair of two of the three fundamental interactions have identical strengths. In all cases, the energies associated with these equal-interaction-strength points are far above that of any conceivable accelerator (although not above that thought to be available just after the big bang). This is interesting, but so what?

What if these three points—the separate energies at which the SU(2) and SU(3), U(1) and SU(3), and U(1) and SU(2) interaction strengths become the same—are in fact *one and the same point?* In other words, what if the strengths of these three fundamental interactions coalesce at a common energy, becoming at that point the single, overarching interaction strength of all the fundamental interactions of nature, barring gravity? This would be compelling evidence that, just as electricity and magnetism were joined together in the 1860s into a unified electromagnetic force, with a common interaction strength given by the value of the electron's charge, the SU(2) and U(1) electroweak force and the SU(3) strong force are, similarly, merely different facets of a single interaction—the grand unified interaction.

Intriguingly enough, this is indeed what the predictions tell us, *more or less.* As shown in figure 10.1, the ever-growing strength of the Abelian U(1) electroweak interaction (you can think of it, if you like, as electromagnetism, although as we've seen in chapter 9, the U(1) interaction is in fact polluted by a small piece of the weak neutral interaction) approaches the decreasing strengths of the non-Abelian SU(2) and SU(3) interactions at roughly the same energy, 10^{17} GeV, at which the SU(2) and SU(3) interaction strengths meet.

Ignoring for the moment the "more or less" and the "roughly," it seems as if this result is nudging us, trying to tell us that our view of the SU(2) plus U(1) electroweak force and SU(3) strong force as *separate* interactions is short sighted. And we, as physicists, know that. The problem is that, to this point, we haven't been able to develop a model that exploits this revelation,

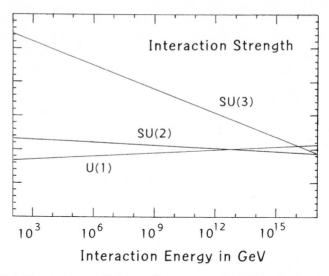

Fig. 10.1. The variation with interaction energy of the U(1), SU(2), and SU(3) charge strengths suggests that the strengths probably coincide at an interaction energy of roughly 10^{16} GeV (10^{16} times the proton mass-energy), or about 10^{13} times the interaction energy of the world's largest particle accelerators. Such a coalescence is the bellwether of a single overarching, unified force that would be the sole source of causation in the natural world.

unifying the existing electroweak and strong nuclear interactions in a way that's both theoretically self-consistent and in agreement with experimental observation.

This is where the hedge—the "more or less" and the "roughly"—may in fact come to our rescue. The U(1), SU(2), and SU(3) interaction strengths do not quite meet at precisely the same interaction energies. They're rather close: the U(1) and SU(2) strengths meet at about 10^{13} GeV, and the U(1) and SU(3) strengths at just above 10^{16} GeV, both a tad below (logarithmically speaking) the interaction energy of 10^{17} GeV at which the SU(2) and SU(3) strengths coincide. The differences between 10^{13}, 10^{16}, and 10^{17} GeV may seem large when quoted in this way, but a glance at figure 10.1 shows that it wouldn't take much of a change in the way the interaction strengths grow and diminish with increasing interaction energy to bring them all together at a common energy. They almost coalesce as it is.

Now, the fact that they don't all come together at quite the same energy could mean that the forces don't unify—the closeness of the three points at which the different pairs of two of the three interaction strengths come to equal each other could just be pure coincidence. However, our prediction of the way in which the interaction strengths change with energy depends on knowing the complete list of ingredients available to concoct the virtual cloud surrounding the object being interacted with. So the failure of the three interaction strengths to coalesce precisely could instead mean that we need to add a little something extra to the Standard Model.

It's quite plausible that there are undiscovered particles, with energies out of reach of current accelerators, that will slightly alter the makeup of the virtual cloud, subtly modifying the way the three strengths change with interaction energy. This small difference in the "evolution" of the interaction strengths with energy might be just what is needed to allow them to coalesce at a single interaction energy, a unification point energy, somewhere between 10^{13} and 10^{17} GeV. And if this is the case, these exotic new particles probably lie just out of reach of current accelerators: the lighter these new particles are, the earlier they change the slopes of the lines in figure 10.1 and the greater chance they have of bringing those lines into coalescence at a common point.

But the notion of unification—of the reduction of the list of independent natural phenomena to a single, all-inclusive underlying agent of causation—is all but an assumption of modern science. Given what we currently observe and measure in the lab, the predicted strengths of the SU(3), SU(2), and U(1) interactions come so close to unifying that it just *must* be the case that they do in fact unify and that there are undiscovered particles that establish this unification lying somewhat beyond the reach of current accelerators.

And, in fact, there is a compelling theoretical framework that provides just the array of particles necessary to establish the unification of the interaction strengths. This framework, which goes by the name of *supersymmetry*, is currently the most promising development in the search for a "grand unified theory" of the electroweak and strong nuclear interactions. The basis of supersymmetry is the notion that each and every field quantum, matter, force, or Higgs, has a supersymmetric counterpart that has properties identical to the original quantum, except for two things: the mass of the supersymmetric counterpart can be much different than its conventional partner; and if the original particle is a fermion, then its supersymmetric counterpart is a boson, and vice versa (note 10.6).

While far from confirmed, supersymmetry does enjoy the support of a number of suggestive notions. If supersymmetry is a correct description of nature, we need to include the supersymmetric partners of the conventional matter and force-field quanta of tables 5.1 and 5.2 in the virtual cloud surrounding any particle being probed by the U(1), SU(2), or SU(3) interactions. As suggested just above, when this is done, it is indeed found that the three interaction strengths coalesce at a single interaction energy (at an interaction energy of between 10^{16} and 10^{17} GeV). In this and other more technical ways, supersymmetry seems naturally disposed to providing a description of nature for which the U(1) and SU(2) electroweak interaction and SU(3) strong interaction are unified into a single overarching interaction.

Another suggestive aspect of supersymmetry is that it requires the Higgs boson (actually, the lightest of the set of at least four Higgs bosons that are necessary to incorporate the phenomenon of mass in supersymmetric theories) to be in the range of 50 to 150 proton masses. As we've seen in the section "Parity Violation and the Electroweak Force" of chapter 9, this is exactly what precise measurements of the ingredient fraction of the Z^0 boson seem to be telling us.

A third suggestive notion is the fact that the large-scale structure of the universe seems to be influenced by more than we see when we look out into space with our most sensitive astronomical equipment. Over the preceding chapters, we've made use of notions from Einstein's special theory of relativity; physicists qualify this theory with the label "special" because there is an additional, even more powerful theory of relativity known as the "general" theory (I mentioned this in passing in chapter 3). Einstein's general theory of relativity tells us how to relate the large-scale properties of the universe (those properties that we observe when we look at the spatial and temporal relationships between the most distant objects we can see) to the amount of mass-energy embedded within it. Our measurements of these most basic structural properties suggest that the universe contains much more mass-energy than we see by adding together the mass of all the stars, galaxies, nebulae, stray light, and so forth, that our astronomical instruments record. So, if we believe the general theory, then much of the universe is composed of "stuff" that we can't see—so-called dark matter. The supersymmetric partner particles are leading candidates for the composition of dark matter.

But all this stands merely as circumstantial evidence—compelling, but not enough to establish the certainty of the universe as being supersymmetric in a court of natural law. The evidence needed to convict—the smoking gun—is the direct observation of one of the particles from this shadow world of field quanta with spins that are complementary to those of ordinary matter. And, try as we may, this evidence has yet to be uncovered by any study dedicated to its discovery.

But given the experiments that we have been able to do so far, this lack of evidence is not necessarily inconsistent with the expectation of super-symmetric theories. According to those theories, the masses of all of these exotic field quanta may well be so large that they are simply beyond the reach of our current panoply of accelerator facilities. These same theories tell us, however, that the *next* generation of experiments, to be performed at the upcoming LHC and the proposed electron-positron Linear Collider, should be much more definitive in either finding direct evidence for su-persymmetry or ruling it out altogether.

And so we again find ourselves anxiously awaiting the results of experi-ments that are to be performed as the next generation of accelerators comes online. Accordingly, this next step on our inward march toward an under-standing of the basis of natural phenomena stands that much more strongly motivated.

Epilogue

It seems inevitable that the study of particle physics will someday reach its practical end. It was barely more than seventy years ago that the American physicist E. O. Lawrence developed the first cyclotron, ushering in the era of accelerator-based particle physics experimentation. Several living parti-cle physicists, one or two of them still active, were young men at the time. With the LHC not coming on until well after 2005, however, the expo-nential growth in accelerator energy enjoyed over those seventy years has fi-nally begun to taper off. Research and development programs are under-way that are beginning to point the way to the next factor of ten or so in accelerator energy beyond the LHC and Linear Collider, but for the first time, fiscal limitations are beginning to play a central role in setting the scope of our imagination.

However, it may well be nature herself and not our own inability to de-vise ever more powerful scientific instruments that ultimately limits our in-terest in pushing forward toward higher and higher energy collisions. The upcoming LHC and proposed Linear Collider will probe the electroweak scale between 100 (10^2) and 1000 (10^3) GeV, an energy range that holds significant promise to provide a wealth of exciting discoveries and measurements.

Beyond this, it may soon be technologically and fiscally possible to reach energies of 10^4 or even 10^5 GeV, but the motivation for doing so will have to arise out of the upcoming studies at the electroweak scale. These studies may well suggest that there's nothing more to be learned from collisions with energy less than that of the grand unified scale, requiring particle accelerators with beams of energy 10^{15} GeV or greater. Even in the best of the old days of rapid technological advancement, no one dreamt of a step forward of *ten orders of magnitude* in accelerator energy. If the next realm of interest does indeed lie at the grand unified scale, then the best we will be able to do is to discern the physics of the electroweak scale (which may well contain important clues about what's going on at the grand unified scale) as precisely and comprehensively as possible and then, effectively, call it quits. It's impossible to predict what the case will actually be as we complete our studies at the electroweak scale, but this scenario is as likely as any.

Whether it's 50, 100, or 500 years from now, the active pursuit of particle physics is destined to wind down eventually. What then will become of all that we have learned?

Perhaps we can gain some insight from the institution of medieval scholarship, which occupied the minds of those given to deeper thinking during a period when political and social conditions did little to foster advances in natural philosophy. While, at least by modern standards, the progress of Western scientific thought was glacially slow during this period, the great strides that had preceded this fractious time were hardly forgotten. A major component of formal medieval thought was given over to the reading and interpretation of the works of the classical philosophers, providing a continuous intellectual thread that preserved these advances until they began to be supplanted in the fourteenth century by the seeds of modern empirical science.

There is within human nature, it seems, a desire to *know*, regardless of whether the attainment of knowledge is accompanied by the thrill of discovery or is won merely by the toil to competence within an established body of thought. I am optimistic that the lessons of particle physics will not fade away as the currency of their pursuit wanes. I suspect that the essential core of this body of knowledge will be preserved over the ages, unless it is itself supplanted by some far-in-the-future and unforeseeable philosophical advance.

The defining characteristic of our modern era is its frenetic drive toward the understanding, domestication, and control of natural causation. We thrill in our scientific insights and revel in the ever-evolving gadgetry that these insights bring forth. There is deep within the psyche of the modern human a sense of forward motion, an expectation of a daily pace of intellectual and technological achievement.

Particle physicists, in this sense, are quintessentially modern. Being responsible for the most fundamental notions of how the universe operates, we push at the leading edge of this defining element of human nature. More so than anything else, then, the eventual cessation of particle physics research may well telegraph the transition to a fundamentally different, postmodern intellectual perspective. Rather than being driven forward by the promise of incessant scientific and technological advance, we will perhaps be forced to look back within ourselves, to make use of our art and artifice to live full and interesting lives within well-defined and no longer rapidly expanding boundaries. Perhaps then, at last, sustainability and the appreciation of the beauty of the natural world will overtake expansion and development as primary societal values.

Or maybe not. But one thing is fairly certain: there's plenty of horizon-expanding left to do. Enough, in fact, that neither you nor I will ever know if the musings of the last few paragraphs are even remotely on target.

In any regard, if you're reading this, then I suspect you're due something along the lines of a congratulation, for it's been a long and action-packed ten chapters. Not that it's been a breeze for me either. The task of writing a book such as this, and perhaps even more dauntingly, unearthing a publisher desperate enough to take it on, should not be underestimated. But here we are, together, celebrating a process that I hope has enriched and enlivened you as much as it has me.

So, I think I'll head downstairs and crack open an ice-cold beverage. Not a "lite" one, mind you, but a robust draft fit for the occasion, along with some rich and salty snacks that will provide additional justification for liquid refreshment. I suspect that none of this will do wonders for my waistline, but then again, if future experimentation indeed demonstrates that the Standard Model is correct, I can safely reinterpret the mass of my well-upholstered frame as being a mirage—the illusory effect of the pervasive drag imposed by the background Higgs field—rather than just the excess weight that my bathroom scale would have me naively believe I've accumulated. And that's good enough for me.

So, you see, particle physics *does* have immediate application to everyday life. Peter Higgs, this one's for you.

Cheers!

Appendix

Exponential Notation

Any scientist will eventually find herself or himself confronted by numbers that are, by everyday standards, either extremely large or extremely small. For example, the number of molecules in a "mole" of any given substance, which is a specific number of molecules that corresponds to an amount of a solid that fits comfortably into your palm, is about 602,000,000,000,000,000,000,000, while the mass of an electron is about 0.00000000000000000000000000000911 kilograms (1 kilogram weighs about 2.2 pounds).

Obviously, we need a way to save ourselves from writing all of these vexing zeros, and our saving grace is known as *exponential notation*. We define the mathematical expression 10^n, where n is some whole number, to mean the number 1 followed by n zeros. So, 10^6 is just 1,000,000, or 1 million. Likewise, 5.2×10^5 is just 520,000, or five-hundred and twenty thousand.

If the number n is negative, then n represents *one greater than* the number of zeros following the decimal point; for instance, 10^{-6} is 0.000001 (one-one-millionth), and 5.2×10^{-5} is 0.000052.

In these terms, then, Avogadro's number (the number of molecules in a mole) is about 6.02×10^{23}, while the mass of the electron in kilograms is about 9.11×10^{-31}. If you like, you can count the zeros in the first paragraph to make sure the typesetter was paying adequate attention.

Notes

Chapter 2   The True Movers and Shakers

2.1   One such cause for optimism, a set of developments that form the field of "string theory," has been described thoroughly and accessibly in a remarkable book, *The Elegant Universe*, by Brian Greene, a theoretical physicist on the faculty at Columbia University.

2.2   If you're unfamiliar with the exponential notation used in the expression for e, please see the appendix on exponential notation for a discussion.

2.3   This notion was modified with the advent of Einstein's general theory of relativity in 1915. In this theory, which now rests on a very solid experimental base, the gravitational "charge" is not mass but "mass-energy," or, even more accurately, "stress energy." The distinction is not important for our discussion.

2.4   For those who already know about the different *flavors* of quarks (up, down, strange, charm, bottom, and top), this may be a bit confusing. Any one of these six flavors of quarks can come in any of the three colors. Technically, there are thus eighteen different types of quarks, although the observable effects of quark color are subtle enough that we usually lump all three quark colors together. So, the term *strange quark* usually refers to the full set of red, blue, and green strange quarks, without regard to the color of any individual strange quark that may be under consideration.

2.5   Technically, it is still possible that all the antimatter is lurking on the "other side" of the universe, but no current theory of cosmology (the science of the birth and evolution of the universe) can explain how such a possibility is consistent with the known properties of the universe, particularly that the universe seems to be expanding from a single point from which it exploded in a big bang 10 to 20 billion years ago.

Chapter 3   The Great Reawakening

3.1   It is no coincidence that Einstein originally formulated special relativity to address apparent logical inconsistencies in Maxwell's theory of electromagnetism that had to do with how observers in different reference frames (i.e., in motion relative to each other) would have to divide up observed electromagnetic forces between their electric and magnetic contributions. In fact, Einstein's original paper on special relativity, published in the German journal *Annalen der Physik*, was entitled "Zur Elektrodynamik bewegter Körper," or "On the Electrodynamics of Moving Bodies." Einstein demonstrated that the logical inconsistencies were not a failing of Maxwell's theory but rather of the common-sense notions of space and time.

3.2 To avoid writing numbers that are too big and cumbersome, we need a shorthand. One thousand eV will be denoted "1 keV" (kilo-electron-volt); one million, "1 MeV" (mega-electron-volt); one billion, "1 GeV" (giga-electron-volt); and one trillion, "1 TeV" (tera-electron-volt).

3.3 Recall that we discussed another unit of energy—the electron-volt (eV). One electron-volt is about 1.6×10^{-19} joules, which is quite small. But individual charged particles (electrons, protons) are pretty small too, so they wouldn't be expected to obtain energies on the everyday scale of the joule.

3.4 A wonderful book, *Mr. Tompkins in Wonderland*, somewhat whimsically speculates about what the world would be like if Planck's constant were much larger—close to one—and the speed of light were much smaller—say, a few meters per second. The book was written by the Russian American physicist George Gamow, who was one of the more prominent contributors to the development of the quantum theory of particle interactions. After a successful international career, Gamow spent the last years of his professional life at the University of Colorado at Boulder. From the robustness of the prose in *Mr. Tompkins*, one can speculate that he enjoyed the skiing.

3.5 Most precisely, if two or more wavelike systems interfere, then the *relative* phases of the interfering systems are important factors in determining the physical properties of the system. However, the phase of the *overall* system—the full many-body system that comprises all the interfering subsystems—has no physical relevance.

3.6 The term *blackbody* refers to the fact that the calculation of the color spectrum of hot objects is most easily done for materials that, when cold, absorb all electromagnetic radiation (light) that hits them, that is, for objects that are perfectly black before they are heated to temperatures at which they begin to glow.

3.7 The equal sign ($=$) in the expression of the uncertainty principle should really be a greater than or equal to sign ($\geq$). $h/(4\pi)$ is really the *minimum* possible value of the uncertainty product; technically, the true value depends on exactly how the localization is achieved. For most physical systems, the uncertainty product is within a factor of two or so of this minimum.

3.8 An object's wave function takes the form of a perfectly uniform wave, extending over all space, *only* if the object has a precisely defined momentum and is free of external influences. If, in particular, the object is acted on by one or more forces (as will be discussed immediately below), the wave function can be rather complicated. This makes sense: Shortly, we'll claim that the wave function encodes within itself all that can possibly be known about the object's physical state. This means that the wave function will in general have to exhibit at least some degree of complexity to encode that information if the object is under the influence of some external force.

3.9 What actually follows is the time-independent one-dimensional Schrödinger equation. The time-dependent part of the Schrödinger equation, the part that tells you how the wave function varies with passing time rather than with location, has been factored out into a separate equation (not shown here), which concerns us less than the time-independent equation presented here. Rest assured, however, that the time-dependent factor of the full Schrödinger equation is much simpler, mathematically, than the time-independent factor presented here, so you are getting your full dose of the quantum theory. We'll come back to the issue of the time-dependence of the wave function in chapter 4, where we'll discuss its peculiar connection to the representation of antimatter.

3.10 Strictly speaking, this property holds if and only if the object is bound—restricted to move in a certain well-defined region in space—by the force represented by the potential energy function $V(x)$. A good example of such an object is an electron that is bound electromagnetically to a proton in a hydrogen atom.

3.11 The concept of energy conservation came along surprisingly late, given its central importance throughout the fields of science. This is because, unlike momentum (also a conserved quantity), energy can come in many different forms—kinetic energy (of both translation and rotation), potential energy (of numerous different forms), heat, radiation, and so forth. The notion of energy conservation seems to have emerged simultaneously with the development of the modern theory of heat in the mid-1800s. Among its earliest proponents were a German physician by the name of Julius Robert Mayer and a British experimentalist named James Prescott Joule, whose professional roots lay in the business of brewing beer.

Chapter 4   The Marriage of Relativity and Quantum Theory

4.1 The modern notion of the "force field" was first introduced by the British physicist Michael Faraday in the 1830s to interpret his revolutionary experiments on the interrelation between the electric and magnetic forces. The story of Faraday's rise from humble beginnings as a laboratory apprentice in London's Royal Institution to the receipt of persistent offers to assume its leadership (all politely refused) is a particularly inspiring one.

4.2 Technically, $k \simeq 8.99 \times 10^9$ newton-meter^2 per coulomb.

4.3 The relation Maxwell derived is simple; if we denote by c the speed with which the disturbance travels through space, then

$$c = \sqrt{\frac{4\pi k}{\mu_0}}$$

where $\pi \simeq 3.14159$ is, as usual, the ratio between the circumference and diameter of a circle, and $\mu_0 = 4\pi \times 10^{-7}$ Tesla-meters per amp is the experimentally measured strength of the magnetic force.

4.4 We also use this term to describe colleagues who continually put forth conjectures that don't make any sense to us.

4.5 In the popular culture of the United States, Feynman was relatively well known for his collections of humorous anecdotes (both scientific and nonscientific) as well as his brinkmanlike demonstration in a congressional hearing that the thermal properties of the material used to make the O-rings was culpable in the midflight explosion of the Challenger space shuttle, a hypothesis later adopted as the most likely explanation for the disaster. Within the culture of professional physicists, Feynman is considered to be one of the most pronounced and free-thinking physicists since the great icons of the early twentieth century.

4.6 Technically, there is one more thing you could ask for: the quantum mechanical phase (see chapter 3) of the scattered wave function. This is necessary to know if you have to combine the results of the calculated process (say, single photon exchange) with the result of the calculation of another process (say, double photon exchange) that has the same input and output particles.

4.7 This coupling goes by another name—the *fine structure constant*. In the early days of atomic physics (before quantum field theory revealed the true meaning of the fine structure constant to be the strength of the coupling between the electron and photon), it was thought to have a value so close to being precisely 1/137 that numerologists started to establish cultish associations with the number 137. However, quantum field theory tells us that the fine structure constant actually depends on the energy of the virtual photon and is only very close to 1/137 at low energy—that is, the energies available to experimenters in those early days of atomic physics. So, there's nothing magical about 1/137. Quantum field theory predicts, and experiments confirm that at virtual photon energies equivalent to about 100 proton masses (remember that mass and energy are related by way of $E = mc^2$) the value of the fine structure constant is about 1/128.

4.8 The boardwalk in Santa Cruz, California, has a particularly nice carousel that turns to the music of a 1900s vintage mechanical calliope. It's well worth the visit if you ever find yourself in town.

4.9 The English physicist Paul A. M. Dirac and the Austrian physicist Wolfgang Pauli are perhaps the most prominent of this crowd, both of whom were in their late twenties when this work was done. Soon after this work, Dirac was awarded the Lucasian Chair of Mathematics at Cambridge University, perhaps the most prestigious academic position on the planet, which has also been held by such greats as Isaac Newton and the renowned astrophysicist Stephen Hawking.

4.10 Don't be discouraged if you've been told that the square of any number is a positive number. This is only true for *real* numbers. In general, solutions to any quantum-mechanical wave equation, relativistic or not, are *complex*. This means that the solutions contain factors of i, the number that when squared (multiplied by itself) somehow gives -1. We'll hear a lot more about complex numbers in coming chapters.

4.11 Raising e, the base of natural logarithms, to an imaginary number that is a multiple of $i = \sqrt{-1}$ (like $i\,\alpha\,Et$) is just a mathematically convenient way to represent functions that oscillate in time or space, that is, sine and cosine functions. So, such complex exponentials are appropriate for representing quantum-mechanical wave functions. By the way, the constant α is simply $1/\hbar$, in case you were wondering.

4.12 There may appear to be a logical inconsistency here. We require energy to be conserved absolutely, in that the sum of the energies of the electron and positron precisely add up to the energy of the virtual photon from which they derive. In the paragraph of the main text just above, however, we relied on the uncertainty principle in a way that may appear to release us from the requirements of energy conservation, so that the fact that real (nonvirtual) photons have zero mass does not restrict the virtual photon from temporarily fluctuating into an electron-positron pair. These two statements are not inconsistent—the reconciliation of these two apparently incompatible statements lies in the recognition that, uncertainty principle or not, we always require energy to be conserved; it's just that, in doing so, the mass of a short-lived state (photon, electron-positron pair, etc.) can temporarily be different from what it would be for the corresponding state with indefinite lifetime whose mass you would measure in the lab. So the mass-energy of the virtual photons is equal to the mass-energy of the virtual electron-positron pair, although neither is equal to the mass-energy of either a real photon or the sum of the electron and positron mass-energies.

4.13 Note that a photon in isolation cannot "fall apart" by getting absorbed by an electron-positron vacuum fluctuation, which would result in the spontaneous conversion of photon into an electron-positron pair. Since both states (the initial photon and the final electron-positron) live arbitrarily long, there is no uncertainty on either's mass-energy, so the differing mass of the initial isolated photon and final electron-positron pair will prohibit (due to energy conservation) the reaction from happening. Instead, a single, isolated photon that is absorbed by an electron-positron vacuum fluctuation will always quickly fluctuate back into an isolated photon. The second photon of figure 4.11 is absolutely essential if the electron and positron are to be made to stick around in the observable final state.

4.14 A laser beam, though intense, is composed of photons that are themselves rather ordinary. The energy of any photon, whether in a laser beam or coming from a child's night-light, is given by the product of Planck's constant and frequency f (or color) of the photon: $E = hf$. So, the individual photons in a laser, no matter how intense the laser may be, are no more energetic than the photons in the night-light. What makes lasers so interesting and powerful is that there are an unspeakably large number of photons concentrated in a small region of space, and, what's more, all with synchronized phase (coherent). However, once the laser beam is bounced off a high-energy electron beam, you have a situation in which *both* the density *and* the energy of the individual photons are tremendous, and interesting things can happen (such as the creation of matter from light).

4.15 An electrically neutral, or unchanged, conductor (such as a chunk of copper) is just a bunch of neutral atoms connected in some sort of regular pattern. What makes the conductor a conductor is that the outermost (most weakly bound) electrons in the atoms become free to wander at will about the conductor. Thus, the conductor contains a large number of free electrons (of order one per atom) even though the conductor as a whole is electrically neutral. These are the electrons responsible for the conduction of electricity when the conductor is hammered into the shape of a wire and connected between the two terminals of a battery.

4.16 For any given particle, the charge strengths associated with the three different forces (electromagnetic, weak, and strong) don't quite coalesce at the same energy unless additional physical laws, beyond those currently known to apply, are introduced. A set of physical laws that seem to lead quite naturally to the fine adjustments necessary to make these three separate extrapolations precisely coalesce are those provided by *supersymmetry*, a discussion of which can be found in Brian Greene's book *The Elegant Universe*. This is an intriguing hint that supersymmetry may underlie the next great leap forward in our fundamental understanding of nature. We'll discuss supersymmetry briefly in Chapter 10.

Chapter 5   Patterns in Nature

5.1 When I was in graduate school at the University of Chicago, I once came home to the sight of a laden cheesecloth dripping its whey into the kitchen sink. When I asked my Czech housemate about the contents of the cloth, she perkily replied: "Quark!" Quark, it turns out, is a creamy, low-fat cheese that is popular in Europe. However, I've never heard one way or the other whether Joyce's Muster Mark ate the stuff.

5.2 A fascinating and detailed recounting of this intense period in the development of particle physics and of the November Revolution (see "The November Revolution" in chapter 5), is contained in a book by Michael Riordan entitled *The Hunting of the Quark*.

5.3 Recall our discussion in chapter 3 of physicists' tendency to quote the mass m of objects in terms of their equivalent energy E using Einstein's famous formula $E = mc^2$, where c is the speed of light.

5.4 Makoto Kobayashi and Toshihide Maskawa have never been recognized for their work with a Nobel Prize because, perhaps, while it is true that the third generation admits matter/antimatter symmetry violation within the Standard Model, it remains to be demonstrated that their approach accounts quantitatively for the observed violating effects. Today, there is an intense global effort to address just this issue, a major component of which are two "B meson factories" (intense intermediate-energy particle accelerators) at SLAC in California and KEK in Japan. Should Kobayashi and Maskawa's conjecture be confirmed, their qualifications for the Nobel would seem strong.

5.5   The recently discovered neutrino oscillations (see the section "Leptons: The (Not-So) Light Ones" in chapter 5) are only possible if there is a similar sort of mixing among the leptons. Much less is known about the nature of this "leptonic mixing." Should the current neutrino oscillation results be borne out, this promises to be a fruitful area of study over the next few decades. The question of what experiments can be done to best understand this phenomenon—and what new technologies can possibly be developed to abet these experiments—is one of the major developmental threads in current experimental particle physics.

5.6   This γ is one and the same as that of "γ radiation" from radioactive nuclei, which results from the release of high-energy photons when an unstable nucleus re-arrays itself into a more stable configuration.

5.7   This is not obvious; to work this out, you really need to know a little about the mathematics of the Lie group that underlies the behavior of the color force. We'll learn a little about Lie groups in chapter 6.

5.8   Note the analogy with colored light: when red, green, and blue light are combined with equal intensity, the result is color-neutral white light. The choice of the term *color* to identify the charge associated with the strong nuclear force is an appropriate one.

5.9   Neutrons in the nuclei of stable atoms live forever because, were they to decay, the sum of the mass-energy of the resulting nucleus plus the electron and antineutrino that go flying off after the decay would be greater than that of the original nucleus with the undecayed neutron, thus violating energy conservation. On the other hand, the process of nuclear beta decay is essentially just the decay of a neutron within the nucleus. The task of understanding which nuclei forbid beta (neutron) decay due to energy conservation, and are thus stable, fell squarely on the shoulders of the nuclear physicists of the 1930s and 1940s, when our basic understanding of nuclear behavior first emerged.

5.10   I recommend the group's pocket-size booklet; write them at Particle Data Group, MS 50R6008; Lawrence Berkeley National Laboratory; One Cyclotron Road; Berkeley, CA 94720, U.S.A. Or visit their Web site at http://pdg .lbl.gov. By the way, I was within days of accepting a job with this group (to clean cages, as I recall) when the opportunity to instead do physics on the redwood-studded beaches of Santa Cruz presented itself.

Chapter 6   Mathematical Patterns

6.1   Here, "Lie" is pronounced "Lee."

6.2   Here's an interesting challenge: can you find another distinct operation (rule for combining the numbers 1 through 4) that still satisfies all the axioms required of a group? If so, then you've discovered a second, distinct group with four elements.

6.3 Whether the superior intellectual capabilities of the human species renders it somehow intrinsically more deserving or worthwhile than all the other animals is a debate into which this author would prefer not to enter, considering that, as a cat "owner," I don't want to offend the true masters of the household.

6.4 The rule for multiplying together two complex numbers $z_1 = a + b \cdot i$ and $z_2 = c + d \cdot i$ is what you might expect, given the rules of algebra: $z_1 \times z_2 = a \cdot c + a \cdot d \cdot i + b \cdot c \cdot i + b \cdot d \cdot i \cdot i = (a \cdot c - b \cdot d) + (a \cdot d + b \cdot c) \cdot i$, since $i \cdot i = i^2 = -1$.

6.5 Technically speaking, the condition $\sqrt{a^2 + b^2} = 1$ is the equation of a circle of radius one in this complex plane. The tip of an arrow forming the radius of this circle reaches every point on the circle as it sweeps through an angle of 360 degrees, so each point on the circle can be labeled by an angle between 0 and 360 degrees.

6.6 Here, we're playing fast and loose about what specific elements of R(3) are actually the generators. The three generators are, strictly speaking, exercise rotations (about the three axes, x, y, and z) by infinitesimal angles—angles θ that are vanishingly small but still not quite zero (such a concept will make sense to those familiar with calculus). For our purposes, it won't do any harm to think of the generators in terms of the three 90-degree exercise rotations.

6.7 More precisely, rotations are those transformations that are both length and shape preserving. Consider two arrows, of sizes s_1 and s_2, that have a common base at the axis of rotation but are separated by an angle θ in the direction that each points. After a rotation, both arrows will have changed direction, but all three quantities—s_1, s_2, and the angle θ that separates the two arrows—will be unchanged. The shape formed by two arrows on a complex two-dimensional plane can be defined analogously to that on a real plane and is similarly preserved in complex rotation.

6.8 For the mathematically inclined, this is most easily done with two-by-two complex matrices for SU(2) and three-by-three real matrices for R(3); the group operation is then simply matrix multiplication.

Chapter 7 The World Within

7.1 To continue this irrelevant but interesting digression: For most of us not steeped in the ethereal world of higher mathematics, it should be said that Karl Weierstrass is widely credited as the father of the mathematical field known as analysis, the formalization and generalization of the techniques of calculus. In the 1850s, as the German Weierstrass was struggling to retain a post as a secondary-school teacher, a young American mathematician by the name of John Pierpont Morgan was garnering offers for academic positions at prestigious German universities. As most of us are aware, J. P. Morgan had a few extracurricular interests, and soon returned to the States to seek his fortune on the bustling streets of Lower Manhattan.

7.2   Think of, for example, the set of all whole numbers (integers), with the two operations being addition and multiplication. The interrelation between the operations is given by the "distributive law": If l, m, and n are three integers, then $l \cdot (m + n) = l \cdot m + l \cdot n$. So, you can see that the concept of a "ring" is an abstraction of a mathematical system somewhat closer to our everyday system of numbers than that of a "group," with its single associated operation.

7.3   More precisely, parity inversion is the operation that takes each point (x, y, z) in three-dimensional space and inverts its coordinates—moves the point to $(-x, -y, -z)$. If you think about this a bit, you'll see that this is not quite the same as a mirror reflection. Nonetheless, mirror reflection and parity inversion have many of the same properties; for instance, both of these are impossible to produce with any combination of the three generating rotations; no amount of successive rotation about the x-, y-, and z-axes will yield the same result as a parity inversion or mirror reflection. In what follows, we will use the terms parity inversion and mirror reflection interchangeably, although, technically, only the true operation of parity inversion is relevant to our discussion. In fact, to be completely accurate, the mathematical representation of the process of mirror reflection is the combination of parity inversion followed by two 180-degree rotations: one about the resulting x-axis and then one about the resulting z-axis.

7.4   For those who remember a little trigonometry, the projection of an object of length L onto a given axis is given by $L \cos \theta$, where θ is the angle between the axis and the projected object.

7.5   More correctly, the spin angular momentum points in the direction about which the particle is spinning clockwise; the axis about which it is spinning counterclockwise is the same line in space but pointing in the opposite direction. We have to pick one direction or the other, and the convention is to choose clockwise. This is the same convention as that of mechanical screws, which turn clockwise as they tighten away from you into a board. The selection of clockwise over counterclockwise has to do with the fact that the majority of human beings are right-handed. If your thumb represents the axis of spin, the fingers of the right hand curl in the clockwise direction as your thumb points away from you.

7.6   Actually, the intrinsic angular momentum associated with a particle of spin s is not $s\hbar$ but, rather, $\sqrt{s \cdot (s + 1)}\hbar$, so for example, a spin-½ particle has an intrinsic angular momentum of $\sqrt{½ \cdot ½}\hbar \approx 0.866\hbar$, slightly greater than ½$\hbar$. This distinction is not important for us, though.

7.7   For the glutton, some more specifics on this. Recall that in nonrelativistic (Schrödinger) wave mechanics, the wave function associates a complex number with every point in space and time, whose size (complex square) is related to the probability of finding the object described by the wave function at that point in space at that particular time. To incorporate both quantum mechanics and relativity, Dirac found he had to have a wave function that associated a point in *two* complex dimensions (a point in the space represented in fig.

7.4) with every point in space and time. The square of the overall distance of this point from the origin still represented the probability of finding the particle at the space-time point of interest. Dirac soon realized, however, that the extra information gained in going from one to two complex dimensions allowed him additionally to encode within the wave function whether the particle's spin was aligned with projection $+\frac{1}{2}\hbar$ or $-\frac{1}{2}\hbar$ relative to any chosen measurement axis — information, as we've seen, that's critical for the description of a spin-$\frac{1}{2}$ particle.

7.8 Both Rutherford and Chadwick received Nobel Prizes, Chadwick in physics in 1935 and Rutherford in chemistry in 1908. Rutherford was somewhat piqued by the fact that the Nobel Committee had turned him into a chemist.

7.9 The term *isospin* is short for *isotopic spin*. The name derives from the analogy with conventional spin, combined with the fact that the operation of exchanging neutrons and protons can be used to generate the wave functions associated with different isotopes of a given element. We won't worry about the physical units of isospin; they're certainly not the same as those of angular momentum.

7.10 The choice of vertical is merely a convention; we could as easily have defined our measurement axis to be horizontal in isospin space in which case protons would be *isospin right* and neutrons *isospin left*. All we require is that, whatever direction the proton's isospin points, the neutron's points the other way.

7.11 What is it then that distinguishes the two states in the middle, making them two distinct particles? Consider, for a moment, a state formed from binding together two particles, a proton (isospin projection $+\frac{1}{2}$) and a neutron (isospin projection $-\frac{1}{2}$). The combined state has to have a total isospin *projection* of $\frac{1}{2} - \frac{1}{2} = 0$ but can have a total isospin (the magnitude of the isospin that's being projected) of either zero or one (to understand this statement, it may help to look at the π^0 state in figure 7.8; the π^0 has a total isospin of 1 but an isospin projection of $I_p = 0$). So that's what distinguishes these two states. While both have an isospin projection of $I_p = 0$, one (the π^0) has a total isospin $I = 1$ and the other (the η) a total isospin $I = 0$.

7.12 Well, this is a bit of an oversimplification. For example, even if the strong force is invariant under rotations in isospin space, it can have a different binding strength for isospin-0 and isospin-1 states that are otherwise identical, such as the π^0 and η of figure 7.8. In addition, the fact that the quark masses are somewhat different means that SU(3) is only an approximate symmetry of the strong force. But these distinctions are beyond us, unnecessary for our discussion, and do not invalidate the statements we make nor the conclusions we draw.

Chapter 8   Physics by Pure Thought

8.1 Those who remember the Pythagorean theorem will have no trouble seeing why this is true. Recall that this theorem, proved to be true over the millennia in a multitude of ways (one, in fact, due to the former American president

Theodore Roosevelt), states that the length of the hypotenuse of a right triangle with sides of length a and b is just $\sqrt{a^2 + b^2}$.

8.2 It should be pointed out here that it was not C. N. Yang and R. L. Mills in 1954 but, rather, the German mathematician Hermann Weyl in 1929, who first recognized the profound implications of local phase-change invariance. However, Weyl did this in the context of the existing theory of electromagnetism, while Yang and Mills were the first to use the principle in an attempt to further our description of the behavior of the natural world.

8.3 In what follows, I will be using nuclear isospin as an analog to the weak-isospin-based theory that is the actual gauge theory of the weak nuclear interaction. The idea of introducing this more accessible analogy is borrowed from I. J. R. Aitchison and A. J. G. Hey's *Gauge Theories in Particle Physics*, Adam Hilger, 1982.

8.4 Those who are arithmetically inclined may want to demonstrate for themselves that the product of two complex numbers with sizes $|z|$, $|z'|$ is a third complex number with size $|z| * |z'|$. The rule for complex number multiplication is what you would expect algebraically:

$$(a + b \cdot i) * (c + d \cdot i) = a * c + (a * d) \cdot i + (b * c) \cdot i + (b * d) \cdot (i * i) =$$
$$a * c - b * d + (a * d + b * c) \cdot i.$$

You'll need to remember that $i * i = i^2 = -1$ and that for a complex number $a + b \cdot i$ of size one, $\sqrt{a^2 + b^2} = 1$.

8.5 In fact, there is one other essential property of the interaction that cannot be predicted solely based on the characteristics of the group SU(2). This is the issue of *parity violation*, the extent to which two interactions that are mirror reflections of each other have the same overall interaction strength g. It's possible for the interaction to have not one but two separate g's: one for our everyday universe, and one for the universe you would end up with were you to reflect our universe through a vast mirror. While the possibility that these two values of g might differ was recognized right off, it was not taken seriously until it was discovered to be the case for the weak nuclear interaction. This discovery was made by T. D. Lee of Columbia University and C. N. Yang of the Institute for Advanced Study at Princeton (they shared the 1957 Nobel Prize); the latter is already familiar to us as one of the originators of gauge theory. The issue of parity violation played a major role in the development of the unified electroweak theory, so we'll get back to it in depth in chapter 9; for the strong nuclear and electromagnetic forces, the value of g is the same for both the regular and mirror-reflected universe. We'll ignore this subtlety for the remainder of this chapter.

8.6 To be perfectly honest, photons do in fact bounce off each other at some level, even if they aren't charged. The way that they do it (I'm about to describe a particular Feynman diagram that represents the process in quantum electrodynamics) is by having one of the photons fluctuate into a virtual electron-positron pair, which then absorbs the second photon. Then, the excited electron-positron pair reemits a photon (a different photon, but how can you tell?), and then finally fluctuates back into a second photon. However, the likelihood of this contrived, yet real, process is so slight that you would only observe it by passing laser beams of most extreme concentration through each other. However, if photons were electrically charged, a process such as that of figure 8.11c (with the W's replaced with photons) would be likely enough that crossed flashlight beams would exhibit observable scattering.

8.7   As mentioned before, don't confuse quark color with quark type, or flavor. A quark of any flavor (d, u, s, c, b, t) can come in any color (R, G, B) and vice versa. Color is a completely new and independent property of quarks.

8.8   In the language that we introduced in chapter 7, we would hypothesize that the three differently colored quarks, being fundamental objects, must fall into the most basic and fundamental manifestation, or representation, of SU(3). This fundamental representation is a pattern of three points into which each of the three colored quarks fall. This pattern is centered around zero (no net color charge), so that when all three colors are combined in equal numbers, the result is just that: zero, or no net color charge.

8.9   This may be a bit confusing because above I claimed, in so many words, that any colored particle, including gluons, can never be observed in isolation. This is true, and what the Large Electron Positron (LEP) experiments really observed was events with four "particle jets," each revealing the presence of a final-state quark or gluon in the diagrams of figure 8.14. Colored particles live on their own for about 10^{-23} seconds, during which time they literally tear other quarks and antiquarks out of the vacuum until everything is paired (or tripled in the case of three-quark baryons) in color-charge neutral particles. Thus, the particle jets that are observed in the detector are fairly well collimated sprays of mesons (pions, kaons, etc.) and baryons (protons, antineutrons, etc.), all of which are long lived enough to travel the meter or two it takes to be observed and measured in today's state-of-the-art particle detectors.

8.10 This does not necessarily mean that absolutely everything in the human experience is attributable to the fundamental physical forces. For example, the nature of human consciousness has not yet yielded itself to the definitive explanation of the natural scientist; although in recent years biologists are beginning to address this most profound issue, the full explication of consciousness remains the subject of a relentless and perhaps unresolvable debate. It does however seem that, whatever human consciousness is or is not, it is not associated with the capacity to influence directly the motion or condition of external objects. Thus, for the elucidation of the nature of human thought and self-recognition, natural scientists must, by and large, defer to the true experts on the subject, who tend to occupy chairs in the philosophy and psychology departments on the other end of campus.

Chapter 9   The Current Paradigm

9.1   Given our current understanding of the weak interaction, it may not seem appropriate to continue to call it the weak nuclear force. What's "nuclear" about two free electrons repelling each other through the weak nuclear interaction? This designation, historically due to the fact that the interaction was first recognized in the context of nuclear beta decay, has been kept nonetheless.

9.2   Why they are so different is not understood, although an interesting line of speculation that has arisen in the past few years suggests that the gravitational charge may not be so small. Instead, it may be the case that most of the gravitational force-field leaks out into spatial dimensions beyond the three that we're used to, making gravity seem weak when it really isn't.

9.3   Aficionados take note. We really should be having this discussion in the context of the Lagrangian, from which the wave equations for the various field quanta can be derived via a variational principle. I have studiously avoided the introduction of the Lagrangian in this book insofar as I feel it to be an unnecessary complication.

9.4   The following presentation of screening and hidden symmetry is inspired by that of I. J. R. Aitchison and A. J. G. Hey's *Gauge Theories in Particle Physics.* The authors, in turn, attribute this approach to explaining the idea of hidden symmetry to the estimable condensed matter physicist Philip W. Anderson.

9.5   Again, any complex number z can be written as $z = a + b * i$, where a and b are ordinary real numbers and i is the mythological square root of -1. Thus, a and $b * i$ are the real and imaginary parts of z, respectively.

9.6   Again, parity inversion is the operation that swaps each of the three coordinates in three-dimensional space with its negative, so that if, for example, one of your atoms is at the point $(x, y, z) = (2, -3.8, 9.1)$ before parity inversion, it will be at the point $(-2, 3.8, -9.1)$ after the inversion operation. Mirror reflection, or at least our perception of it, only reverses the horizontal (x) coordinate, leaving the height (y) and depth (z) coordinates unchanged. In fact, mirror reflection is really the combination of *two* operations on space: parity inversion, as described above, followed by a rotation of 180 degrees about the x-axis. Regardless, mirror reflection, like parity inversion, cannot be achieved by any sequence of ordinary rotations. So, if some physical law is not invariant under parity inversion, it won't be invariant under mirror reflection, and vice versa. Thus, in our discussion, we'll ignore the distinction between mirror reflection and parity inversion and treat them as if they're one and the same.

9.7   The notion of the physical quantity of parity is potentially confusing. The parity of a system is a property of the system in our own unreflected universe. The parity of the system simply catalogs what would happen to it if the system were to be mirror-reflected. If the value of the system's wave function changes sign (goes from positive to negative and vice versa) when we mirror-reflect, the parity of the system is *odd*. If the wave function is unchanged when we mirror-reflect, the parity of the system is *even*. Mathematically, we would say that

$\psi(-x, -y, -z) = \psi(x, y, z)$ for even parity, and $\psi(-x, -y, -z) = -\psi(x, y, z)$ for odd parity.

9.8 This C. N. Yang is one and the same as the Yang of Yang and Mills, the duo who brought us local phase invariance and the seeds of gauge theory.

9.9 For the curious, the argument goes as follows. Since the spin axes of the cobalt-60 nuclei are all aligned and pointing north like a bunch of compass needles, imagine yourself viewing the cobalt sample from the south, so that all the nuclear spins are clockwise. Now, nickel-60, the nucleus left over after the beta decay, has a net spin of $4\hbar$, one $\hbar$ less than that of cobalt-60. So, to conserve angular momentum, the $\frac{1}{2}\hbar$ electron and antineutrino spins must both be clockwise, so that combined they have the $\hbar$ clockwise spin needed to make up the difference between cobalt-60 and nickel-60. So, if the clockwise-spinning electron recedes from you (to the north in our compass analogy) it's right-handed; if it comes toward you (south) it's left-handed. Therefore, as claimed, right-handed particles tend to go north and left-handed particles south.

9.10 Technically, this prejudice is accommodated within quantum field theory through the space-time property of the minimal interaction vertex of the electron and weak-interaction field quantum.

9.11 The team also included Leon Lederman, co-recipient of the 1988 Nobel Prize in Physics for demonstrating that the electron- and muon-type neutrino are distinct particles. Lederman also served a long stretch as the director of the Fermi National Accelerator Laboratory outside of Chicago and is the author of the popular book *The God Particle*.

9.12 If you are one of the victims of congenital left-handedness, you may well be standing up in your chair and shouting something like: "Aha! I knew it! We really *are* better than them! The very universe itself prefers left-handedness!" Being left-handed myself, I will happily concede you this point. By the way, the complementary processes for antimatter, achieved by exchanging all particles for their antimatter counterparts, give the same results—maximal parity violation—although for antiparticles, it's right-handedness that's exclusively preferred. Note that our universe is composed of matter, not antimatter, so the first part of this footnote still pertains.

9.13 In the year 2000 *Review of Particle Properties*, the compendium of particle physics data put out annually by the Berkeley Particle Data Group, one finds that the least accurately measured of these quantities, the Z^0 boson mass, is now known to a precision of about 20 parts per million. From that book, we find that $\alpha = 0.00729735253 \pm 0.00000000003$, $G_F = 0.0000116639 \pm 0.0000000001$ per GeV^2, and $M_Z c^2 = 91.188 \pm 0.002$ GeV, where in each case the second number is the extent of experimental uncertainty on the measured value. (We multiply the Z^0-boson mass M_Z by the square of the speed of light so that we can quote its mass-energy rather than its mass, which is the convention in particle physics.) The resulting value of $\sin^2\theta_W$ calculated above is quoted to the number of decimal places that are meaningful, given the precision of these measurements.

9.14 It may occur to you that, for all this to work, we have to know the exact amount of right-handedness preference of the B^0. In fact, we do know this because we know that the photon has exactly zero preference between right- and left-handed particles. Since the amount of W^0 in the photon is also specified by the number $\sin^2\theta_W$, we know exactly how much B^0 right-handedness preference it takes to cancel precisely the complete left-handedness preference of the W^0.

9.15 A word on what is meant by *experimental uncertainty*. The convention is that the true value should fall with about 68% probability within the range bounded by the measured value minus the uncertainty and the measured value plus the uncertainty. Then, assuming Gaussian, or bell-curve statistics (a good assumption in many cases), the probability increases to about 95% for a range of plus or minus twice the uncertainty and 99.75% for plus or minus three times the uncertainty. The probability quickly approaches one as the range increases beyond three times the quoted uncertainty.

9.16 Remember—these calculations are done in the context of quantum field theory, which is an extension of old-fashioned quantum mechanics. In quantum mechanics, waves can interfere constructively or destructively. In the latter case (destructive interference), the addition of a new process, a new source of quantum-mechanical waves, will actually *reduce* the calculated collision probability. Such is the case for the additional modes of collision admitted by the introduction of the Higgs boson.

Chapter 10　Into the Unknown

10.1 Recently, a number of readable books have been published that present some of the latest thinking in the field of cosmology and the large-scale structure of the universe. Joel Primack, a colleague of mine at the University of California at Santa Cruz and a renowned cosmologist, has recommended the following. For well-formed text enhanced with beautiful color plates, *The Little Book of the Big Bang* by Craig Hogan and *One Universe: At Home in the Cosmos* by Neil de Grasse Tyson, Charles Tsun-Chu Liu, and Robert Irion are sure to please. A more descriptive, but nonetheless concise, book is Trinh Thuan's *The Birth of the Universe*, while a more comprehensive text is provided by Timothy Ferris' *The Whole Shebang*. Finally, all of the recent books by the eminent British cosmologist Martin Rees (*Just Six Numbers, Before the Beginning*, and *Our Cosmic Habitat*) provide good background on the ideas behind the modern-day theory of cosmology and its implications.

10.2 A detailed and insightful account of the rise and fall of the Superconducting Supercollider, tentatively entitled *Tunnel Visions—The Rise and Fall of the Superconducting Supercollider*, and coauthored by physicist and science historian Michael Riordan, should soon be available.

10.3 Charged particles that travel on circular paths, such as those in a circular colliding beam accelerator, must be continually reaccelerated due to the fact that the magnetic forces that keep them going in a circle cause them to radiate their energy away. For electrons and positrons, which are about 2,000 times lighter than protons, it becomes prohibitive to continually restore this lost energy for beam energies greater than one-tenth of a TeV or so. An alternative approach, pioneered at the Stanford Linear Accelerator Center (SLAC) at Stanford University, is to collide beams from two opposing *linear* (rather than circular) accelerators; hence, the name Linear Collider. With technology in hand today, accelerator physicists believe that, for about the cost of the LHC, a 1 TeV linear electron-positron collider can be built. Alternative Linear Collider acceleration schemes under development at this time hold out the hope that the Linear Collider may eventually reach collision energies in excess of 5 TeV.

10.4 Accordingly, the first Web site in the United States was developed at the Stanford Linear Accelerator Center.

10.5 Recall that the weak force is not really weak, in that the weak-isospin charge of the fundamental particles, which sets the strength of the weak interaction, is roughly the same size as that of the electromagnetic and strong-force charges. Again, the weak force appears weak because its associated force-field quanta — the W^+, W^-, and Z^0 — each have a mass of about 100 times that of the proton. Because of this mass and Heisenberg's uncertainty principle, the range of influence of the weak interaction is very limited, making the interaction appear weak even though its true strength (as measured by the magnitude of the weak-isospin charge) is in fact slightly greater than that of the electromagnetic interaction.

10.6 Recall that *bosons* have spin that is an integer multiple of $\hbar$, while *fermions* have spin that is a half-integer multiple of $\hbar$, i.e., the spin of a fermion is an integer multiple of $\hbar$ plus $\frac{1}{2}\hbar$.

Index

Abelian (commutative) group, 142. *See also* non-Abelian group

abstract (mathematical) space, 183

accelerator. *See* particle accelerator

alpha (α) radiation, 187

Ambler, Ernest, 312

amplitude (of waves), 26

Anderson, Carl, 14, 73, 99

Anderson, Philip W., 295, 367n9.4

angular momentum, 70–71, 176–77, 194, 207; and atomic nuclei, 178; conservation of (*see* conservation: of angular momentum); quantization of, 177–81; units of, 70, 177

antifundamental representation of SU(3), 200–201

antimatter, 14, 73–77, 102, 122–24; and stability, 128

antiproton, 14, 117

associative law, 141

asymptotic freedom, 266–67, 274–75

atomic spectra, 23–24, 31

background fields, 294–97

bare electron charge, 82, 263, 341

Barkov, Lev, 286

baryon, 132, 134–35, 267; Λ, Σ, Δ, and N, 134, 192–94; Ω^+, 204

Berners-Lee, Tim, 337

beta decay, 14, 99, 253, 367n9.1; of cobalt-60, 312, 368n9.9

blackbody, 356n3.6; radiation, 31

B Meson Factory experiments, 16

Bohr, Neils, 51

Bose, Satyendra Nath, 71

boson, 71, 178; and supersymmetry, 347

bottom (b) quark, 117, 129

B^0 quantum, 284–86, 288–89, 300, 315–19, 369n9.14; and parity violation, 316

broken symmetry. *See* hidden symmetry

Brookhaven National Laboratory, 15, 100, 105, 204

Bryn Mawr College, 170

bubble chamber, 93

Budker Institute, 286

Cartesian coordinate system, 154, 176

Casimir, Hendrik, 86

Casimir effect, 86

causal connection, 218

CDF experiment, 117

CERN, 14, 103, 112, 120, 252, 273, 320, 327–28, 337

Chadwick, James, 187, 364n7.8

Chamberlain, Owen, 14

charged current interaction, 108–9, 115–16, 251

charmed (c) quark, 105–16; decay 126, 129

cheating term, 223–24, 239–43, 255–56, 270, 277

meson, 131, 135–37, 267; K (*see* kaon); π (*see* pion)
MeV, 356n3.2
Michelson, Albert Abraham, 18, 19
Mills, R. L., 216
minimal interaction vertex, 65–70, 229, 247, 256–57, 261, 270–72, 278, 343; gluon-only or triple-gluon, 272; Higgs-W and Higgs-Z, 201; space-time property of (*see* coupling: space-time property of); strength (*see* coupling: strength)
mirror nucleus, 188
mirror reflection, 175, 363n7.3. *See also* parity: violation
mixing: leptonic, 361n5.5; quark, 113–14, 116, 124, 128
Mod(4), 144
momentum, 34, 44; units of, 37
Mr. Tompkins in Wonderland, 356n3.4
multiplets, 203
muon, 99, 119, 244; decay, 126, 249; lifetime, 246
muon neutrino. *See* neutrino, muon

Nambu, Yoichiro, 266, 294
Neddermeyer, Seth, 99
negative energy solutions, 73–75
neutral current interaction. *See* neutral weak interaction
neutral weak interaction, 110, 251–54; and gauge theory, 248; and neutrino scattering, 112, 252
neutrino, 14, 16, 17, 100–105, 108, 125; electron, 100, 109, 118, 124; interactions, 111–15; mass, 104–5, 124; muon, 100, 124, 108–9, 111, 116, 244; oscillation, 104–5, 124, 361n5.5; tau, 103, 116, 124
neutron: lifetime, 133, 361n5.9; quark content, 133
Newton, Isaac, 51, 139, 339, 358n4.9
Nobel Prize, 4, 15, 17, 33, 35, 36, 68, 69,

74, 94, 97, 98, 99, 100, 105, 107, 108, 120, 133, 204, 286, 315, 317, 334, 364n7.8
Noether, Emmy, 169–70
Noether's theorem, 170–74, 187, 191, 206, 228, 246, 281
non-Abelian group, 158, 240, 254; and gauge theory, 254–61; implications for physical processes, 256–60, 342–43; and SU(3), 272
November Revolution, 105
nuclear beta decay. *See* beta decay
nuclear isospin, 137, 187–195, 232, 364n7.9; conservation of, 192; definition of, 189; of proton and neutron, 189; space (*see* isospin space)
nucleon, 12, 188

parity, 175, 306–9, 367n9.7; conservation, 308–9; even and odd, 307; intrinsic, 198–99; inversion, 175, 363n7.3; and spin, 309–15; violation, 309–16, 365n8.5; —maximal, 314–15
particle: exchange forces (*see* exchange forces); jet (*see* jet); stability of, 125–28; zoo, 96, 130–38
particle accelerator, 21, 36, 93; applications of, 337; complementarity, 105, 324
Particle Data Group, 135, 361n5.10
particle physics detectors. *See* detectors
Pauli, Wolfgang, 99, 358n4.9
Perl, Martin, 99
PETRA, 120
phase: and Lie groups, 231–38; quantum-mechanical, 29–30, 209–12, 245, 255, 276, 358n4.6; of waves, 25, 26–29, 356n3.5
photoelectric effect, 32–33
photon, 33, 48, 54, 119, 121, 125, 228–90; and Abelian gauge groups, 257; and effective mass, 293–96; and laser beams, 359n4.14; and parity violation,